Advances in Industrial Control

Springer
London
Berlin
Heidelberg
New York
Barcelona
Budapest
Hong Kong
Milan
Paris
Santa Clara
Singapore
Tokyo

Other titles published in this Series:

Microcomputer-Based Adaptive Control Applied to Thyristor-Driven D-C Motors
Ulrich Keuchel and Richard M. Stephan

Expert Aided Control System Design
Colin Tebbutt

Modeling and Advanced Control for Process Industries, Applications to Paper Making Processes
Ming Rao, Qijun Xia and Yiquan Ying

Modelling and Simulation of Power Generation Plants
A.W. Ordys, A.W. Pike, M.A. Johnson, R.M. Katebi and M.J. Grimble

Model Predictive Control in the Process Industry
E.F. Camacho and C. Bordons

H_∞ Aerospace Control Design: A VSTOL Flight Application
R.A. Hyde

Neural Network Engineering in Dynamic Control Systems
Edited by Kenneth Hunt, George Irwin and Kevin Warwick

Neuro-Control and its Applications
Sigeru Omatu, Marzuki Khalid and Rubiyah Yusof

Energy Efficient Train Control
P.G. Howlett and P.J. Pudney

Hierarchical Power Systems Control: Its Value in a Changing Industry
Marija D. Ilic and Shell Liu

System Identification and Robust Control
Steen Tøffner-Clausen

Genetic Algorithms for Control and Signal Processing
K.F. Man, K.S. Tang, S. Kwong and W.A. Halang

Advanced Control of Solar Plants
E.F. Camacho, M. Berenguel and F.R. Rubio

Control of Modern Integrated Power Systems
E. Mariani and S.S. Murthy

Advanced Load Dispatch for Power Systems: Principles, Practices and Economies
E. Mariani and S.S. Murthy

Supervision and Control for Industrial Processes
Björn Sohlberg

Modelling and Identification in Robotics
Krzysztof Kozlowski

Pietro Carlo Cacciabue

Modelling and Simulation of Human Behaviour in System Control

With 113 Figures

Springer

Dr Pietro Carlo Cacciabue
European Commission Joint Research Centre
Institute for Systems, Information and Safety
21020 ISPRA
Varese
Italy

British Library Cataloguing in Publication Data
Cacciabue, Pietro, C.
Modelling and simulation of human behaviour in system control. - (Advances in industrial control)
1.Human behaviour - Computer simulation 2.Control theory
3.Human-machine systems
I.Title
629.9

Library of Congress Cataloging-in-Publication Data
Cacciabue, Pietro, C.
Modelling and simulation of human behaviour in system control / Pietro Carlo Cacciabue.
p. cm. -- (Advances in industrial control)

ISBN-13: 978-1-4471-1569-4 e-ISBN-13: 978-1-4471-1567-0
DOI: 10.1007/978-1-4471-1567-0

1. Automat control- -Mathematical models. 2. Automatic control- -Computer simulation. 3. Human-machine systems- -Mathematical models. 4. Human-machine systems- -Computer simulation. I. Title. II. Series.
TJ213.C2 1998 97-51768
629.8- -dc21 CIP

Softcover reprint of the hardcover 1st edition 1998

Cover design by Lino Parisi
Typesetting: Camera ready by author
69/3830-543210 Printed on acid-free paper

Advances in Industrial Control

Professor Dr -Ing M. Thoma
Westermannweg 7
D-30419 Hannover
Germany

Professor H. Kimura
Department of Mathematical Engineering and Information Physics
Faculty of Engineering
The University of Tokyo
7-3-1 Hongo
Bunkyo Ku
Tokyo 113
Japan

Professor A.J. Laub
College of Engineering - Dean's Office
University of California
One Shields Avenue
Davis
California 95616-5294
United States of America

Professor J.B. Moore
Department of Systems Engineering
The Australian National University
Research School of Physical Sciences
GPO Box 4
Canberra
ACT 2601
Australia

Dr M.K. Masten
Texas Instruments
2309 Northcrest
Plano
TX 75075
United States of America

Professor Ton Backx
AspenTech Europe B.V.
De Waal 32
NL-5684 PH Best
The Netherlands

To Catherine, my wife
Nicola, Nadia and Annick,
our children

SERIES EDITORS' FOREWORD

The series *Advances in Industrial Control* aims to report and encourage technology transfer in control engineering. The rapid development of control technology impacts all areas of the control discipline. New theory, new controllers, actuators, sensors, new industrial processes, computing methods, new applications, new philosophies..., new challenges. Much of the development work resides in industrial reports, feasibility study papers and the reports of advanced collaborative projects. The series offers an opportunity for researchers to present an extended exposition of such new work in all aspects of industrial control for wider and rapid dissemination.

The potentially devastating effect of an operator making the wrong decision in the control of a highly automated system or process is well known. However as even more large-scale automated systems become likely, for example automated highways for cars, it is increasingly important to be able to assess the safety of these mixed or joint systems. Carlo Cacciabue's monograph on the modelling and simulation of these mixed processes of technological systems and human operators is extremely timely. The monograph provides an up-to-date and systematic presentation of the basic concepts and tools needed. This comprehensive coverage of the subject also includes a review of the last twenty years of research effort in the field. The monograph culminates in two applications of the joint system methods to two examples concerned with safety in nuclear power plant.

It is pleasing to have this monograph in the Series because it is on a very unusual but important systems topic in control engineering. We hope that Carlo Cacciabue's careful exposition of mixed technological systems and human factors will stimulate new avenues of research and new monograph contributions on the subject.

M.J. Grimble and M.A. Johnson
Industrial Control Centre
Glasgow, Scotland, UK

FOREWORD

The Simulation Of Joint Systems

The idea of using simulations both to describe systems and to represent their dynamic properties in such a way that they can be calculated by a computer is far from new. The foundations of this field were laid in the early 1960s when Kristen Nygaard and Ole-Johan Dahl began to develop the SIMULA language at the Norwegian Computing Centre in Oslo. The beginnings were slow, since in the 1960s both computers and programming languages were relatively novel entities. But in the years since then, and in particular since the mid-1980s, the use of simulations has grown at a phenomenal rate. This is mainly due to the fact that computing power has become a common commodity that is available to anyone. Calculations and simulations, such as Computational Fluid Dynamics, that would have required the power of a mainframe less than a decade ago can today be performed by a standard Personal Computer at a fraction of the cost - both for the hardware and the software! The use of simulations has therefore grown both in depth and in breadth. The depth or degree of realism and complexity of simulations has steadily been improved and constantly seems to challenge the capacity of even the most powerful computers. But more importantly, the breadth of simulation has grown even faster. Whereas simulations initially were used for a small number of technically well-developed domains, such as weather forecasting and operations analysis, they are now present in almost every kind of human endeavour and have, in particular, stepped out of the world of mathematics to the more open territory of visualisation and human-machine interaction.

The greater availability of the computer as a universal scientific tool has been matched by the development of powerful software tools that have made it less of a daunting task to attack a new problem. Indeed, specialised tools now exist within many scientific fields that enable people with technical rather than computing expertise to develop their own simulation models. Simulations are usually thought of as part of the technical or engineering sciences, but they are in fact used commonly in design, training, advertising, political decision making, company planning, etc. We come across the use of simulations in the daily and scientific press for subjects as diverse as global warming, quasars and black holes, car crashes, urban pollution, global weather phenomena such as El Niño, stock market developments, simulated warfare, synthesis of chemicals and compounds, earthquakes, traffic patterns, etc. The use of simulation has clearly become a common tool for practically every problem.

The attraction of simulations is that they are easier and cheaper to use than the real system, e.g. simulation of car crashes. In many cases a simulation may also be the only way to investigate something either because the target phenomena may be out of reach, such as quasars, impossible to control, such as global warming or an earthquake, or simply because they do not exist yet, such as a new car or a drug. The common basis for a simulation is a model of a phenomenon in a formal language that is programmed so that the calculations can be performed automatically. In this way it is possible to see what the consequences of changes in the model parameters will be - at least according to the assumptions built into the model.

Simulations have understandably been used mostly for phenomena that could be modelled in precise, and preferably mathematical, terms such as technological systems or other phenomena that broadly speaking are subject to the laws of nature, cf. the examples given above. As the use of simulations have grown in breadth, we have been faced with the challenge of simulating what one might call interactive systems, i.e., systems where the overall performance is determined by how the sub-systems interact with each other, specifically on how they respond to the behaviour of other sub-systems. This is, of course, particularly important when one or more of the sub-systems are living or biological systems rather than technological systems, especially if the biological system also is a psychological one - i.e., a human being. Examples of that could be the behaviour of cars on a highway, the behaviour of people in a train station, the behaviour of the stock market traders, and the behaviour of a person charged with controlling a dynamic process or machine. As the simulations of technological systems (using the term broadly) have become better and better, the challenge to model mixed technological and psychological systems has grown. In technical terms such systems are often referred to as joint systems, to emphasise the importance of considering the characteristics of the system as a whole rather than attributes of the parts. The simulation of joint systems, in fact, represents not one but two challenges: the first challenge is to model the interaction between psychological and technological systems, while the second is to model the psychological system itself.

The modelling of the interaction is a challenge because the simulation must go beyond a straightforward calculation of the internal state changes in each system. The interaction is an active exchange of data or information, and in the case of psychological systems the description of how input is noted and captured is not a trivial matter. The simulation of the interaction also raises the spectre of synchronisation in real time, hence of defining what real time is for each sub-system. The modelling of the psychological system, commonly referred to as operator modelling, is an even more daunting task. Despite the multitude of systems that have been built to simulate human cognition - specifically within the field of artificial intelligence and cognitive science - the results have been relatively meagre. This is because there is a crucial difference between simulating the reasoning of a person in splendid isolation (e.g. solving a problem or making a diagnosis) and simulating how a person interacts with a dynamic process - involving real-time reasoning, anticipation and perception, planning, etc.

Despite these double challenges, the simulation of joint psychological-technological systems is necessary if we are to understand and manage the complexity of human interaction with technology - which is central to the role of the human in the industrialised society and to maintaining control of a precarious equilibrium. Specific uses of joint system simulation have already emerged in areas as diverse as risk analysis, ergonomic evaluation, team performance, and communication, but the field as a whole lacks a coherent theory or approach. The present book is thus an important contribution to this rapidly developing field, because it presents the issues that must be addressed in a systematic and highly comprehensive manner. In particular, it provides an excellent treatment of the principal issues in designing simulations of interactive systems - psychological as well as technological. It describes the problems that are inherent in developing a cognitive model from a practical rather than psychological perspective, and thereby avoids much of the conceptual detritus that often makes cognitive psychology and cognitive science unnecessarily convoluted. The book is not only a highly readable introduction for the technically minded practitioner but also an excellent starting point for venturing further into the exciting world of joint system simulations.

Erik Hollnagel
Professor, Human-Machine Interaction
University of Linköping, Sweden

AUTHOR'S PREFACE

Si parva licet componere magnis
(Virgil, Georgics, IV, 176)

In September 1985, in San Miniato (Italy), a NATO workshop was held on the subject of "Intelligent Decision Support in Process Environments" (Hollnagel, Mancini and Woods, 1986).

The workshop was held in an old convent located in the magnificent landscape of Tuscany, where the participants stayed for two weeks.

The organisers managed to gather together the major scientists and researchers in the area of human factors and cognitive sciences of those days. Amongst them were Veronique De Kaiser, Keith Duncan, Baruch Fischhoff, Len Goodstein, Erik Hollnagel, Gunnar Johannsen, Jacques Leplat, Beppe Mancini, Neville Moray, Don Norman, Jens Rasmussen, James Reason, Penelope Sanderson, Henk Stassen, Willem Wagenaar, Dave Woods and others.

This list, although incomplete, shows that that event was extraordinary.

The impact that such a collection of experts had on a "young" engineer (I was then in my early thirties) as well as on some other participants eager to learn and open to new "subjects", has been enormous. To add to this fortune, the lectures of specialists like Reason, Moray, Hollnagel, Norman, or Rasmussen could be further developed in the informal discussions that went on outside the formal presentations.

Since then, I have been very much interested and engaged in the study and development of human factors.

However, my engineering background has also influenced the way in which I have approached the problem. In particular, one question has always governed my thoughts. I can express it like this:

"If engineers and mathematicians have not yet managed to represent completely, by analytical solution, "simple" phenomena like boiling or heat transfer, why should we be worried about accepting simplified and approximate numerical solutions representing human behaviour, which is generally recognised as the most difficult machine of the universe ?"

To me the *models* and principles expressed by psychologists, describing human cognition, were nothing more than the conservation equations developed by physicists about natural phenomena. Therefore, one could think of following the same process utilised by engineers in transforming those conservation equations into numerical forms, so as to "calculate", or better *simulate* human behaviour.

I soon discovered, alas, as usual, that this was not a really novel idea.

It is however an exciting idea that generates interest, enthusiasm and expectations that can foster a lifetimes work of research.

In this book I have reported the most interesting findings of this research in the years that go from that time in San Miniato to current days.

Short Summary

This book is aimed at supporting safety analysts, usually engineers, but also ergonomists and psychologists, who have the goal of designing and evaluating the behaviour of the "joint system", formed by the human and machine, paying special attention to cognitive aspects and technology.

Although humans and machines have "always" existed, this discipline is relatively recent. The study of humans has traditionally been the subject of philosophy or psychology, while the analysis of machines has been the reign of engineering and physics. The novelty lies in the attempt to marry these domains of science, so as to generate means and tools that may be applied in practice.

This issue has already been dealt with in the past by very well known and experienced scientist. In particular, a number of important monographs and books have been published, to which I have referred to throughout this book. They represent pioneering and fundamental works that are an essential element of the background of any human factors specialist. These books are: *Human Reliability Analysis: Context and Control* (Hollnagel, 1993a), *Information processes and human-machine interaction. An approach to cognitive engineering* (Rasmussen 1986), *Human error* (Reason, 1990), *Systems Engineering Models of Human-Machine Interaction* (Rouse, 1980), and *Telerobotics, Automation and Human Supervisory Control* (Sheridan, 1992).

In this book, four main Chapters are developed:

Chapter 1. The first Chapter focuses on the theoretical basis of simulation of human-machine interaction. It develops a number of definitions, and attempts to identify the boundaries of the discipline under discussion.
A reference Framework for modelling and simulation of human behaviour in system control is discussed, and the most important requirements and specifications are developed.

Chapter 2. The second Chapter presents the engineering methods and Cognitive Task Analysis techniques that may be applied for the identification of algorithms, data and parameters that are utilised for the development of numerical simulations.

A first example of the application of these methods is shown with the development of a simulation of an operator behaviour by *qualitative physics*.

Chapter 3. This Chapter presents a compendium of models and simulations of cognition that cover the last 20 years of research in this area. It offers the reader the possibility to identify the theoretical paradigm and numerical implementation that may suit his/her specific problem. The *basic guidelines* for developing a simulation of cognition are singled out.

A second example of application is shown, where one of the simulations discussed is applied in a detailed study of human-machine interaction.

Chapter 4. The fourth and last Chapter is initially dedicated to the description of Machine models and simulations and to the discussion on the Interaction Model, i.e., a model that manages the simulation of the whole human-machine system.

Finally, two very extensive applications are presented concerning the simulation of a human-machine interaction for the safety study of two real complex systems typical of a nuclear power plant.

These applications are based on two very different simulations of human behaviour, so as to show how the principles, requirements and specifications described in the preceding Chapters may actually be applied and exploited, for different objectives.

ACKNOWLEDGEMENTS

Writing a monograph is by nature an endeavour of intellectual, scientific and creative character that gratifies primarily its author, and is the product of years of research and study.

In reality, a long journey in the world of sciences is never made in splendid isolation.

This is even more true when there is more than one discipline involved in the study, as in this case, where *engineering*, in its role of analysis of complex technical systems, is combined with *psychology*, in its role of analysis of cognitive behaviour.

In carrying out this effort, in which the duality and variety of these two sciences, apparently impossible to conciliate, result in a single study which I hope will make a contribution to the development of technology, I have been sustained and supported by many experts, researchers, scientists and students.

Together, we have developed ideas, models, techniques, and many applications.

In this book I have included the most significant.

The theoretical and conceptual basis of this book have been developed and are strongly influenced by the discussions, debates and interactions with one person in particular: Erik Hollnagel.

I am sincerely grateful to Erik for his friendship, which was born many years ago, when, during a conference, *a certain gentleman*, sitting at the back of the audience stood up and supported me in such a fine and cultured manner that I immediately warmed to him. And I was right! Since then, I have always gained from each of our meetings and exchanges of views. Our friendship and honesty of attitude has matured over the years and helped us in difficult moments. Thank you, Erik!

Many of the students who graduated or developed their PhD studies with me are still successfully working in the area of *Human Factors*.

Mauro Pedrali is one of them.

I am grateful to Mauro for his patience and help in revising the first manuscript, for supporting me in better defining concepts and basic ideas and in ensuring consistency throughout the Chapters.

With some students and young researchers, such as Hugues Bersini, Alberto Codazzi, Stefano Mancini and others, we have developed and integrated models and simulations of cognition and technical systems.

I am grateful to all of them, and, in particular, to: Alberto Pace for his support in developing the study of the qualitative physics model of Chapter 2; Michel Masson and Françoise Decortis for the case study of the application of COSIMO discussed in Chapter 3; Andrea Carpignano and Claudia Vivalda, for the test case on the Chemical and Volume Control System presented in Chapter 4; Giacomo Cojazzi, for the Auxiliary Feed-Water System also discussed in Chapter 4.

I have also appreciated very much the support of some colleagues, and in particular, Monica Bacchi, Lino Parisi and Cornelia. Spitzer, who have dedicated their precious time to reading my manuscript, to trying to understand and apply the concepts described in the book and to discussing with me. Their support is greatly acknowledged, as it has been instrumental for the development of the book.

I am particularly grateful to Lino Parisi who has given his full support and much time and effort to preparing some of the most valuable complementary, but essential, features of the book. For example the cover page was designed by Lino.

Finally, even if I have been told several times not to do so, I have to express my deepest thanks to my wife, Catherine.

She has read and suffered with me every single word, expression, and sentence of this book. Even when my English was more Italian than anything else she has grasped the concept, diligently modified the grammar structure, and turned around the sentences until everything was acceptable. And when she was not happy, she would start again ..... All this work has been done in addition to the patience of waiting and looking after me, when I spent hours and days totally absorbed by my endeavour.

This book belongs to her as much as it belongs to me.

CONTENTS

Chapter 3. Models and Simulations of Cognition

Chapter 4. Modelling Machine and Interaction

CHAPTER 1
A FRAMEWORK FOR HUMAN-MACHINE INTERACTION SIMULATION

1.1 Introduction

In order to develop a framework that provides guidelines and support for the analysis of human-machine interaction, it is important to start by illustrating some basic concepts and definitions that characterise the domain of our work. Therefore, we will begin by outlining what is meant by a "classical" architecture human-machine-interaction and by simulation of cognition. In particular, we will discuss the meaning and differences between the concepts of human behaviour models and simulations.

We will define types of simulations, types of analyses and, domains of application of human-machine interaction. The domains of application are particularly important, as they characterise the boundaries and the objectives of the studies carried out by the simulations.

We will examine the content of human-machine system simulation and how models are actually transformed into computerised, numerical simulations. Special attention will be devoted to practical implementations of cognitive simulations, as machine models are better known and much more developed.

A new architecture and framework of human-machine interaction will then be devised, which is better adapted to the mechanisms and environment of a modern system control. In particular, as the dynamic nature of human-machine interaction is considered a fundamental process, a specific model will be defined that governs interaction and handles this key aspect. This model plays the role of a "pivot" between the simulations of humans, machines and environment.

We will then tackle the problem of data, parameters and taxonomies sustaining such simulations of humans and machines. A framework will be discussed for structuring data and taxonomies in accordance with the human-machine architecture of reference. This data architecture manages the correlation and transfer of information to and from the Human-Machine Interaction Architecture.

Finally, on the basis of all the previous analyses and discussions, we will be able to outline the basic requirements necessary for the development of comprehensive architectures of human-machine simulation.

As this first Chapter is dedicated to the identification and definition of the basic principles and methods that govern human-machine interaction, we will start by defining the concept of *Human Factors*, which is the general frame of reference that embraces all the subjects discussed in this book.

> ***Human Factors*** **may be defined as the technology concerned with the analysis and optimisation of the relationship between people and their activities by the integration of human sciences and system engineering in systematic applications and working environment frameworks.**

This definition, that "per-se" can give rise to endless discussions, needs to be elaborated further, at least for what concerns the crucial issues that are debated in this book. Firstly, by this definition, *Human Factors* extends the concept of *Ergonomics*, as the science of (human at) work, beyond the workplace and behavioural performance to the cognitive and social aspects involved in human activity (Edwards, 1988).

Human Factors is conceived here as a "technology", emphasising upon its practical nature rather than its disciplinary character. In this sense, the difference between human factors and human sciences is the same that exists between engineering and physics: physics and human sciences look at the basic principles and fundamental criteria that govern their locus of interest, while engineering and human factors concentrate on the implementation in the real world and working environment of these principles and criteria.

This distinction is particularly important, as it is recursively called upon for distinguishing different subject matters, such as the concepts of *model* and *simulation*, and *micro-* and *macro cognition* discussed in other parts of this book.

1.1.1 Human-Machine System

A Human-Machine Interaction (HMI) process always occurs in a realistic context which is characterised by the plant or machine under control, by the socio-technical working environment and by the operator(s) in direct contact with the machine.

The plant interacts with the human operator through its interfaces, i.e., display panels, indicators, decision support tools. The socio-technical working conditions, also called context and environment, influence the human behaviour. These conditions comprise: a) the actual environment in which operations take place, which implies factors such as noise, space, light, temperature etc.; b) other operators co-operating directly or collaborating at a distance with the decision maker; and c) the whole social context represented by the management, the company, the society and the cultural climate.

The plant interfaces and socio-technical working conditions are the main sources of stimuli for the operator. They affect the allocation of resources and knowledge base. They may modify the unfolding of the reasoning and cognitive processes as well as the performance of manual or control actions by, for example, causing errors or inappropriate behaviour or by altering the amount of knowledge accessible to the human in a given circumstance.

A formal description of the human-machine interaction is then necessary in order to lay down the basis for a generic simulation framework. Following Rouse (1980), we will start by defining the (human-machine) *system* as the "overall combination of human and machine" and we will depict the *system* as a simple loop connecting its two main components, as shown in Figure 1.1.

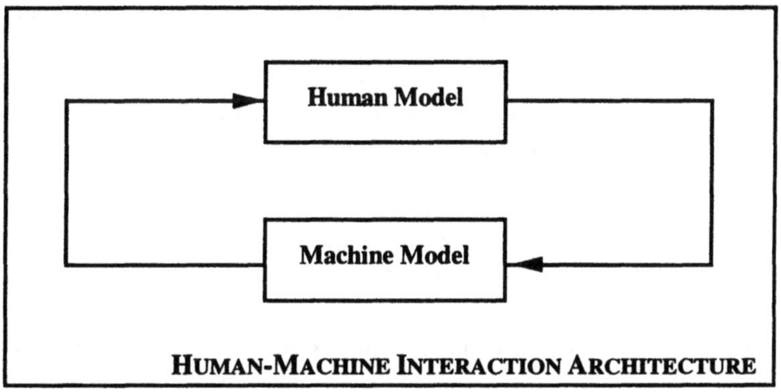

Figure 1.1. Human-Machine "System", (adapted from Rouse, 1980).

This structure is called *System Simulation Architecture* (SSA), or *Human-Machine Interaction Architecture* (HMIA). It represents a simple starting configuration of human-machine interaction, which needs to be further developed in order to obtain a more elaborate simulation framework, able to account for other aspects of the interaction, such as dynamic features, human errors, component failures, working environment, cultural effects, etc. Moreover, this Human-Machine Interaction Architecture does not show explicitly the structure of data and parameters (Data Architecture), which are instrumental for the development of "good" models and for setting the limits of validity with respect to real domains.

More comprehensive definitions of (human-machine) *system* have been proposed by other authors and researchers in the domain of safety, as the notion of the combined human-machine element has further expanded and has become crucial for the development of all systematic safety theories (Miller, 1965; Kuhlmann, 1986; Sheridan, 1992).

For example, Miller (1965) has defined a "safety system" as "*....the integration of the skills and resources, specifically organised to achieve accident prevention, over the life cycle of an air vehicle system .....*".

A more complete definition of the *Human-Machine System* can be found in the document MIL-STD-882B (DOD, 1984), which defines the "system" as:

"A composite, at any level of complexity, of personnel, procedures, materials, tools, equipment, facilities and software. The elements of this composite are used together in the intended operational or support environment to perform a given task or achieve a specific production, support, or mission requirement".

This definition is deemed very appropriate, as it considers more closely the effects of socio-technical environment discussed above. Moreover, it looks explicitly at crucial aspects derived from the use of modern computer technology for the control and management of plants and machines, in normal operations and emergency or transient conditions.

In more generic term, however, the terminology "system" is associated with the whole of a plant or with hardware and software components of a plant, and, thus, it represents a synonym of "machine". Sometimes, the terminology "system" is used for identifying the context and environment.

For the remaining of this book, either the term "*system*", when referring to the overall combination of humans and machines, will be clearly pointed out, or, preferably, the expression "*Human-Machine System*" will be used. The generic expression "system" will simply indicate the whole or part of a plant.

As we will see later on, in a more complex representation of human-machine system, the working environment and context need to be more precisely identified and described by their own data and dynamic simulation, so as to provide a time dependent feedback on human and machine performance.

Moreover, as we will always be considering human activities with reference to a plant and a working context, the interactions between human and machine will be also simply called *human interactions*. The machine attribute in this definition is implicit and it has been dropped, as it covers all interactions of the human operator with other humans as well as with the plant and its control system.

1.1.2 Models and Simulations

A clear distinction exists between the concepts of *model* and *simulation*. These two terms, quite often used as synonyms, characterise two complementary ways to represent a process, a system or any physical or cognitive phenomenon that can be subjected to a formal description.

A *model* is essentially a theoretical account of a process or a system based on a number of hypotheses, conservation principles, and simplifying assumptions, which can take the form of analytical functions or differential equations or lexicographic expressions.

The major goal of a model is to describe in detail a process or a system in all their different features. A model is not concerned with the existence of a solution method to deal with its formulation, nor with the availability of data and parameters for quantifying outputs. A "good" model may be limited to an abstract or conceptual representation of a system or process, which is able to capture the essence of the fundamental phenomena and behaviours.

A *simulation* is an expression or implementation of a model in a form that is controllable or executable by numerical application or computation.

The major goal of a simulation is to "reduce" the complexity of the analytical model by a numerical or proceduralised method able to confine the theory in a manageable, usually computerised, form. A simulation is thus subordinate to the existence of a theoretical model. Moreover, a simulation must consider the availability of data and parameters in order to quantify the numerical representation of the model. This enhances further the process of simplification of a theoretical model. A simulation is by definition a numerical replica of a system or process.

Ideally, "analytical solutions" resolve the dichotomy between models and simulations, as they represent the complete and closed "mathematical" solution of the equations of the model. But analytical solutions seldom exist and are available only for very simple cases, which never represent the reality. Consequently, analytical solutions are not applicable in practice.

Simplified numerical simulations or functional descriptions are the unavoidable and manageable instances of theoretical analytical models. In this sense, any simulation contains an underlying theoretical model.

A common general purpose of models and simulations is the explanation of passed events as well as the prediction of time evolution within certain margins of precision, given boundary and initial conditions, and parameters and data. But, while models are focused on the theoretical aspects and description of these predictions and explanations, simulations are concerned with the numerical transformation of the theories.

As an example, in the domain of thermo-hydraulics, any plant, such as heat exchangers, power plants, engines etc., can be modelled by the well known mass and energy conservation principles and Navier-Stokes equations. These equations capture the entire theoretical behaviour of the components of a plant. However, the actual response of such plants can be calculated only after an appropriate elaboration of these equations by simplification, numerical reduction, definition of parameters and data and possible computerisation for extended calculation purposes. This is a typical process of transformation of a theoretical model into a computerised simulation.

It is now clear why models and simulations are considered two complementary ways to represent reality. This difference will be maintained all throughout this book, as to distinguish whether a modelling paradigm is formulated and kept at high conceptual level or it is also further developed into some numerical or logical expression, able to calculate or predict behaviour.

Focusing on human performance, the problem of separation between models and simulations is enhanced even further by the objectively difficult task of trying to bound human behaviour within the limits of a structured representation (theory) and then by the need to transform such theory into a set of computerised numerical algorithms (simulation).

Letting to one side the philosophical problem of modelling human behaviour in abstract theoretical terms, which is deeply rooted in the culture and the history of humankind, the specific endeavour of developing simulations of human behaviour is relatively recent. This contributes further to the immaturity of the task of simulating human behaviour with respect to machine simulation, which is supported by well consolidated mathematical theories, physical principles, and widely developed numerical methods.

1.1.3 Modelling Modern Working Contexts

The most important product of modern technology is certainly the development and use of *automation* in plant management and control.

The concept of *automation* has been defined and studied by many authors in relation to design and safety issues, as well as to the role assigned to the human in relation to it (Moray, Ferrell, and Rouse, 1990; Rouse, 1990; Sheridan, 1992; Billings, 1997). The following definition will be adopted in this book:

> ***Automation* is the way in which processes of production and operational tasks are carried out, performed and controlled by self-operating machines and/or electronic devices.**

Automation and the expansion of computerised systems have enabled the performance of an enormous number of tasks, with a level of accuracy unthinkable for human sensory abilities.

Even if automation is always designed to work co-operatively with the human operators, its development has strongly affected the human interaction for two correlated aspects: the *tasks* and the *role* of operators.

The *tasks* of operators are mainly concerned with managing the process control system, by setting up the envelope of operations and the plant configuration. In so doing, operators have to exploit their knowledge of procedures, of control means and goals, and of the plant's topographical layout, functional and physical behaviour. Automation takes care of carrying out accurately, precisely, timely and faithfully the requests and plans of operators.

The *role* of human operators is then becoming mainly one of supervisors and monitors of procedures carried out automatically, once these have been set up by the operators themselves (Sheridan, 1992).

In such configuration of human-machine system, it is evident that the tasks and role assigned to the human operators have become much more complex and demanding from the cognitive viewpoint and risky from the human reliability side, as the consequences of a "human error" can be unrecoverable and catastrophic (Nagel, 1988).

Operators are removed from the direct impact with the phenomena and interact with the control systems, by means of visual display units, graphical presentations, decision support means and process management systems. All these means help to perform diagnosis, planning and execution of procedures, and a new control strategy of partnership between humans and machines is created (Ralli, 1993).

This discussion leads to two main conclusions:

- Firstly, it is impossible to separate the effects of the human "supervisor" from the plant dynamics, and, thus, it is necessary to integrate the human component
  a) in early stages of design,
  b) in safety assessment,
  c) in training, and
  d) in accident investigations.
- Secondly, the human-machine interactions and plant transient evolutions are affected essentially by the cognitive and knowledge processes of operators, rather than manual control actions and behavioural responses.

In this scenario of modern human-machine systems, modelling human behaviour entails modelling primarily cognitive processes and system dynamic evolution. The following principle may then be applied for the development of *human models* in a Human-Machine Interaction Architecture:

In a modelling architecture of human-machine system, the human element has to be accounted for by a simulation of the cognitive processes and control actions, i.e., the mental and manual activities, resulting from the time dependent interaction of humans and machines.

While the basic need for simulating the man-plus-the-controlled-element was enunciated by McRuer (1965) and colleagues already in the 60s, the demand to focus on cognitive processes is more recent. Following the development of computer technologies of the 70s, the first formulations of theoretical models of cognition appeared, inspired by the metaphor of the operator as *Information Processing System* (Neisser, 1967; Newell and Simon, 1972).

Prior to those years, the most substantial theories on human behaviour did not consider predictive models outside the observable domain (behaviourism). The relevant human activities were mostly associated with manual control and direct interaction with the physical phenomena. The inclusion of the human element in the control loop of a process was thus a simple exercise of mathematical consideration of action delays and parameters estimation. A short history of the development of models paradigms and simulations will be developed in the next Chapter.

In more recent years, the progress of technology towards supervisory control and automation has required the formulation of much more complex models of human reasoning and decision making process, able to account primarily for cognitive activities rather than manual control, as is enunciated in the above principle.

In the next section the Human-Machine Interaction Architecture will be analysed with reference to various types of applications that can be made for practical and structured simulations of human-machine interactions.

1.2 Types and Applications of HMI Simulation

1.2.1 Types of Cognitive Simulation and Types of Analysis

There are many reasons why a simulation of cognition is used rather than, for example, a study of what happens in practice. Obvious reasons are: access to a simulation may be easier than access to a workplace; simulation covers a wider range of situations and events than can be observed; simulation can be controlled and restarted "at will" (for example using snapshots and breakpoints or partial backtracking); and simulation provides a structured way to analyse events.

Sometimes, simulations (of cognition) are divided into two main types: *qualitative* and *quantitative*. *Qualitative* simulations give an account of the cognitive processes, by describing the structure, the links and the logical and dynamic evolution. These descriptions can be used to explain how human-machine interactions occur, but they are not computational methods, even if they can be considered as the first formal framework of theoretical models. Qualitative simulations can be used for predicting expected behaviours, in some well defined specific cases, where machine performance is also simulated to the same level of precision.

Quantitative simulations contain, in addition, the computational part of a simulation and can be utilised to make numerical estimates of human behaviour. Clearly, this demands that machine and environment evolution are calculated in parallel to the human behaviour simulation.

It is obvious that a quantitative approach rests on a qualitative analysis and that the use of algorithms and numerical estimates is only one way to "substitute" reasoning with numbers, which allow easier comparison as they are single dimension quantities (Hollnagel, 1993a, 1997).

From this discussion it results that quantitative and qualitative simulations do not present fundamental differences. In particular, predictive simulations can be either qualitative or quantitative, and the richness of their content is based primarily on the theory and method behind them.

It should therefore be more appropriate to talk about quantitative *numerical* and quantitative *descriptive* methods, instead of the too simplistic distinction qualitative vs. quantitative methods.

In a wider context of human-machine interaction studies, *numerical* and *descriptive* simulations of cognition support both *prospective* and *retrospective* types of analyses (Figure 1.2).

> ***Retrospective* analyses consist in the assessment of events involving human interaction, such as accidents, incidents, or "near-misses", with the objective of detailed search for the fundamental reasons, facts and causes ("*root causes*") that have promoted and fostered certain human behaviour.**

A retrospective analysis needs the data on the event to be investigated, as well as an appropriate formalised scheme describing human behaviour.

In essence, retrospective analyses must be supported by a sound descriptive model and simulation of cognition. In practice, retrospective analysis tend to be predominantly qualitative.

However, the possibility to perform quantitative numerical estimation and analysis of past events should not be ruled out, as it represents an important exercise for safety and for the definition of preventive measures.

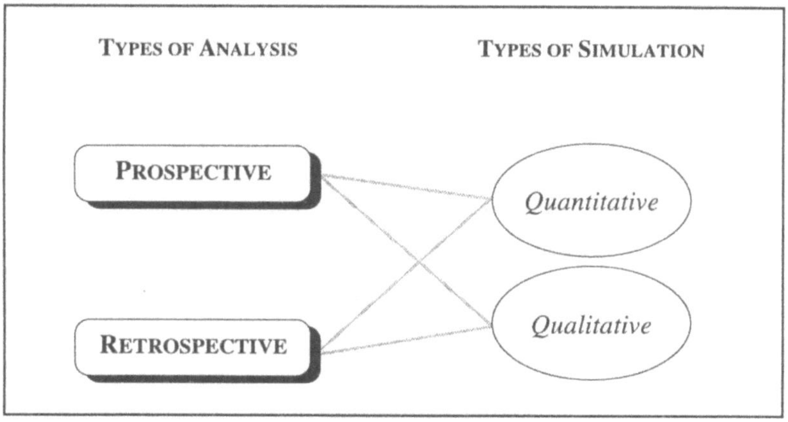

Figure 1.2 Types of simulation and types of analysis.

***Prospective* analyses enable to predict and evaluate the consequences of human-machine interaction, given an initiating event and boundary configuration of the system.**

Prospective studies allow the design and assessment of the safety measures and limits of the plant associated with human behaviour. Prospective analyses demand the evaluation of data, parameters and sometimes probabilities associated with each expected human control and decision process, including the evaluation of recovery actions.

A prospective analysis is thus based on hypotheses and assumptions and it speculates on possible consequences by means of predictive estimations and simulations.

However, as results from the above discussion on qualitative and quantitative methods, not all prospective analyses need to be performed by a numerical computerised simulation of human-machine interaction, but, depending on the requirements and objectives of the analysis, qualitative approaches may be perfectly suitable.

The essential invariant property is:

The same theoretical paradigm (model) of cognition is utilised for both prospective and retrospective analyses.

This important prerequisite is mandatory for the development of coherent methods for prospective and retrospective studies.

As a matter of fact, if the models used for prospective and retrospective analysis differ substantially, it could easily occur that the findings of retrospective studies may not support predictive (prospective) analysis and that the results of prospective studies may not match the findings of the (retrospective) analyses of real events.

1.2.2 Types of Application

Simulations of cognition are used in four different principal types of application: *Design, Safety Assessment, Training,* and *Accident Analysis* (Table 1.1). These four different types of application will now be discussed in detail.

Table 1.1. Types and applications of cognitive simulations

Application	Specific Application	Type of analysis	Type of simulation
Design	*Interfaces* *Procedures*	Prospective	numerical/descriptive
Safety Assessment	*Design Basis* *Probabilistic Assess.*	Prospective	numerical/descriptive
Training	*Classroom* *Simulator*	Prospective/ Retrospective	numerical/descriptive
Accident Analysis	*Root Cause Studies*	Retrospective	numerical/descriptive

Design

Simulations of cognition for *Design* purposes aim at the development and evaluation of *Procedures* and *Interfaces* (Stokes and Wickens, 1988; Degani and Wiener, 1994a).

In particular, using a human-machine interaction simulation, different procedures and interfaces can be designed, compared and tested for a large variety of initial and transitory conditions derived from plant malfunctions, emergencies, as well as normal operations.

The study of procedures and interfaces, for diverse human behaviours, is a typical application of prospective analysis using quantitative numerical simulations.

However, in some cases, the safety state of a system is evaluated by comparing the effectiveness of different interfaces and/or control procedures to the same initiating event or accident scenario. Then, a computerised numerical analysis of events may not be needed, and a structured representation of the human interactions is sufficient in combination with a clear definition of boundary and initial conditions.

Therefore, for design purposes only prospective analyses are performed, applying both numerical and descriptive types of simulation.

Safety Assessment

A *Safety Assessment* study can be performed from two quite different perspectives: *Design Basis* (DB) and *Probabilistic Safety Assessment* (PSA).

Both types of analysis demand the evaluation of the human-machine interaction. The *Probabilistic Safety Assessment* of a plant evaluates the probabilities associated with certain accidents, their consequences and frequencies of occurrence, in relation to a predefined selection of initiating events. PSA, as basic methodology for systematic safety analysis needs, primarily, the combination of structured representation of events with: a) classes or levels of erroneous behaviour; b) reliability data and error/failure probabilities; and c) reliability methods for human and hardware/software systems (Hollnagel, 1993a; Parry, 1994).

These requirements are also conditioned by the fact that, in a PSA study, the human reliability analysis needs to be combined with the plant reliability assessment according to the rules of probability and reliability analysis. Moreover, a model of the plant is needed for the description of the machine responses to human actions.

PSA is essentially a prospective type of analysis, but the types of simulation that support PSA methodology may be very simple, depending on the depth and the accuracy required by the specific study. In some cases, no accurate models are required and a simple descriptive (but predictive!) type of simulation may be sufficient. In other cases, refined models and numerical predictive simulations may be required of plant and human behaviour, which also include the effect of the organisation and working context.

In any case, for PSA purposes, the probabilities associated with human erroneous actions are needed to perform the evaluation of human interaction processes and to determine the probability of success/failure of a task or procedure.

Design Basis accident analyses consist in the safety study of specific accidents. The boundary and initial conditions are pre-specified by the designer and are believed to represent the set of worse possible accidental scenarios, which encompass all other conceivable accident configurations. The safety measures and protection devices of a systems are designed and dimensioned on the basis of the results of design basis accident studies.

Design Basis accident studies imply the evaluation of all engineered safety devices, standards, procedures and training, including human interactions and plant performances.

Design Basis studies are prospective types of analysis. They need the support of: a) models and simulations of human-machine interaction, including cognitive behaviours and organisational factors; b) simulations of plant, control and instrumentation systems; and c) appropriate data and parameters, and classification schemes.

The results of design basis accident studies are frequently numerical and are used to tune the performance of the system with respect to its safety limits.

Training

Training human factors insight has nowadays become common practice for highly specialised operators, such as nuclear power plant operators, pilots, air traffic controllers, etc.

In some cases, such as in civilian and military aviation, the training procedures and methods in human factors are already formalised by regulatory bodies and authorities and are integral part of the overall curriculum of expertise development.

In particular, two specific types of human factors training are considered: *Classroom* and *Simulator*. The *classroom* human factors training consists in introducing the concepts of human behaviour, human-human interaction, and human-machine interaction in very specialised discussions and lectures, as part of the standard and recurrent training (Wiener, Kanki, and Helmreich, 1993; Degani and Wiener, 1994b).

The *simulator* training is carried out during practical, hands-on, sessions at a "full-scale replica" simulator. Operators are trained in these sessions with the objective to develop their "technical" skill in controlling and supervising the machine during abnormal conditions, but also to manage, and exploit human competence and potentialities at their best, specially when working as a team.

In both these cases, i.e., classroom or simulator training, the instructor or facilitator, must clearly master a model and a structured representation (simulation) of human behaviour in order to describe, review and characterise different human performances (Johnston, McDonald, and Fuller, 1994).

Moreover, specially in the classroom work, some complex scenarios are discussed and evaluated with the trainees with the objective to evaluate their ability to predict system response and to maintain situation awareness. Also in these cases the reference model and simulation of human behaviour is applied for predicting and evaluating likely human performances.

A model of this kind must be well structured and descriptive in nature, as to offer a solid paradigm of reference to the instructor. The most frequent application of such a model in training occurs when the instructor has to explain punctual performances and patterns of behaviour. The same model and simulation can also be applied to guide predictive analysis.

For these reasons, simulations of cognition may be utilised in training to perform both prospective and retrospective types of analysis.

Accident Analysis

Accident Analysis and investigation are oriented to the identification of the *root-causes* of an accident, either related to human errors and mishaps or to system failures and malfunctions, and to the definition of preventive measures against future events of the same type.

An accident analysis aims at establishing the correlation between causes-effects-consequences that may be recurrent within the system at all levels, and have contributed to the sequence of events.

From the human analysis viewpoint, a method for accident analysis requires a methodological framework that comprises models of cognitive processes and organisations (Broadbent, Baddeley, and Reason, 1989; Senders and Moray, 1991; Maurino and colleagues, 1995; Pedrali, Cojazzi, and Cacciabue, 1995). These models lead to classification schemes, or taxonomies, that allow the categorisation of observed behaviours (Pedrali, 1996).

A structured representation (simulation) of the dynamics of events and of human-machine interactions, is needed where dependencies, contextual occurrences and logical links are accounted for.

In any case, accident studies and investigations are retrospective types of study. They do not need quantitative predictive calculations, as all necessary data and information are retrieved from available recordings of real occurrences. However, their analyses can resolve into quantitative estimates of the causes and consequences of the events studied.

Summary: Applications of HMI Simulation

Following the above discussion, it is interesting to compare the general connections existing between types of applications, types of analysis and types of simulation (Figure 1.3) and to review some of the most important issues discussed in this section.

Firstly, it has been argued that the distinction between descriptive and numerical approaches is only apparent, as they are both important and necessary representations, in their own way, of the human-machine system. What matters is that descriptive/numerical simulations are instances of theoretical frameworks, that are able to support retrospective and prospective types of analysis.

Secondly, the importance of applying the same theoretical model for prospective and retrospective analysis has been discussed, so as to ensure complete correspondence and maximum feedback from the outcome of one type of analysis to the other.

Thirdly, it is important to realise that, for retrospective and, specially, for prospective types of analysis, a sound basis of realistic and consolidated body of data is essential. This is particularly important when a computerised numerical analysis is required.

Finally, there must be a balanced level of detail between the simulations of human behaviour and of machine performance. In many cases, a simple model of machine is sufficient for the goals of the study and consequently a non-complex simulation of cognition may be also applied.

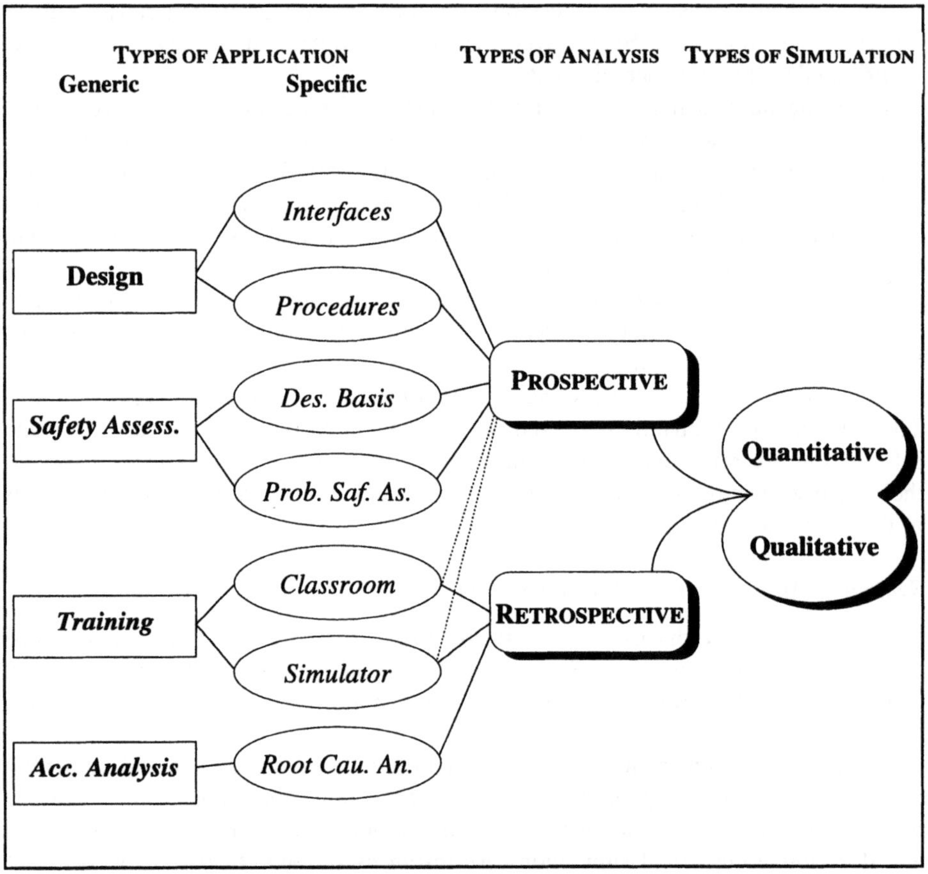

Figure 1.3. Types of applications, types of analysis and types of simulation of human-machine systems.

According to the foregoing discussion, Figure 1.3 shows the equivalence of qualitative and quantitative types of simulation for prospective and retrospective types of analysis.

Moreover, it can help in identifying the type of simulation and analysis that should be developed or exploited when a certain application is required.

As an example, a probabilistic safety assessment of a human-machine system needs to be performed by a prospective type of analysis and either a qualitative or a quantitative simulation can be used, according to the detailed specifications of the problem under study.

We will return in much more detail, in the forthcoming Chapters, on the subject of types of cognitive simulation, types of analysis and on generic and specific applications for design purposes, safety, training, and accident investigation.

In the next section we will focus on the elements of the Human-Machine Interaction Architecture. In particular we will describe the models of human cognition and plant behaviour and the working environment, as basic constituents of high-tech control contexts.

1.3 Elements of HMI Simulation

In this section we will concentrate on the paradigm of human behaviour as the information processing metaphor. Particular attention will be dedicated to the new cognitive features of the human operator, and to the formulation of an architecture of human-machine system simulation able to accommodate and fully support such new features.

We have already considered in the previous section the main applications of human-machine simulations, the generic types of human behaviour simulations that exist, and the types of analyses that can be performed.

We will now tackle the following questions, specifically related to the content of cognitive simulation:

- *What* is meant by simulation of human cognition in the context of modern system control environment.
- *What* are the elements of human-machine system simulation.
- *How* models are actually transformed into computerised, numerical simulations.

This analysis will lead to the definition of a set of requirements, specifications and conditions that characterise a new basic framework for the simulation of human-machine systems.

1.3.1 Micro-Cognition, Macro-Cognition and Simulation

The study of cognition can be approached from many different perspectives. It can be viewed as an objective "per-se", without any specific linkage or application to practical working domain.

In a simulation context, such as the one under development here, one is mostly concerned with the effects that cognitive processes produce in terms of plant control and other perceptual-motor skills.

This distinction is very important, as it implies two different lines of development, which can be identified with the *micro-cognition* and *macro-cognition* (Cacciabue and Hollnagel, 1995):

> ***Micro-cognition*** **denotes the academic pursuit of modelling psychological phenomena.**
>
> ***Macro-cognition*** **lies, instead, in an engineering context, where the primary purpose is to construct a representation of a phenomenon or system. This is then used to calculate or predict how the system will develop.**

Micro-cognition

Micro-cognition is intended as the detailed theoretical account of how cognition takes place in the human mind. The emphasis is on the scientific use of models, which are employed as a means for achieving a deeper understanding of a phenomenon. A model explains the phenomenon in question and provides an account of the mechanisms or functions, i.e., the causal or functional architecture, that underlie the phenomenon.

The focus is on the "mechanisms of intelligence", rather than the way the human mind works, which identifies the objectives of Artificial Intelligence (AI). Micro-cognition is concerned with building theories for specific phenomena and with correlating such theories with available empirical and experimental evidence.

Typical examples of micro-cognition are studies of human memory, problem solving in confined environments, learning and forgetting in specific tasks, language understanding, and so on. Many problems that are investigated are substantially realistic as they correspond to real-life situations in generic terms: any empirical observation of human-machine interaction can be abstracted in terms of cognitive processes.

Therefore, for any event studied in terms of micro-cognition the emphasis is on experimental control, on predictability by a well formalised paradigm and model or theory that go in depth into cognitive mechanisms or processes.

Theories of micro-cognition are the underlying mechanisms in most of the early AI systems, such as Epam (Feigenbaum and Simon, 1984) and the General Problem Solver (Newell and Simon, 1963). They then evolved in models such as the Human Information Processor (Card, Moran and Newell, 1983), and more recently have reached their apogee in the Unified Theory of Cognition - SOAR (Newell, 1990).

Macro-cognition

Macro-cognition refers to the study of the role of cognition in real tasks, that is in interaction with the environment. Macro-cognition only rarely looks at phenomena that take place exclusively within the human mind or without overt interaction. It is thus more concerned with human performance under actual working conditions than with controlled experiments.

Macro-cognition emphasises the external validity, the regularity across conditions and goes in breadth as to cover the widest possible spectrum of predictable behaviour under well defined environmental conditions.

Typical examples of macro-cognition are diagnosing accidental/incidental conditions, controlling an industrial process, operating an aircraft, writing a program, designing, planning a mission, and so on.

The simulation is a correct translation of essential functional relationships and dependencies into a form which enables controlled manipulation of the independent parameters (including the environment). An engineering model does not necessarily constitute an explanation. Conceptually speaking, it can either be a black box, a finite state automaton, or a convenient representation of input-output correlation.

Examples are neural networks, mathematical models (differential equations), linear regression models, supervisory control models, Kalman filters, etc.

Some phenomena may, in principle, belong to both categories. Examples are problem solving, decision making, communication, information retrieval, and so on. But if they are treated as macro-cognition the interest is more on *how* they are performed and how well they serve to achieve their goals, rather than on the details of *what* goes on in the mind while they are performed.

The simulation of macro-cognition is therefore radically different from the simulation of micro-cognition. As yet, only a few cases of macro-cognition are found in literature and some of them will be described and compared in Chapter 3 (Woods, Roth and People, 1987; Cacciabue and colleagues, 1992a; Amalberti and Deblon, 1992; Takano, Sasou and Yoshimura, 1995).

Simulation of Cognition

In the context of this book we will concentrate on *simulation* of (macro) cognition, in the sense that simulation is always developed with a specific practical purpose in mind.

However, we will not advocate the behaviourist "black box" approach for which one can analyse the behaviour of a system by measuring or observing simply input and outputs (stimuli and response) and, from these, infer transfer relations used to generalise. We will also be concerned with the mechanisms that are inside the box, only as far as they are able to explain, describe and, eventually, predict the resulting behaviour.

The concern is therefore to maintain a sufficient correspondence between the known or observed regularities of the cognitive processes and the outcome of the simulation. In other words, the simulations of interest here are concerned with the description of actual human behaviour, which includes behavioural and cognitive processes as they are described by a theoretical model.

A definition of *simulation of cognition*, also referred as cognitive simulation, that extends the definition already given and suits the above discussion, including the principle on simulation of human-machine system enunciated in the previous section, has been given by Cacciabue and Hollnagel (1995).

This definition is here reported for completeness:

> **"The simulation of cognition can be defined as the replication, by means of computer programs, of the performance of a person (or a group of persons) in a selected set of situations. The simulation must stipulate, in a pre-defined mode of representation, the way in which the person (or persons) will respond to given events. The minimum requirement to the simulation is that it produces the response the person would give. In addition the simulation may also produce a trace of the changing internal mental states of the person"** (p. 58).

1.3.2 Theoretical Content of HMI Simulation

Having defined what is meant by simulation of cognition, this section deals with the identification of the main constituents of a human-machine system.

In practice we will try to discuss the detail content of the two basic elements of the *Human-Machine Interaction Architecture* described in Figure 1.1. Before entering this elaborate analysis, it is important to re-examine the scope and limits of application of simulations of cognition, in the perspective of macro-cognition and human information processing metaphor.

Hollnagel's "Minimal Modelling Manifesto"

The *Information Processing System*, chosen as reference analogy for human behaviour modelling, implies that specific attention is devoted to the stages or mental operations that occur between the stimuli and responses.

In modelling these specific activities it is important to remember that our objectives are focused on what has been defined as *macro-cognition*, as to mean that the target of the simulation is the "variety of human performance" that matches the observed experience and empirical studies. The simulation of cognition should aim at minimising the theoretical effort when describing the essential phenomena and processes relative to the domain under study, rather than attempting to model all types of theoretically conceivable and possible performances. The validity and scope of the model can be evaluated by the degree of agreement between the predicted and actual performance given certain contextual working conditions.

This principle was found by Hollnagel (1993a) to satisfy very well the law of Requisite Variety of Ashby (1956), which states that "*only variety in R* (Regulator) *can force down the variety due to D* (Domain under control); **only variety can destroy variety**" (p. 207).

At first glance, this law may seem to imply exactly the opposite to what we are trying to demonstrate. But, taking a closer look, one has to consider that, here, our target (i.e., the Domain under control) is the cognitive simulation, while the Regulator is the human performance, as it can be ascertained from experience. The main requirement for controlling variety in the simulation is to balance the observed variety of human behaviour: by increasing the amount of observed human performances, we decrease the variety of possible hypothetical cognitive patterns to be simulated!

This principle was further developed by Conant and Ashby (1970) in the following rule: "*every good regulator of a system must be a model of that system*".

From these consideration Hollnagel (1993b) developed a "Minimal Modelling Manifesto", which reads as follows (p. 379):

> **"A Minimal Model is a representation of the main principles of control and regulation that are established for a domain - as well as for the capabilities and limitations of the controlling system".**

According to this definition, a minimal model is specific for a given domain, such as aviation or nuclear reactors, and for a specific application, e.g., during normal operations and emergency conditions.

However, as strong similarities and commonalties can be identified between different applications and domains, such as, for example, the exploitation of automation and supervisory control, the development of *"common" minimal model* can, in some cases, be envisaged.

Elements of Human Model

We will now try to identify a number of minimal basic cognitive processes and functions of cognition which are able to sustain a simulation of cognition, in accordance with the principle of the minimal modelling manifesto.

Primarily, the simulation of interest here shows two elements that are generally accepted. These are: *Stimuli* and *Responses.*

Stimuli are generated by the machine under control or by the working environment and the context affecting the human operators, and *Responses* are the control or manual actions carried out by humans in response to stimuli and/or following a cognitive process.

A first difference with behaviourist approaches can be identified, as we do not necessarily imply that human response is directly engendered by a physical event occurring in the plant, but it may rather be derived from another reasoning process or planning developed at a different time, and not necessarily connected to the ongoing physical process or plant behaviour.

Moreover, the concept of *stimulus* is quite substantially enlarged, as to include not only the cues coming from the plant but also the vast amount of information, non-technical signs and "background noise" due to the entire working environment. Examples of these types of stimuli are external events, organisational and cultural factors, and plant generic operating conditions.

The main characteristic of the information process paradigm of human behaviour is the attention given to modelling functions or mental processes which occur between stimuli and responses. The substantial differences between cognitive and behaviourist approaches reside in these models. Following the principle of minimal modelling manifesto, there is a minimum number of functions and cognitive processes that need to be included in a simulation approach to describe actual human behaviour.

In particular, four main cognitive functions are considered sufficient. These four functions are: *Perception, Interpretation, Planning* and *Execution* (*PIPE*) (Figure 1.4).

In addition to these four main cognitive functions, two critical cognitive processes of allocation of knowledge and resources are considered as reference: *Memory/Knowledge Base* (*KB*) and *Allocation of Resources* (*AoR*).

These functions and processes are not chosen at random, but they reflect the consensus on the characteristics of human cognition that has developed over many years (Sheridan and Ferrell, 1974; Pew and Baron, 1983; Rasmussen, 1986; Sheridan, 1985; Hollnagel, 1993a, 1993b) and can be considered a logical extension of the notion of the perceptual cycle described by Neisser (1976).

We will now look in more detail at each cognitive function and process that support the model, as they are represented in Figure 1.4.

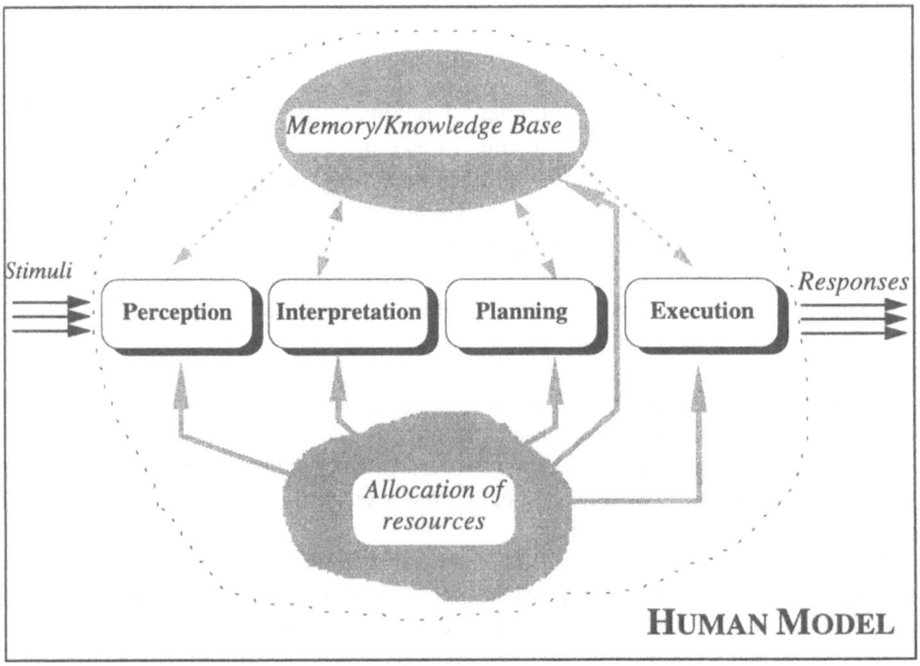

Figure 1.4. A model of Information Processing.

The first function of cognition is *Perception* of information. Perception is related to the content of information transferred to the human sensory activity, as referring to the normal sensory receptors (vision, hearing, taste, touch and smell). This sensorial function is to be combined with a contribution due to expectations which represent the cognitive support to the search for and acquisition of information for further processing.

The second cognitive function, *Interpretation*, consists in the elaboration of perceived information, in the form of identification of cues from the environment and comparison with the already acquired and formalised knowledge.

The third function, *Planning*, implies a choice or a development of a plan for control or manual action to be carried out, as result of the previous steps of cognitive processes.

Finally, the fourth function, *Execution*, entails the implementation of the decision, which may take the form of manual responses or control actions, but may also simply be the beginning of a new cognitive process.

As for the two fundamental cognitive processes of allocation of knowledge and resources the following short descriptions can be given. The first process, *Memory/Knowledge Base*, embodies previous experience, knowledge acquired during training and practice, and theoretical know-how. The type of knowledge contained in the Memory/Knowledge Base is made of rules, beliefs, procedures, physical principles etc. and supports the four cognitive functions.

Allocation of Resources is a process even more crucial for the whole cognitive model, as it describes how the resources available to the human are distributed throughout the model and how they affect the cognitive functions. This process is influenced by the context and environment and aims only at representing a sort of meta-cognition, with no other complex function (Amalberti, 1996; Wickens, 1984; Wickens and Flach, 1988). *AoR* affects the human model in two ways:

1. Firstly, it defines the level of operability of the cognitive functions and their relative sequence.
2. Secondly, it affects the amount and type of knowledge available for the cognitive processes.

Mechanisms of Human Model

In order to better define the elements of a human-machine system, let us consider more closely the main cognitive functions and processes defined in the previous section.

In particular, it may be noticed that in Figure 1.4 no connections between the four *PIPE* functions have been shown and that no discussion has been carried out concerning the *mechanisms* that generate human behaviour and govern the interaction of knowledge base, cognitive functions, and cognitive processes.

This has been done on purpose, as these mechanisms are very important and play a crucial role in the architecture of the model and should not be mixed with the elements that compose the cognitive model itself, i.e., the *PIPE* cognitive functions and the *KB* and *AoR* cognitive processes.

As an example, a cognitive mechanism is usually performed over several non-procedural recurrent loops of reasoning, interpretation and observation. The simulation of cognition should, thus, be able to reproduce such mechanism by the cyclical interaction of cognitive functions and, also, by generating non-sequential links between any two functions of cognition (non-procedural thinking).

Therefore, as a further step of development of the human model, the definition of the interconnections amongst cognitive functions can be formalised as shown in Figure 1.5.

The links between the fundamental cognitive processes of *Allocation of Resources* and *Memory/Knowledge Base* are clearly defined. In particular, as already mentioned in the previous section, the Allocation of Resources influences the way to perform any of the other cognitive processes and functions.

Memory and *Knowledge Base* are only activated by the two cognitive functions of *Interpretation* and *Planning*, as these are considered "high level" mental activities, while *Perception* and *Execution* are mainly related to sensorial and "low level" mental activities. However, *Memory/Knowledge Base* can feed data and information to all four cognitive functions in order to support their performances.

The *Stimuli* and *Responses* appear as the sensory motor connection of the operator with the machine.

The effects of the *Context* and *Environment* are considered on the Allocation of Resources and on the cognitive functions. In this way the socio-technical context can modify the whole cognitive process and can be linked to specific behaviour like human performance.

The connections amongst the four cognitive functions of *Perception, Interpretation, Planning* and *Execution* are established to maintain the cyclical nature of cognition, by which information perceived from external stimuli is combined with reasoning over past events and with anticipation about plant future response, to produce further perceptions, reasoning, planning etc. This process is simulated by the repetition of the loop of the 4 *PIPE* cognitive functions until the execution of an action (*response*) is generated.

Another way to see this process is also to consider that the output of the *Execution* function is either an action (*response*) or the initiation of a new loop of perception of data and management of information.

In the modelling architecture, also the sequential and by-pass processes between functions are represented. For example, the direct link between *Perception* and *Execution* is clearly shown. These shortcuts between the functions of cognition may be due to the effect of a specific allocation of resources or may be the consequence of the knowledge available in the memory storage.

It is important to observe that no direct links have been postulated between *Perception* and *Planning* or *Interpretation* and *Execution*. This is because it is assumed that the two functions of *Interpretation* and *Planning* are strictly related to each other and, together, interact with *Memory/Knowledge Base* to give rise to the highest level of mental process.

In other words, the human model assumes that, once the interpretation function takes place a (mini-)mental loop is started that eventually results in an execution, but only after a process of planning and interaction with the memory and knowledge base.

The function *Execution* can therefore be reached in two ways: either by direct connection with the perception of information or by a cognitive process of interpretation and reasoning, which involve also the memory and knowledge base.

The *Memory/Knowledge Base*, once activated by *Interpretation* or *Planning*, can affect or act on any of the four functions of cognition. This means that once human memory and knowledge are explored for further processing, any function of cognition can be activated as result of this process.

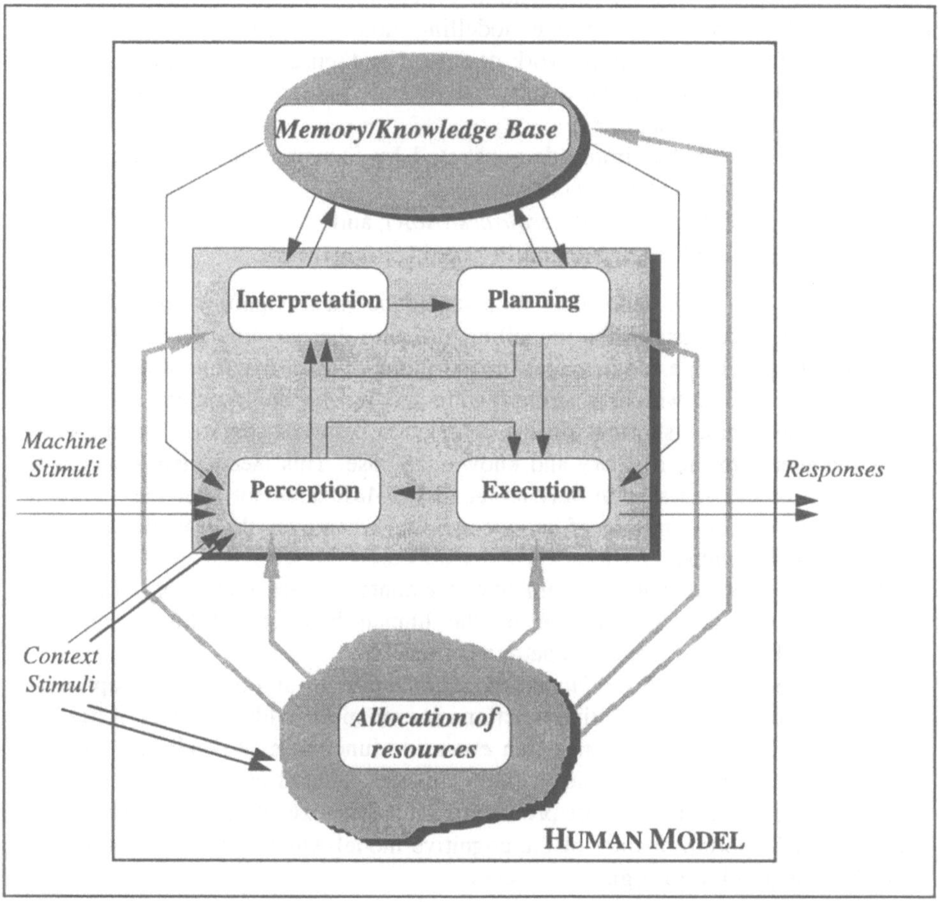

Figure 1.5. Structure of Human Model in a Simulation Architecture.

In summary, the basic requirement for the development of a human model is the consideration of the four cognitive functions Perception, Interpretation, Planning and Execution and the two basic processes Knowledge Base/Memory and Allocation of Resources. This type of human model can be integrated in an architecture of human-machine system simulation, where data are exchanged between machine and human models in the form of stimuli, responses and contextual effects.

The modelling elements and structure described in this section, and outlined in Figure 1.5, constitute the *Reference Model of Cognition (RMC)*, or *human model of reference*, that will be considered all throughout this book.

Human Model of Reference and IPS Metaphor

The *Reference Model of Cognition* discussed in the previous sections was developed from the standpoint that the Information Processing System (IPS) metaphor is a general paradigm on which cognitive modelling should be based.

However, the modelling framework described in Figure 1.5 encompass cognitive functions, processes and mechanisms that extend quite considerably beyond the boundaries of the IPS metaphor.

This argument can be simply demonstrated by focusing on two aspects of the human model described above, namely:

- The process of *Allocation of Resources* (*AoR*); and
- The mechanism of connection of the cognitive functions.

The process of *AoR* represents in essence the meta-cognition element of the human model. It consists of the modelling structure that governs the distribution of the available knowledge and manages the unfolding of cognitive functions.

The *AoR* process, which is strongly affected by the environment and context conditions, determines shortcuts in the connection between cognitive functions and limits the access to the memory and knowledge base. This mechanism matches the information processing system only as far as the *AoR* does not alter the "nominal" sequence of functions and use of knowledge, i.e., when everything works according to design specifications.

However, as soon as the environment or the context affect the process of *AoR*, the equivalence with the IPS is lost, as the human behaviour follows paths and mechanisms that do not match the patterns of an IPS.

The difference between the *human model of reference* and the IPS metaphor due to the connections of the cognitive functions is even more evident. In fact, it has been argued that, in some circumstances, an *execution* function may result in a complete or partial repetition of the cognitive loop. This mechanisms can be repeated several times, with different paths, before producing a manifest execution of an action.

This is an important feature of the cognitive model which is not contemplated in an IPS and is peculiar to cognitive processes.

Although the IPS metaphor remains a milestone in the evolution of cognitive modelling and we will continue to refer to it as a reference paradigm of human behaviour, it can be concluded that the *Reference Model of Cognition* considered here extends beyond the concepts of the IPS and differs from it for important features typical of cognitive systems but not so characteristic for technical systems.

We return now to the discussion on the Human-Machine Simulation Architecture.

Elements and Mechanisms of Machine Model

We will now focus on the machine element of the simulation architecture of Figure 1.1.

The methods most commonly applied to describe the behaviour of plants and working environments at theoretical level are balance equations, conservation principles and logical correlations (models). Numerical or analytical algorithms are then applied for the development of simulations, that transform the models in computerised applications, exploiting the power of approximate techniques and computer technology.

In particular, the *Plant* performance (Figure 1.6) is modelled in consideration of: a) the evolution of its physical and state variables; b) the functioning states of its components; and c) the control actions (*Responses*) performed by the operator.

The operator actions are carried out through the *Actuators* available for direct control.

Following operator actions and plant physical evolution, a new set of physical quantities, states of components and relative indicators readings are calculated by the machine simulation. These are the primary source of information available to the operator, and are the main, but not the only, *stimuli* directed to the perception function of the human model.

Other types of information about plant behaviour that reach the operator are graphical representations of main circuits and components, suggestions through decision support, alarms, procedures, and so on. The whole body of information is available to the operators through the *Interfaces*.

In summary, the plant behaviour is modelled and simulated by the evolution of its basic physical variables, actuators and interfaces.

Another essential element of the machine model is the *context* and working *environment*. This needs to be modelled and simulated by logical and analytical correlations, which account for a variety of contributors:

- The presence of other operators in the control room, or in other distant locations, which affect the allocation of resources and may generate blunders.
- The effect, on cognitive processes and functions, of personality factors, management, training, real experience and social context as a whole. This set of factors affect primarily the process of resources allocation.
- The effect of external factors and external events, like noise, humidity, temperature and physical phenomena, which may limit or alter human performance.

Usually, when a Machine Model is developed with the objective of performing predictive analyses, the major effort is dedicated to the model of the plant and its interfaces and control system. However, as the focus of human-machine studies are cognitive human performances, the need has now risen to consider the Context and Environment, affecting human behaviour, on the same level of detail as the plant evolution.

As there is no real feedback of human actions on the context and environment, its simulation can be performed almost independently from the mainstream calculation of human and plant transient evolution. What should be preserved is the synchronisation between human and machine interactions, so that logical and realistic simulation is performed.

In most cases, the methods and algorithms that are utilised for the simulation of the context are very simple and are limited to the description of major events and phenomena affecting human, and more rarely, plant behaviour.

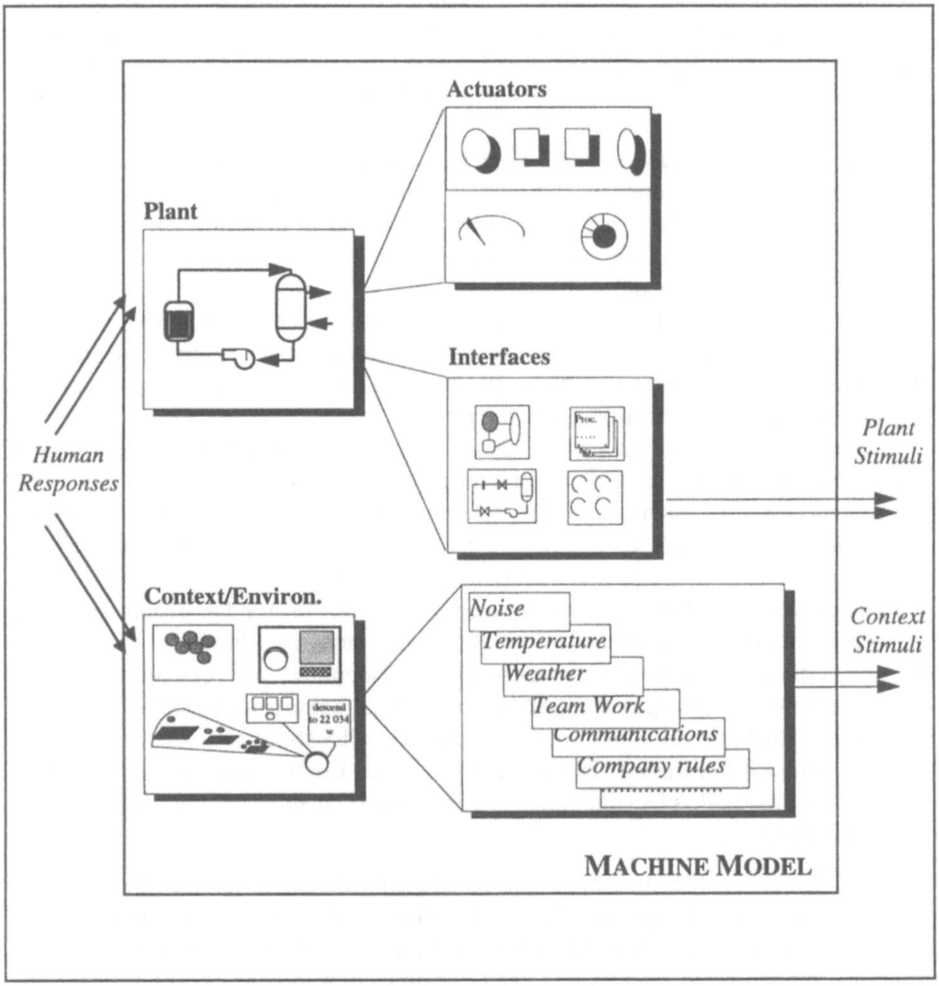

Figure 1.6. Structure of Machine Model in a Simulation Architecture.

In summary, the main requirement that need to be satisfied in order to develop a Machine Model, able to interact with a cognitive human model, is the inclusion in the modelling framework of:

- "classical" deterministic models of functional, physical and state variables;
- models of actuators and indicators generating interfaces and responses; and
- models of context and environment for creating external conditions that may affect human behaviour and plant performance.

1.3.3 Practical Implementation of HMI Simulation

The set of cognitive functions and processes that sustain the *reference model of cognition* described in Figure 1.5 is quite standard and, as already mentioned, it follows for many aspects the most known paradigms developed during the last two decades by many authors.

The important question concerns **how** these cognitive functions and processes are transformed into computerised numerical expressions. Many methods and techniques are available. As an example, Rasmussen (1986) gives a catalogue of techniques and approaches with reference to his well known Skill-, Rule-, and Knowledge-base (SRK) framework.

These approaches cover a very wide range of formal logic and numerical theories, like information theory, signal detection, manual control, and decision theory, optimal control theory, fuzzy sets and artificial intelligence, scripts and plans. These methods are very useful in that they allow the development of cognitive functions and processes into some formats that can be dealt with by numbers and logical expressions. Therefore, they lead to the formulation of simulation forms that can be handled by a computer, once transformed into a software program using an appropriate programming language. In this way, quantitative numerical predictions can be obtained.

However, in a modelling perspective, the support of cognitive psychology is needed for the final development of a simulation. Cognitive psychology can give an answer to questions like "How do we evaluate which are the quantities actually perceived by an operator ?" or "How does one deal with different contrasting information ?" and "How does the model complete the evaluation of hypotheses and goes on with the execution of a control action ?"

In practice, by supplying the theoretical background that gives an answer to these questions, cognitive psychology contributes to the formulation of the connections mechanisms between different cognitive functions and processes. In more general terms, cognitive psychology comes in support in closing the gap between modelling methods and simulation formats.

Examples of useful theories of cognitive behaviour are Reason's "Cognitive underspecification theory" (Reason, 1990), Hollnagel's "Contextual control theory" (Hollnagel, 1993a), and Searle's "Intention and action theory" (Searle, 1983). These theories allow the identification of "conservation principles of cognition" which can play the role of invariant and thus can be considered as mechanisms of general application. They are also defined "*computational 'primitives' of the cognitive system*" (Reason, 1990, p. 103).

The primitives of cognition act as *filters* for the modelling process, in order to discriminate and select the activities that are actually carried out during the simulation of human-machine interaction.

We will make large use of these theories in developing practical means to "calculate" crucial parts of the human behaviour. These means are the specifications that need to be applied when developing simulations, and they will be discussed in more detail in the next section.

Simulation of Cognitive Functions

We will now consider how the various theories, that can be applied in formalising a cognitive model, are simplified and adapted for the development of simulations.

In particular, we will focus on the theoretical construct of the human model of Figure 1.5, and on the four cognitive functions and basic processes, that support the model.

Figure 1.7 shows the connections existing between the cognitive functions and the primitives of cognition or psychological hypotheses that sustain each function, and the computational means that can be used for the numerical development of the simulation.

<u>The Function Perception</u>

The function *Perception* implies that some, or all, the information and cues produced by the models of plant and context are actually perceived by the human sensory activity.

A mechanism has to be devised to enable to distinguish between cues that are actually perceived and cues that are not. One criterion is the definition of *salience*, which is a parameter that regulates the information content attributed to each "cue of the world", i.e., the cues coming from the plant, the context and the working environment. These cues are calculated by the "machine" element of the human-machine system.

There are two types of salience: the *physical salience*, which relates to the physical appearance of cues, and the *cognitive salience*, which refers to their cognitive relevance.

Cognitive salience is related to the knowledge base and previous experience of humans and to the history of the human-machine interaction under analysis, i.e., the immediate past and present trend of the transient under study.

Physical salience depends on the sensory relevance of signals or information, such as an indicator or an alarm of safety state of the system. The physical salience of a cue contributes to the relevance of the stimulus and, thus, one can say that the salience characterises the *stimulus* received from the plant information system.

The overall value of salience is a function of two variables, i.e., the cognitive and the physical salience, and can be calculated by fuzzy membership functions or by "weighted" arithmetical expressions.

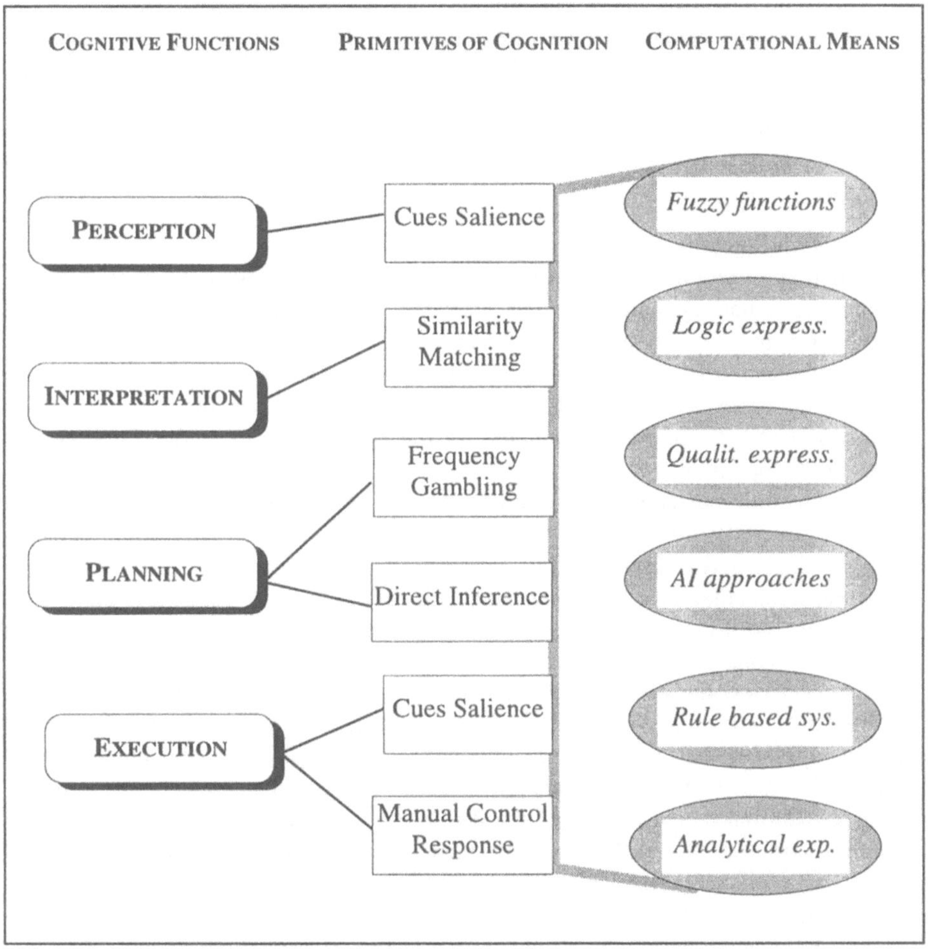

Figure 1.7. Cognitive Functions, Primitives of Cognition and Computational Means.

<u>The Function Interpretation</u>

The function *Interpretation* requires that the perceived information is matched with the knowledge existing in the operator memory/knowledge base.

The quantity and quality of the perceived cues contribute to the matching mechanisms. The more pertinent and precise are these cues, the less numerous and more accurately supported will be the hypotheses selected from the knowledge base. In any case, interpretation is a process that matches perceived cues with the existing knowledge.

In practice, full matching between perceived and known world is very rare and usually, either more than one diagnosis results from the interpretation process or, sometimes, no diagnosis can be performed.

This matching mechanism can be simulated, for example, by the cognitive primitive of "Similarity Matching" (Reason, 1990), even if other matching mechanisms can be found in the literature (Helander, 1988).

The implementation of matching mechanisms can be performed by applying production rules or fuzzy membership functions or analytical and logical expressions.

The Function Planning

The function *Planning* deals with cases of conflict or ambiguity between different matching alternatives and with the selection of the sequence of actions and interventions to carry out in order to satisfy the aims of a task.

A mechanism of selection proposed by Reason (1990) follows the cognitive primitive of "Frequency Gambling", which tends to favour the similarity with situations more frequently encountered in the past and, thus, more familiar and known to the operator.

This selection mechanism, being a primitive of cognition, governs many different aspects of everyday life, and may be applied when a selection is possible between different alternatives. However, it can also be used when no alternatives are available, as no diagnosis has been made. In this case the "Frequency Gambling" primitive would favour the plan of actions most familiar to the operator in the wider context of the ongoing human-machine interaction and control process.

In some cases, the planning process demands more than one selection between different alternatives. A complex reasoning and decision process may then be carried out, by applying basic principles, knowledge about physical phenomena, plant functional and structural behaviour. This process may be called *Direct Inference*.

Direct Inference requires the interaction between the cognitive function of planning and the memory/knowledge base. This process may be lengthy and repetitive, before achieving a plan of responses and actions.

Planning and choice can be simulated by analytical and fuzzy comparison between knowledge base content and results of the interpretation process. Reason's cognitive primitive of "Frequency Gambling" can be implemented by "tagging" knowledge "chunks" and "frames" according to characteristic aspects of the humans being simulated, such as experience, amount and type of knowledge, culture, etc.

In the case of *Direct Inference*, more complex computational methods are necessary, like logic and fuzzy theories, rule based systems and artificial intelligence approaches. The latter are very popular nowadays and make use of production rules and, usually, object oriented programming languages.

The Function Execution

The function *Execution* is the immediate result of the detection-interpretation-planning process.

This consists in the *response* of the human model, formalised either as a specific control/manual action, or as the development of new hypotheses relative to the state and performance of the plant.

In particular, as diagnosis and planning are dynamic processes, carried out over extended periods of time, several observations and change of planning may occur before any action is taken.

Consequently, a typical output of execution is the evaluation of hypothesis. This requires the repetition of the loop and the research of further information, with new values for the cognitive salience associated to certain cues. In this case, the objective of the operator is the further confirmation of the diagnosis and task resulted from the previously performed planning process or the acquisition of new information for further planning of actions and interventions.

Therefore, the level of confidence in one particular explanation or hypothesis can be dynamically modified and adjusted according to the interaction loop of human and machine.

The execution function can be simulated by rule based mechanisms, which are particularly qualified to represent control procedures and specific actions, carried out in the form of *responses* of the human model. Fuzzy membership functions and analytic expressions may be utilised for the repetition of the loops of collection of new information and confirmation of hypotheses.

Summary

Figure 1.7 summarises the foregoing discussion. In particular, two specific considerations can be made:

1. Firstly, each cognitive function is supported by a well identified set of primitives of cognition that fulfil its requirements. As an example, Direct Inference is applied for developing *Planning*, while Similarity Matching can be used for *Interpretation*.
2. Secondly, the computational means that can be implemented for the development of models into simulations are of generic application. They cover a wide range of methods, from fuzzy functions to logic, analytical and qualitative expressions, to artificial intelligence approaches and rule based systems.

The spectrum of primitives of cognition and computational means presented in Figure 1.7 is not necessarily exhaustive of all possible theoretical constructs and analytical methods that can be applied for developing a human simulation.

However, what is important here is that it has been shown that there exist in literature a number of well developed methods and numerical formalisms, which are able to transform a theory and a model of human behaviour into a computerised set of instructions and algorithms, by which it is possible to describe and predict such behaviour, in response to the stimuli of a machine and environment.

Simulation of Knowledge Base and Allocation of Resources

Knowledge Base (*KB*) and *Allocation of Resources* (*AoR*) are crucially important cognitive processes in the economy of the cognitive model. They define the complete set of available knowledge, the amount and form of knowledge, which is actually utilised in a dynamic context condition, and they may affect the dynamic unfolding of the cognitive functions during a human-machine interaction.

Figure 1.8 shows the primitives of cognition and the computational means most commonly adopted for modelling and simulating these two cognitive processes.

Memory/Knowledge Base

The *Memory/Knowledge Base* is the combination of limited, serial and computationally powerful working memory and a virtually limitless knowledge base.

The latter is a repository of structured expertise gained by the operator during everyday work, formal and practical training. The knowledge base contains: a) some of the main procedures, or at least the main conditions to identify and initiate control or emergency procedures; b) the behaviour of specific parts of the control system; c) the response and basic behaviour of many individual components; d) the physical relations between variables and plant parameters; and, finally, e) a vast number of heuristics or rules derived from training and empirical/practical experience.

As the content of knowledge base is vast, it can affect and feed back information to any of the four cognitive functions.

The computational forms of the content of the knowledge base are production rules, logical, analytical and numerical algorithms, fuzzy rules and membership functions, qualitative descriptions.

Allocation of Resources

The *Allocation of Resources* is crucially important as it governs the forms in which the knowledge base is actually utilised by the cognitive functions and how these interact dynamically with each other and with the plant under control.

The process of allocation of resources influences all the other cognitive processes and functions of the model. This process is at the core of the whole simulation, as it can affect the output of a simulation according to the type of control modes applied by the human model. Different types of control strategies allow to simulate many control attitudes and control means.

In defining the control modes of the simulation, it is very important to ascertain as closely as possible the effects of the context and working environment. These conditions, together with personal attitudes, generate different control modes and implicitly show the effect of the environment on the human behaviour.

The effects of context and environment on the mechanisms of allocation of resources are described and computerised through rules and numerical algorithms that compare different control strategies with context/environment variables. In this way the actual sequence of functions and the available knowledge can be simulated.

The same computational algorithms utilised for memory/knowledge base, such as fuzzy functions, logic and qualitative expressions etc., can be applied for developing simulation of the process *Allocation of Resources*.

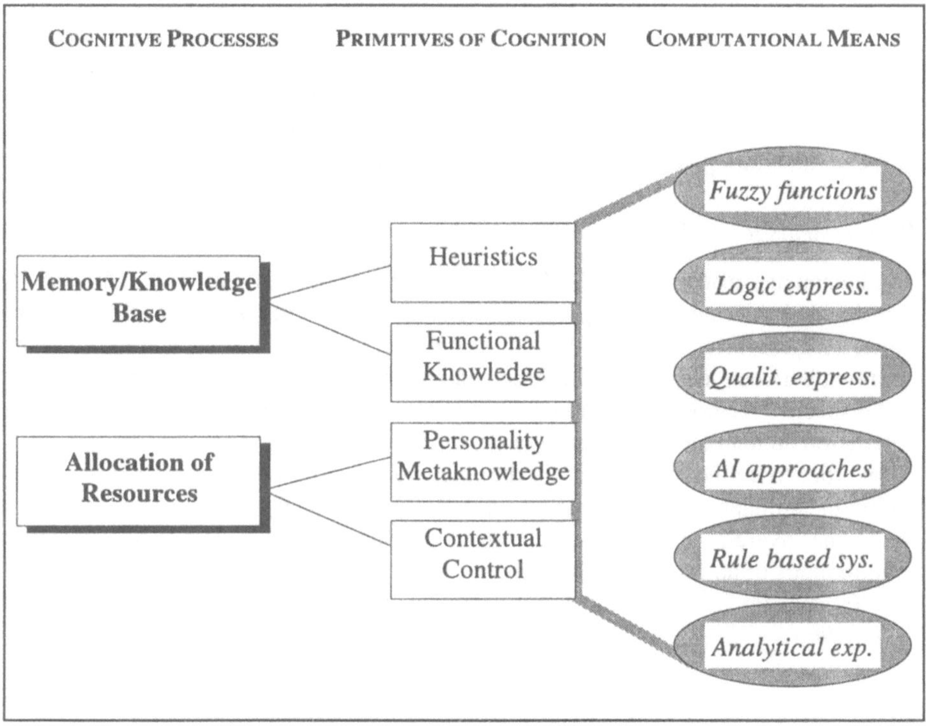

Figure 1.8. Cognitive Processes, Primitives of Cognition and Computational Means.

1.4 An Architecture for HMI Simulation

In the preceding sections we have analysed in detail the two constituents of the Human-Machine Interaction Architecture, as defined by Rouse (1980) (Figure 1.1). The techniques and algorithms to transform models into simulations have been presented and discussed.

The complexity of these models demands that an appropriate mechanism of interaction is devised, that deals with the issue of managing the exchange of data, the dynamic synchronisation of simulation and the logical and stochastic behaviour of the constituents of the human-machine system.

This means that the Human-Machine Interaction Architecture proposed by Rouse (Figure 1.1) may be questioned, because of its extreme simplicity. In particular, the need to develop a specific model for managing the human-machine interaction has to be considered.

1.4.1 Interaction Model and HMI Simulation

The first two issues to discuss, in relation to the Human-Machine Interaction Architecture described in Figure 1.1, are: 1) the reasons why a simple direct connection between human and machine behaviour models is not sufficient anymore; and 2) which characteristics should show a more elaborated and specific constituent of the architecture that manages the interaction between human and machine models.

Requirements for an Interaction Model

The human and machine models generate a variety of distinct outputs that are strictly correlated, but require an appropriate filtering before they are combined to each other. This characteristic of the human-machine system implies that a straightforward coupling between human and machine behaviour simulations may not be established and a specific method has to de developed. This new element of the human-machine simulation architecture is the *Interaction Model.*

In order to define more precisely the role and objectives of the Interaction Model, within the simulation architecture, a number of requirements can be identified which make the overall simulation more effective and practical. The most important requirements of an Interaction Model are:

1. To manage the time evolution and synchronisation of the human-machine system.
2. To handle the exchange of data between the human and machine (plant and working environment) models.
3. To manage the stochastic behaviour of human and machine and the time dependent occurrence of errors and malfunctions

Firstly, the dynamic aspects and time dependence of interactions represent an element of complexity for the overall architecture. The time integration algorithms need to account for the effects of human action, plant response and environment evolution, which may not always occur in sequence, and usually evolve in parallel with different time constants. The overall output depends on the time management procedure and the accuracy of the simulations:

- A *first requirement* for the Interaction Model is thus to be able to manage the time evolution of the human-machine system. This requirement is identified as: *Time Management.*

Secondly, the configuration of human-machine systems is usually made of a simulation of the plant under control and a simulation of single human behaviour, or sometimes of a team. Rarely, a complete model of the working environment is simulated in parallel to the models of human and plant. As we showed earlier, when discussing the structure of the machine model, the environment evolves almost independently of the interaction between operators and plant.

It is then up to the model of interaction to trigger off occurrences in the environment, that may affect human or plant behaviour, and interface them with the rest of the system simulation.

To further clarify this point, it is sufficient to consider the example of the architecture of a flight simulator. Advanced training simulators consider two piloting positions and flight control systems, but the interaction with the environment, like air traffic controllers or changing weather, are not directly accounted for by the simulation. Usually, the instructor plays the role of the air traffic controller or injects changing atmospheric conditions through the side control panel, specifically dedicated to generate extraneous events or activity external to the cockpit.

The effects of this type of working conditions and external environment can be accounted for by the interaction model, usually by simple algorithms, very specific to the technical domain under study:

- A *second requirement* for the Interaction Model is thus to handle the interface between the models of working environment humans and plants, by considering the logical connections that exist between interconnected events. This requirement is identified as: *Logical Human-Machine Interaction.*

A third crucial feature of the Human-Machine Interaction Architecture under development is the ability to describe and predict interactions in presence of human errors, component failures and malfunctions. This requirement is related to the application of the Human-Machine Interaction Architecture for accident analysis and safety assessment studies.

In this case, the interaction model must contain a structured set of logical and probabilistic algorithms that support stochastic consideration and likelihood of failure and error occurrences.

This model should be able, in principle, to combine statistical theories with deterministic considerations about the causes and times of occurrences. Moreover, it needs to be fed with the appropriate data about failure probabilities, failure types and failure modes.

In this way the interaction model is in a position to manage the human-machine interaction by generating malfunctions of components and human erroneous manifestations as time evolves and as the working environment and context dynamically develop:

- This third fundamental *requirement* can then be summarised in the need for the Interaction Model to manage the stochastic behaviour of human and machine in dynamic conditions. This requirement is identified as : *Dynamic Reliability.*

The result of this discussion is that an appropriate *Interaction Model* becomes necessary, and needs to be formally included in the loop of the human-machine simulation architecture.

New Human-Machine Interaction Architecture

If one takes into consideration these requirements and combines them with the preceding discussion about the characteristics and requirements of human and machine behaviour models, a new architecture of human-machine system simulation can be developed.

Figure 1.9 shows this architecture, which exhibits some important differences from the original one of Rouse.

These are:

1. In the proposed new architecture a specific Interaction Model manages the connection between Human and Machine Models.
2. The human and machine behaviour models are no longer linked in sequential mode, but iterations and loops may be generated during time cycles. In this way, the two models can evolve independently from each other for some time periods, before actively interacting.

All connections between Human and Machine Models are managed by the Interaction Model.

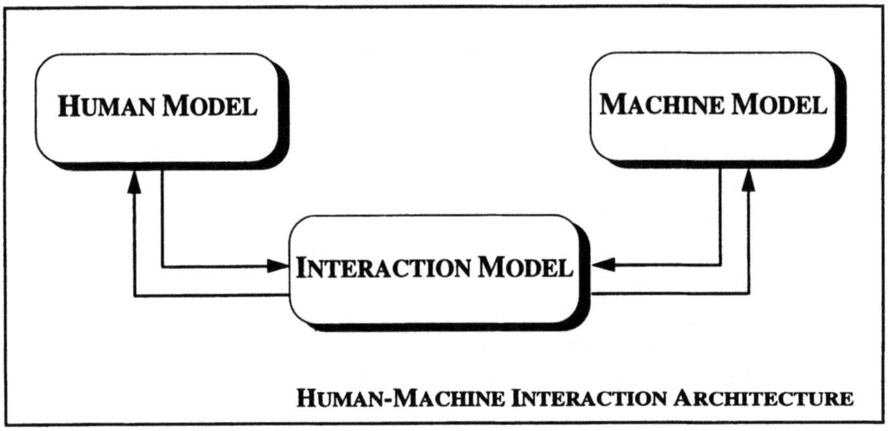

Figure 1.9. New Architecture of Human-Machine System Interaction.

This representation expands quite substantially Rouse's architecture and is able to represent in a more structured format the complexity of modern human-machine systems. In the remainder of this book we will always refer to this scheme for representing a Human-Machine Interaction Architecture.

1.4.2 Interaction Model and Data Management

Elements of Interaction Model

The three main requirements of the Interaction Model, discussed in the preceding section, can be satisfied by a variety of theoretical expressions and numerical algorithms (Figure 1.10).

Primarily, the *Time Management* can be dealt with by considering independent time steps for human and machine simulations and by developing specific models for the interaction and for the exchange of data. In this way, appropriate synchronisation can be achieved between the models of human and machine behaviour, while independent loops of evolution can be calculated.

The computational means that can be used to implement time management are, in most cases, logical and analytical expressions.

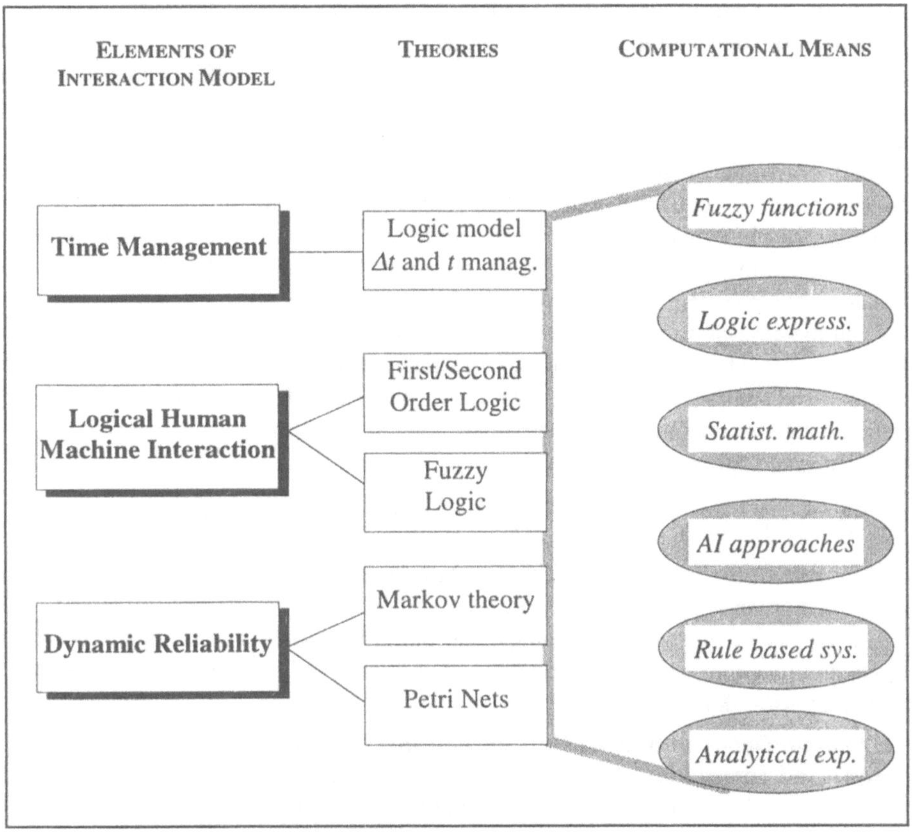

Figure 1.10. Elements of Interaction Model and computational means.

The second important element of the Interaction Model is the handling of *Logical Human-Machine Interaction.* The Interaction Model should filter all information that is produced by the human and machine models, by means of its algorithms specially developed for this purpose.

In so doing, not all information calculated by one model is actually made accessible to the other, as occurs in reality when plant data may not be perceived or recognised by the operator for many different reasons, or human actions may not be transformed in active plant operations, because of failures, malfunctions or unavailable components or actuators.

The theories that sustain this part of the interaction model are based on first or higher order logic, and, specially, on fuzzy logic, which is particular suitable for the representation of qualitative and "quasi-numerical" thinking (Dubois and Prade 1980; Van Orman Quine, 1959). The appropriate numerical and mathematical expressions to be used for implementing these theories are *fuzzy functions*, *analytical expressions* or *rules* and *Artificial Intelligence algorithms*.

Finally, the third requirement implies that the Interaction Model is able to handle false or incomplete information or system failures and human errors. To this purpose, a model of *Dynamic Reliability* can be applied in the interaction model.

The theories that sustain this part of the interaction model are based on statistics and reliability, in consideration of time dependencies, and logical behaviour. In particular, Markov theory and Petri nets can be applied in order to describe the time dependencies and the multiple correlations existing between the elements of the human-machine system (Bharucha-Reid, 1960; Peterson, 1981).

As in the case of cognitive functions and processes, the computational means and algorithms by which the elements of the interaction model are transformed into computerised numerical formats are quite generic. The most commonly used methods are fuzzy functions, statistical, logical and analytical expressions, AI approaches and rule based systems.

Data Management and Transfer

A major role of the Interaction Model is to manage the flow of information and data within the human-machine architecture.

The variety of input-data, variables, and parameters that are exchanged between human and machine models, can be defined as "data" and can be organised into two main categories, defined as follows:

> ***Dynamic Data*** **combine the human and machine behaviour models while time and interaction develop. These data consist in the "variables and parameters" calculated and exchanged by the models, and in the "failure and error modes and types" resulting from the dynamic interaction.**
>
> ***Static Data*** **consist in the "input-data, variables and parameters" for the machine and human models, and in the "correlations" for failure and error causes, modes and types due to human-machine and socio-technical context, independently of the dynamic interaction.**

Dynamic data and *Static data* are utilised by the algorithms of the Interaction Model for determining which information is transferred from one model to the other, at each time step of the transient analysis (Figure 1.11). In this way the interaction model manages the unfolding of the human-machine system interaction.

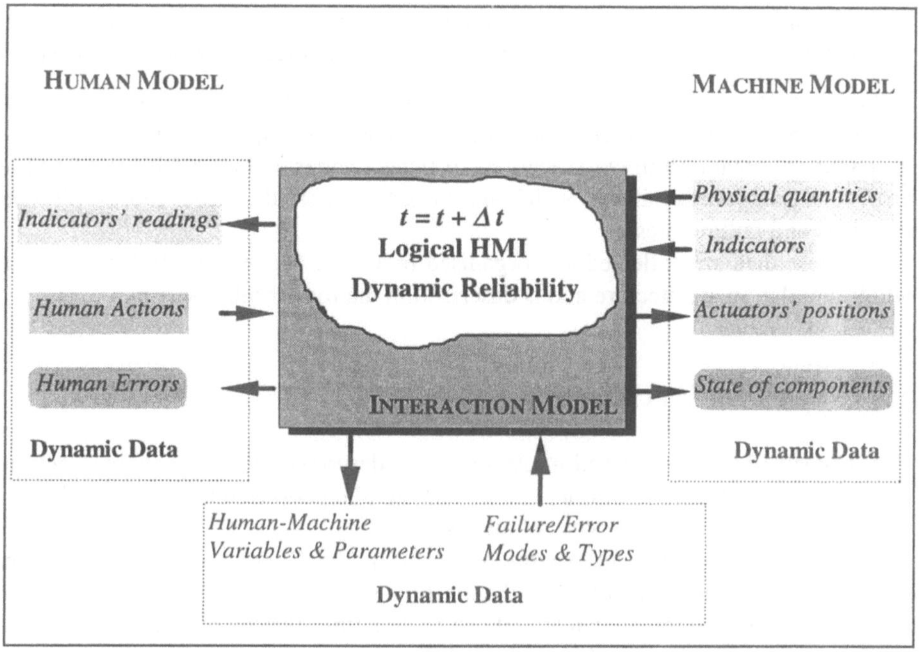

Figure 1.11. Interaction Model and Data.

More in detail, the governing mechanisms and "data" that are exchanged from one model to the other through the Interaction Model, can be described as follows:

- *Physical quantities* and *Indicators* values are dynamic variables calculated by the Machine Model.
  They are transformed, by the Interaction Model, into *Indicators' readings* that are fed to the Human Model for further calculation.
  The transformation into Indicators' readings is performed by the algorithms of the Interaction Model, according to the *Failure Modes* and *Types* of indicators (dynamic data) obtained combining *Human-Machine Variables & Parameters* (dynamic data) with failure correlations of indicators defined at the beginning of the analysis (static data).
- *Human Actions* are dynamic variables calculated by the Human Model.
  They are used by the Interaction Model to generate the *Actuators' Positions.* Once again, the transformation of human actions into actual positions of actuators depend on the algorithms of the Interaction Model, and on the failure modes and types of actuators obtained combining dynamic data and input correlations (static data) relative to failures and malfunctions of actuators.

- *Human Errors* and *State of components* are calculated by the Interaction Model, accounting for the failure/error modes and types, identified combining the human-machine variables and parameters with input failure/error correlations.

In order to manage the exchange of data between the models of human and machine, the interaction model requires a large amount of input quantities.

These are organised in classifications of types, modes and probabilities of human errors and component failures, which are related to external/internal causes, physical variables and failure rates.

All these data are collected and organised in a Data Architecture that sustains the whole simulation architecture and is described in the forthcoming sections.

1.5 A Data Architecture

The Human-Machine Interaction Architecture described in the previous sections enables us to evaluate the transient behaviour of a human-machine system, either for prospective or retrospective types of analysis.

In this section we will discuss the question of "data" with specific focus on the simulation of human behaviour. For the machine simulation, the data issue is similar, and the analogy with the human model can be straightforwardly made.

The topic will be discussed from three different perspectives:

1. Firstly, we will concentrate on data that are correlated with the theoretical models and paradigms, so as to sustain the simulation architecture. This is done by defining classification schemes, or taxonomies, of behaviour, i.e., by identifying *which data are needed* for the evaluation of interactions.
2. Secondly, the question of *how data are obtained* is tackled. A number of possible ways are explored for identifying data in relation to real working environment and plant performances.
3. Thirdly, the architecture that combines data with human-machine system simulation is discussed, with the aim of clarifying *how data correlate* to models.

1.5.1 Taxonomies

In the context of a human-machine interaction study, the cognitive model acts as a paradigm of reference for the simulation of human behaviour and for the manager of data and parameters concerning human behaviour.

A cognitive model is sustained by a reference classification, or *taxonomy*, of data and parameters.

> **A *taxonomy* is a set of categories according to which data are structured and collected in a systemic way. A taxonomy must be sufficiently generic and flexible as to adapt to the different types of applications and analyses to be performed.**

For example, in some cases, the taxonomy must be extended as to comprise socio-technical and environmental effects on human and plant behaviour. In other cases, a less complicated structure of data may be sufficient.

In any case, a taxonomy is a formalised representation of a plant or human behaviour, which refers to a conceptual model, and allows to structure data in useable formats.

In the domain of human factors, many taxonomies and classifications for the organisation of data on human behaviour exist. The most known classifications have been developed in the context of safety assessment and probabilistic safety studies. In particular, the classifications of Swain and Guttmann (1983), or Rasmussen and colleagues (1981) and Rouse and Rouse (1983) have been developed in relation to the first human models based on the "information processing paradigm." More recently, the taxonomies of Bagnara and colleagues (1989), Reason (1989), Hollnagel (1991), Hollnagel and Marsden (1996) have been developed with greater concern for the cognitive aspects and for the socio-technical context.

Usually, as taxonomies are developed with reference to a theoretical model, they are very generic and, only after an appropriate customisation, can be applied to a specific domain. This implies that all generic expressions and items of a taxonomy must be further adapted to physical dimensions, time constraints and working characteristics of the domain in which the taxonomy is being applied.

As an example, data and parameters concerning organisational factors affecting behaviour are generically defined according to well established theoretical models, such as those developed by Reason (1990), Westrum (1995), or Hofstede (1991). Then, using practical experience of a specific working environment and socio-technical context, a more focused and limited number of parameters, data and indicators of organisational factors can be defined and used for real safety studies (Pedrali, 1993; Hudson and colleagues, 1994; Maurino and colleagues, 1995).

In the domain of machine simulation there exists an even greater spectrum of classifications for the organisation of data on hardware components and physical performances. Data about plant performance, physical parameters, failure rates and constants have been studied over long periods of time and can, nowadays, be found in literature as well as in easily accessible databases.

In the case of human behaviour, a totally different situation exists. In general, only theoretical methodologies are available which describe how to organise, structure and collect data (Bagnara and colleagues, 1989; Hollnagel and Marsden, 1996) and, apart from few exceptions, no databases exist that directly support the proposed methodology.

Consequently, a substantial amount of work is required to develop a body of data ("database"), which follows the requirements of such taxonomies. Moreover, these data are always very specific to the working environment for which they are developed. Finally, a particular ability in experimental design and some experience of field studies are essential to collect and evaluated these types of data.

The quality of the results of the human-machine study rests on the accurate performance of the field study and on the definition of a coherent and consolidated set of data and parameters, as much as on the method and theoretical constructs on which the empirical studies are based.

This is the reason why an equivalent effort has to be made in both directions, i.e., the theoretical and the experimental work, in order to develop a method that can work in practice and that is supported by realistic consolidated data.

The reason for the scarcity of databases in the area of human behaviour is primarily related to the need of being very specific to each working environment. Contrary to hardware components, it is hardly possible to extrapolate human behaviour from one working domain to another. In some cases, an even more detailed level of refinement is needed, as many aspects of cognitive behaviour depend on the particular social and contextual working environment.

In substance, each safety study or analysis of real human-machine systems needs a very specific set of supporting data.

For these reasons, it is very difficult do develop generic database on human behaviour.

This requirement, however, is not always respected, and there exist methods which are associated with built-in generic databases dedicated to human performance. These methods, such as THERP (Technique for Human Error Rate Probability) of Swain and Guttmann (1983) are particularly successful, independently of the quality of the associated analytical approach or database.

The reason of such success rests on the fact that they offer a solid support of data for the numerical evaluation of human performance. This is indeed the primary requirement of the end users, i.e., designers, safety analysts and assessors, and sometimes the aspects of quality assessment of the database is underestimated.

The problem of data has to be seriously considered, because the quality of the outcome of a study can be substantially invalidated by the lack of reliability or appropriateness of the initial data, even when very detailed models and extensive methods of investigation are used.

The development of an appropriate and validated database and taxonomy requires at least the same amount of attention that is dedicated to the development of the models and simulations sustaining the interaction.

1.5.2 Data and Parameters

The taxonomies of human-machine behaviour describe the structure in which data are organised and should be collected for use in performing retrospective and/or prospective analyses. Different sources of information and means of investigation are available to the analyst for collecting and defining the data required to fill in the database.

A close similarity exists between different sources of data for human and machine models. We will analyse firstly the way in which data are obtained and generated for human models and then we will consider the analogy with the sources of data for machine models.

Data for Human Models

Input-data, constant variables, parameters and correlations between human, machine and context are part of the so called "Static Data", according to the definition given above.

These data are necessary for human models and can be gathered and collected in a number of ways (Figure 1.12):

- Data are recorded from actual instances during human interaction processes. Examples of these data are: Voice Recorders, audio-video recordings of training sessions in simulators etc. It is important to point out that these data are collected in total absence of external observers.
- Data can be collected in the working environment and within the organisation by direct observation of normal operations, e.g., in control rooms, or during flights, or during maintenance operations. This type of data can also be collected from laboratory experiments. The main difference between this type of data and the previous one lies in the presence of the observer in the field of study.
- Data can be obtained by interviews and questionnaires at all levels of the organisation.
- Data are contained in mandatory and voluntary reports on accidents, incidents and near misses (Reynard and colleagues, 1986; Taylor and Lucas, 1991).

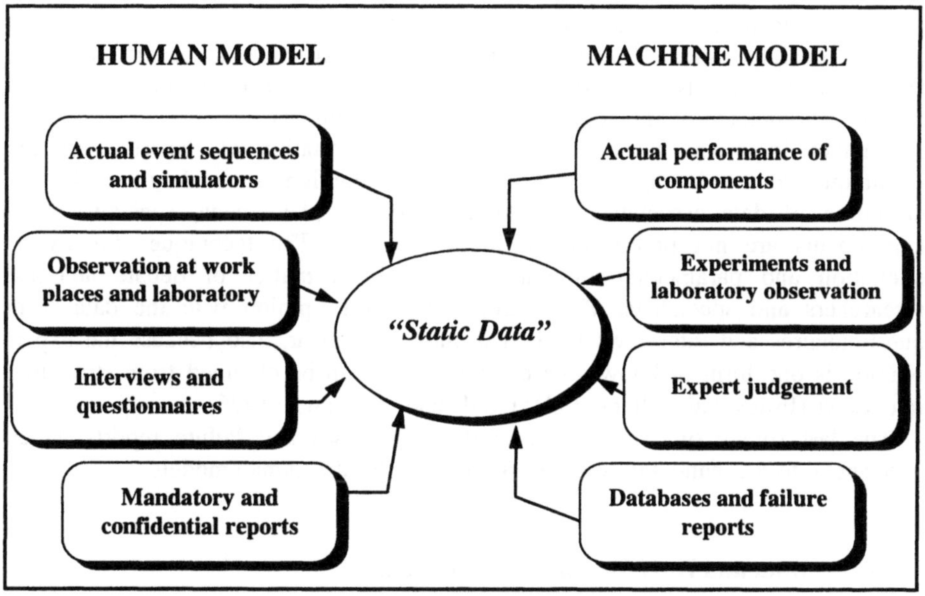

Figure 1.12. Input Data and Parameters for Human and Machine Models.

The first type of data, i.e., the actual recording of real situations, have to be utilised as they are. All the human factors implications associated with the event under observation are derived by methodological considerations, especially those related to socio-technical context and work organisation. These data are the most genuine, as they are not affected by the "disturbing" presence of an observer and the "altering" boundary conditions due to the experimental setting.

The other 3 types of data differ substantially from the first type as they are all structured a-priori, i.e., before they are collected, in accordance with a reference taxonomy.

If a good correspondence has been established between models and taxonomy, the formats of these data are such that they can be immediately used in the simulations for generating predictions or for describing behaviour, with no further manipulation.

Data for Machine Models

As already mentioned, a close analogy exists between data collected for human models and for machine models. In particular, the number and type of sources of data and parameters for machine and human models are similar (Figure 1.12).

The observation of behaviour of humans in real working environment correspond to recording of actual performance of components in their real working context. As data and performance characteristics may result different than those given by the constructor, this process is an important element in the validation and testing behaviour of components. Not many data are collected in this way.

The vast majority of data and parameters concerning hardware components and physical characteristics of plants are derived from experimental observations and laboratory experiments. These data are the same as those for human models derived from observations of workplaces and laboratory experiments.

Another important source of data is the exploitation of expert's opinion. Techniques for combining expert judgements and opinions are widely used for the definition of data associated with highly complex phenomena, specially when experiments are not practicable or are too costly. The techniques for expert elicitation and parameters combination are still a matter of debate amongst researchers and specialists. The analogy of expert opinion with the data from questionnaires and interviews for human models is clear, also because the expert opinion is one form of knowledge elicitation that can be obtained from structured interviews (Firley and Hellens, 1991; Guida and colleagues, 1996).

The last source of data are the existing databases and failure reports which correspond to the mandatory reports on accidents for the human models.

1.5.3 Data and HMI Simulation Architectures

The body of *data*, collected or derived from a specific working domain, and the *taxonomies*, that help in structuring and limiting the scope of collection, are the two basic elements of the Data Architecture.

This architecture sustains and correlates with the Human-Machine Interaction Architecture for the evaluation of human-machine interaction. The following discussion, which was partly developed in connection with and limited to the Interaction Model, will be only shortly recalled.

The Data Architecture manages both "static" and "dynamic" data. *Static data* are utilised to define the initial conditions of models and for correlating the behaviour and performance of human, machine and context.

Dynamic data are managed and elaborated by the Data Architecture in two correlated steps:

1. Human-machine "dynamic data", calculated by the simulations of interaction, are used by the taxonomy for searching for specific modes and types of component failures or human errors.
2. These are returned to the simulation architecture for further calculations.

This exchange of data between the two architectures continues for the whole analysis and is managed by the Interaction Model, as was anticipated in the previous section (Figure 1.11).

This is a further demonstration of the importance of the Interaction Model in the whole economy of the simulation of human-machine interaction.

1.6 A Framework for HMI Simulation

In the last two sections of this chapter we will review and summarise firstly the most important characteristics of the framework for human-machine system simulation developed so far and then we will outline the basic requirements and specifications that need to be fulfilled in order to transform the theoretical framework into a set of computerised methods.

From the discussions carried out in the previous sections we can finally develop a framework that can suitably describe modern human-machine systems, in consideration for contextual factors, data management, reliability and logical behaviour of humans, software and hardware components.

1.6.1 Elements of Framework

The basic elements of such framework are a Human-Machine Interaction Architecture and a Data Architecture.

The Human-Machine Interaction Architecture has the objective of simulating the interaction process and is structured in three modelling elements, namely (Figure 1.13): the Human Model, the Machine Model and the Interaction Model. These three models are able to describe the processes of interaction between human and machine, the errors and failures that may occur during the interactions and the likely outcomes and sequences of events.

The Data Architecture has the objective of managing the data that support the simulations. The amount of data required is particularly extensive, especially for what concerns human factors, as they are domain, plant, company and culture related, and must be organised in appropriate taxonomies that match the models utilised for the simulation of the human-machine interaction.

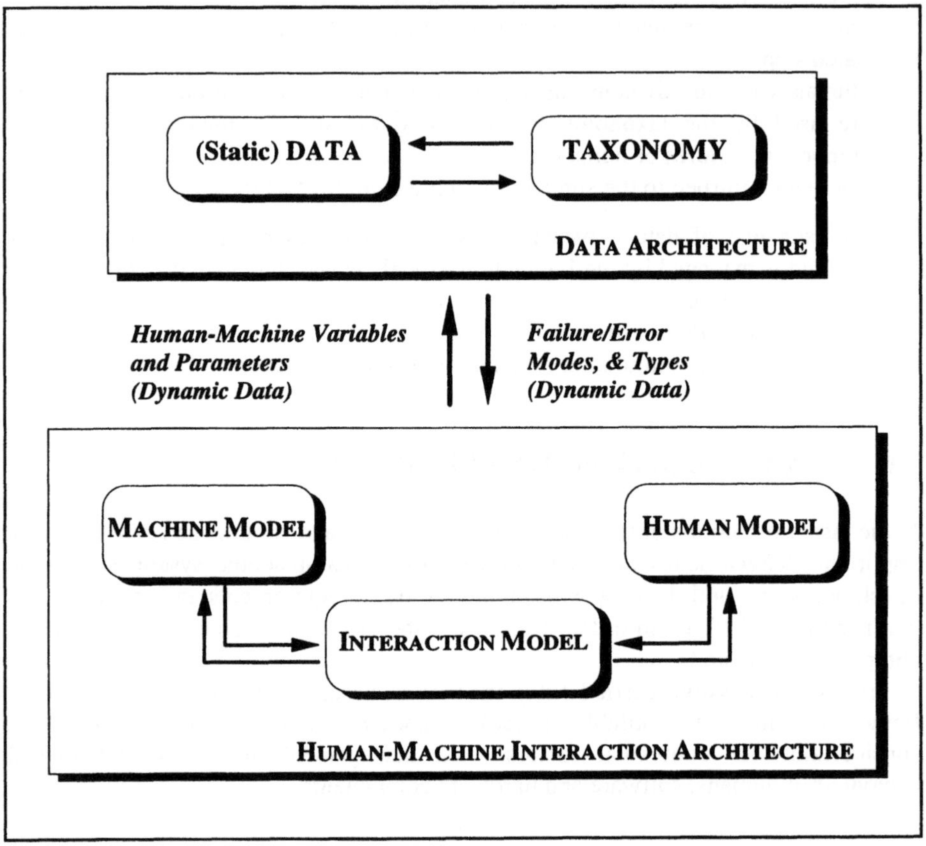

Figure 1.13. Data Architecture and Human-Machine Interaction Architecture.

The predictive capabilities of a simulation of human-machine interaction depend on the quality of the models and mechanisms, as well as on the soundness of the taxonomies and databases.

The features that characterise the proposed framework with respect to the more "classical" architectures, such as the one proposed by Rouse, can be summarised in the following 4 main points:

1. The models of the Human-Machine Interaction Architecture have been changed and extended: the Human Model considers cognitive processes, in addition to sensory-motor activities, and the Machine Model includes the environmental and contextual factors affecting the overall human and machine behaviour.
2. An Interaction Model has been included for the management and exchange of data and information between the models of human and machine behaviour, with the specific objective of controlling the synchronisation of the models, the logical unfolding of events and occurrences, and the stochastic and reliability factors affecting human and machine behaviour.
3. The connection between human and machine behaviour models has been modified to include the possibility that iterative and non-sequential processes occur in addition to the mechanisms of direct link between models.
4. A specific architecture of data management has been devised, which enables to structure and organise data in classification schemes conforming to the theory and formalism of each simulation.

1.6.2 Mechanisms of Framework

The mechanisms that govern the connection between the various elements of the framework are also very important, especially for what concerns the time evolution and time step management.

In Figure 1.14 the general framework for human-machine system simulation is shown, outlining the operational aspects in a loop of interaction and time step evolution, i.e., the processes of exchange of data, parameters and variables.

The exchange of data between the architectures of human-machine simulation and data management concerns mainly the search within a database and associated taxonomy of failure, causes, modes and types. This process runs in parallel to and makes use of the human-machine dynamic variables (*dynamic data*) which are calculated during the simulation loop.

The models of human, machine and interaction exchange data and calculate updated quantities in a structured fashion, according to the rules of the Interaction Model.

In particular the following remarks can be made:

- The Machine Model uses as *input:*
  a) The *physical variables* calculated during the previous time step;
  b) The *state* of components calculated by the algorithms of the Interaction Model, using the data and correlations from the Data Architecture; and
  c) The *actuators' positions* set by the Human Model, but also filtered by the Interaction Model and Data Architecture, to account for possible failures.

  The results (*output*) of the calculation for each time step are:
  a) The updated *physical variables* (at $t+\Delta t$) and
  b) The *indicators* values calculated by the machine model.

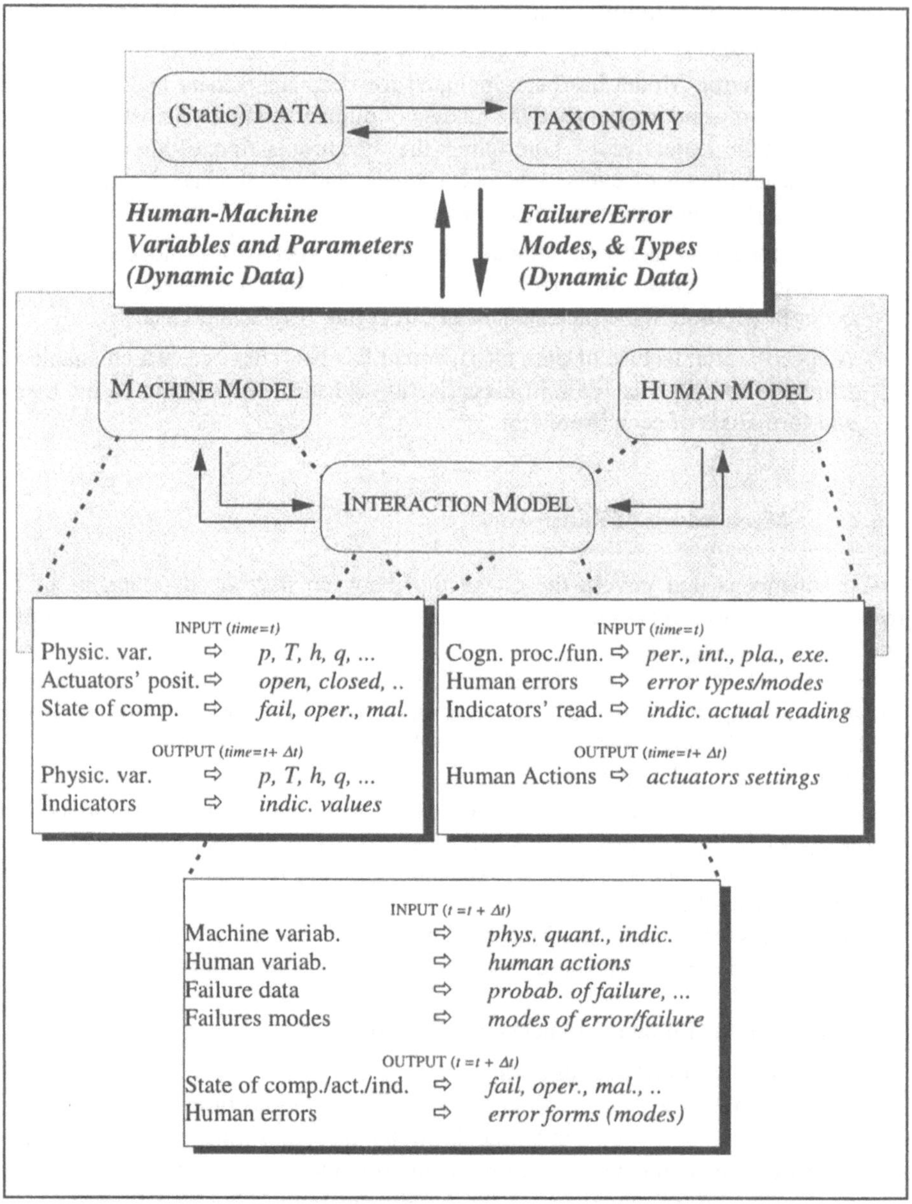

Figure 1.14. Operational aspects of Human-Machine System Simulation and Data Architectures.

- The Human Model uses as *input*:
  a) The information about the *cognitive processes* and *functions* from the calculation of the previous cognitive cycle;
  b) The *human errors,* derived, as for the state of components, from the Interaction Model and the data retrieved from the Data Architecture; and
  c) The *indicators' readings,* calculated by the Machine Model, but also filtered by the Interaction Model and Data Architecture for possible malfunctions.

  The *output* of the Human Model consists in:
  a) The *human actions* carried out on the plant control system by setting the actuators to certain positions.

- The Interaction Model receives as *input:*
  a) The human-machine the variables calculated by the machine and human models (*dynamic data*); and
  b) The *modes* and *types* of component *failures* and human *errors* derived from the correlations of (static) failure/error data and taxonomy contained in the Data Architecture.

  The *output* of the Interaction Model consist in:
  a) The actual *state of components* (failure modes);
  b) The actual *actuators positions;*
  c) The actual *indicators' readings;* and
  d) The actual *human errors* (error modes).

This discussion concludes the description and analysis the Human-Machine Interaction and Data Architectures, and offers a reference framework for the development of integral simulation approaches dedicated to the study of human-machine interactions in real plants and work settings.

In the forthcoming Chapters of the book we will review the variety means for putting this theoretical framework into practice and we will develop to a number of specific applications.

1.7 Summary Requirements and Specifications

The requirements and specifications necessary for developing a model and simulation of human-machine interaction have already been identified in the previous section during the discussion on the different elements of the reference framework.

These requirements and specifications will now be shown in a compact form which enables to identify immediately what should be available in terms of theoretical models, simulation algorithms, data and classifications, once the needs or goals of a specific application are defined.

1.7.1 Areas of application

The first milestone of our theoretical architecture consists in the evaluation of the reasons that lead to the development of a simulation approach.

There are a number of areas of applications that demand the use of human models and simulation approaches in order to make more complete and realistic analyses of complex human-machine systems. These areas of applications are of particular concern for designers, engineers, safety analysts, regulators and managers.

Four main types of applications have been identified, namely: *Design, Safety Assessment, Training,* and *Accident Analysis* (Figure 1.15).

Analysing in more detail the needs of these four types of applications, it has been realised that they require the development of two types of analytical methods: *Prospective* and *Retrospective Analysis.*

The discussion on the types of simulation that sustain prospective and retrospective analysis has demonstrated that there are practically no differences between qualitative and quantitative methods. They are two ways to represent the same reality, and their richness and validity rest on the content of the underlying theories rather than in the form in which results are presented.

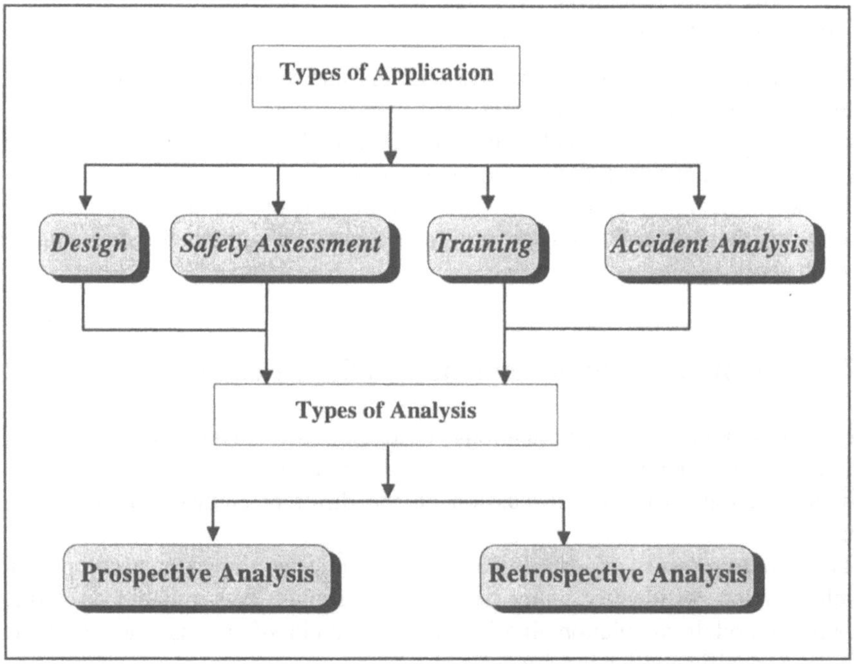

Figure 1.15. Application needs and method requirements.

1.7.2 Models, Simulations and Data

We have argued that a viable and powerful way to solve the problem of performing extended and accurate studies of human-machine interaction, both in retrospective and prospective, is in the use of *simulation*.

The development of a simulation framework demands the consideration of a number of architectures for dealing with and managing models and data. The analysis and discussion of models, mechanisms, data and means of implementation for the simulation architectures and framework have led to the identification of requirements and specifications (Figure 1.16).

In practice, three models need to be included in the simulation architecture, and these must be associated with a detailed treatment of data, parameters and variables, organised in a taxonomy in the data architecture. This implies that requirements and specifications concern four main elements: *Human Model, Machine Model, Interaction Model*, and *Data.*

Human Model

The model and simulation of the human behaviour can be carried out by modelling the basic *Cognitive Functions*, the *Memory and Knowledge Base*, and the *Allocation of Resources*.

These have been discussed, and a number of possible means of implementation in computerised numerical or logical expressions have been identified.

A set of potential specifications have also been formulated: Primitives of cognition; Analytical and Logical expressions; Artificial Intelligence approaches; Rule based systems, etc.

In practice, these are only means of implementation of a model and, thus, have to be adapted to a number of diverse and empirical considerations, as, for example, specific aspects of the model to be simulated, computer type and power available, preferences and experience of the model developer, etc.

Machine Model

The requirements for the development of a machine model and simulation are quite classical and are commonly fulfilled when studying physical phenomena.

They can be grouped in three main categories of models, able to represent the features of a plant and working context that intervene in a human-machine dynamic interaction. In practice, the machine model must be able to evaluate the performance and response of: *Physical variables; Actuators and indicators;* and *Context and environment*.

In the Human-Machine Interaction Architecture under study the requirements of the machine model are broader that what is usually demanded to a plant simulation. The behaviour of actuators and indicators is not always carried out in deterministic simulation, nor are the logical interactions between physical variables and indicators and actuators, or between actuators and indicators themselves. Moreover, the model of the context and environment is seldom included in the scenario of human-machine interaction, while, in the architecture considered here, it plays a major role in affecting human behaviour and plant performances.

As in the case of the human model, a number of means of implementation are identified for the machine model. These are only some of the possible specifications that can be applied to transform the models into numerical and logical expression, as for example: numerical methods, analytical, logical and polynomial expressions, etc.

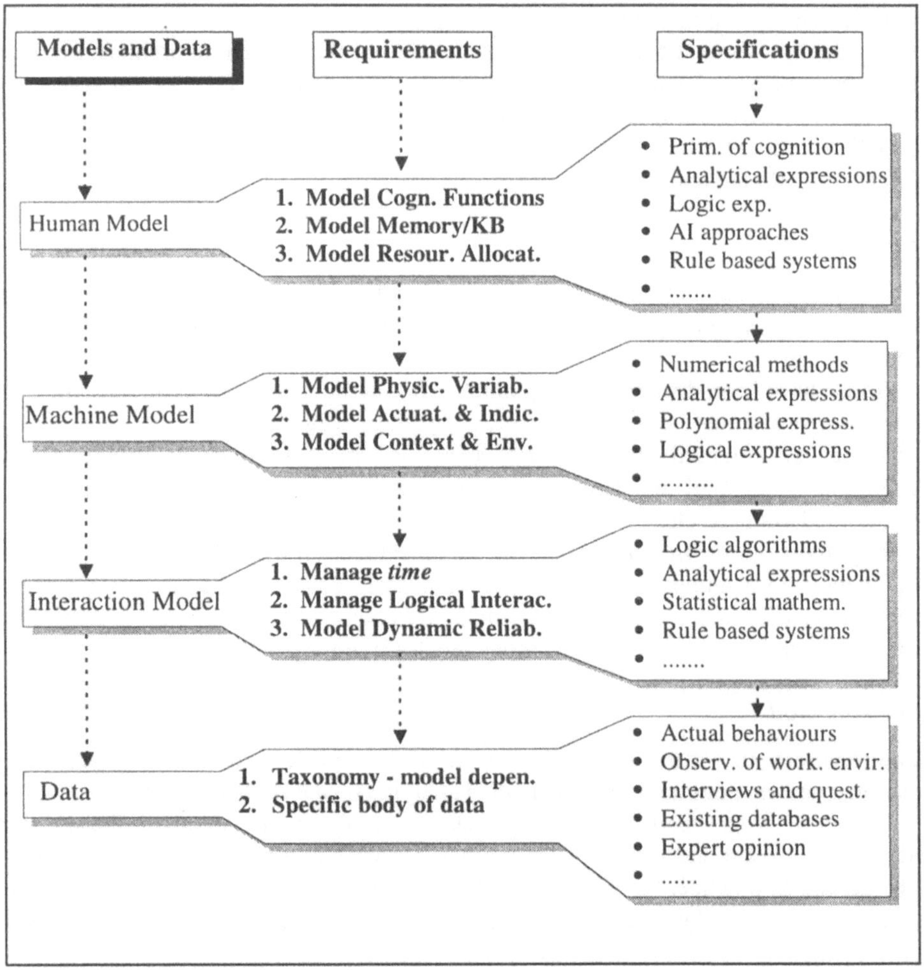

Figure 1.16. Requirements and Specifications for models, simulations and data.

Interaction Model

The interaction model has to concentrate on the management of the simulation process. Therefore, the basic requirements for its development are related to the control of the time evolution and of the exchange of data between the models of human and machine.

In particular, these two models have to be properly synchronised, even when different individual time steps are used for numerical or logical reasons. This implies that the first two requirements of the interaction model are the *Management of time* and of the *Management of Logical Human-Machine Interactions.*

This second requirement is very relevant from the methodological viewpoint, as it contains the need to formalise the sequence of "calls" to the models and their interaction during the performance of a case study.

The third requirement consists in the application of a model for the reliability and systematic safety analysis of the human-machine system. This model should specifically account for dynamic aspects.

The interaction model represents the real element of conjunction between the databases, containing parameters, coefficients, failure rates, etc., and the simulations, where these data are included in the solution algorithms.

The implementation of the interaction model can be carried out following different specifications. As in the cases of human and machine models, the use of logical and numerical algorithms or analytical expressions can be very useful, especially for managing simple correlations of failure and error modes and types. For more accurate quantitative calculations, statistical methods or rule based approaches should be utilised, so as to implement the reliability theories necessary to calculate probabilities and frequencies of occurrences and uncertainty bounds.

Data

As we discussed in the previous sections, the human-machine system simulation can perform properly if it is supported by a structured body of data.

These data should be organised according to the reality of the working environment under study and to the models that are used for simulation or analysis. Consequently the requirements for data are essentially two: a *taxonomy* and a *specific body of data.*

The development of a *Taxonomy* for human-machine interaction enables to classify and organise behavioural parameters with contextual and modelling conditions.

A *specific body of data* supports the simulations of human, plant and environment.

The specifications by which data are defined and collected demand the application of empirical and field observations methods, such as observation of actual behaviour and of working environments, interviews and questionnaires, etc.

The use of existing databases and of expert opinion are also valuable ways to develop a body of data necessary for a human-machine interaction study.

1.7.3 Types of Analysis

The last important aspect of human-machine simulation to be summarised is the need to formalise properly the type of analysis to be performed.

This methodological requirement ensures coherence between different types of analysis and enables the performance of different types of studies with correlated models and simulations.

In other words, the fundamental characteristic of a Human-Machine Interaction Architecture is that prospective and retrospective types of analysis are carried out with reference to paradigms and models of human and machine behaviour that are coherent with each other.

In this way, basic data and parameters can be exchanged from one analysis to the other: retrospective analyses can generate data to be utilised in prospective studies; and prospective analyses can reproduce the sequences of events occurred in real cases (Figure 1.17).

To reach this aim, the models adopted for prospective and retrospective analysis need to be identical, while the structure and the combination of human-machine interaction differs according to the type of analysis being carried out.

This is the meaning of the first requirement for both prospective and retrospective analyses, which states that models should be preserve for retrospective and prospective analysis

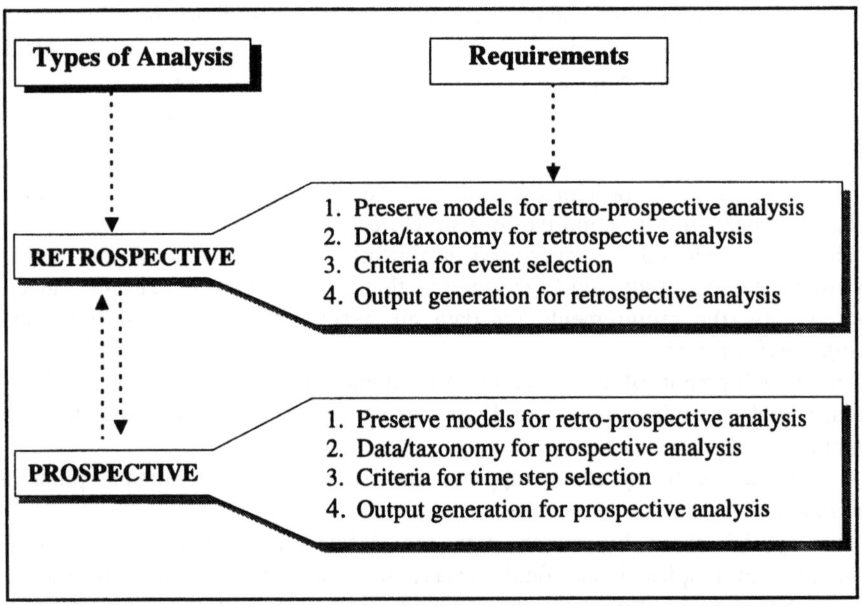

Figure 1.17. Requirements for prospective and retrospective types of analysis.

The second requirement concerns the management of the data and taxonomy for the performance of analyses.

While the sources of data are equivalent for the two types of analysis and are developed according to common criteria, the ways in which they are utilised and analysed differ according to the specific type of analysis, whether prospective or retrospective.

The third requirement for prospective and retrospective analysis concerns the time management, and enhances the fact that both prospective and retrospective analysis need to contain a strategic approach to deal with dynamic interactions.

In case of retrospective analysis, as the specific sequence is totally known, the crucial issue is the definition of the events to be studied in detail for the identification of root causes of failures and/or human erroneous behaviour. Therefore, the third requirement for retrospective studies is: *criteria for event selection.*

In case of prospective analysis the problem is the management of the time sequence, so that the human actions and machine evolution are logically co-ordinated and synchronised. Hence, the third requirement for prospective studies is: *criteria for time step selection.*

Finally, the last requirement is the formulation of the output of the analysis. This can be stated in generic terms as *output generation.*

However, the analyst must be able to develop an appropriate form for presentation of the output according to whether the study being carried out is to be combined with other analyses, as in the case of a probabilistic safety study, or is the only outcome of the study, as in the case of accident investigations.

These are the reasons why a distinction has been made for the requirements of prospective and retrospective analysis.

1.7.4 Tables of Requirements and Specifications

The summary of all requirements and specifications for the models, simulations and data can be found in table 1.2 and 1.3.

These are the tabularised versions of Figure 1.16 and 1.17 and reflect completely the content of the above summary arguments.

Table 1.2 contains the summary of requirements and specifications for models, simulations and data.

Table 1.3 contains a summarised format of the requirements for prospective and retrospective types of analysis .

Table 1.2. Summary of requirements and specifications for models, simulations and data.

MODELS/DATA	REQUIREMENTS	SPECIFICATIONS
HUMAN MODEL	1. Model Cogn. Functions 2. Model Memory/KB 3. Model Resour. Allocation	• Primitive of cognition • Analytical expressions • Logic expressions • AI approaches • Rule based systems • ......
MACHINE MODEL	1. Model Physic. Variables 2. Model Actuat. & Indic. 3. Model Context & Env.	• Numerical methods • Analytical expressions • Polynomial express. • Logic exp. • ....
INTERACTION MODEL	1. Manage *time* 2. Manage Logical Interac. 3. Model Dynamic Reliab.	• Logic algorithms • Analytical expressions • Statistical mathem. • Rule based systems • ....
DATA	1. Taxonomy - model dep. 2. Specific body of Data	• Actual behaviours • Observ. of work. envir. • Interviews and quest. • Databases • Num. expressions • ......

Table 1.3. Requirements for prospective and retrospective types of analysis.

TYPE OF ANALYSIS	REQUIREMENTS
PROSPECTIVE	1. Combine models for prospective analysis 2. Data/taxonomy management 3. Criteria for time step selection 4. Output generation
RETROSPECTIVE	1. Combine models for retrospective analysis 2. Data/taxonomy management 3. Criteria for event selection 4. Output generation

1.7.5 Conclusions

In this first Chapter, a Framework for Human-Machine Interaction Simulation has been developed and discussed at length from a theoretical and methodological standpoint:

- All the elements of the Data and Human-Machine Interaction Architectures have been discussed and correlated.
- The mechanisms for combining and synchronising the dynamic interaction of models have been analysed.
- The approaches that support the transformation of models into working simulations have been identified.

In the following Chapters a number of techniques, models, and simulations of cognition and human-machine interaction will be reviewed, according to the requirements and specifications described above.

The applications and sample case studies will be performed at different levels of detail and complexity, so as to show practical implementations of the reference framework and architectures described in this Chapter 1.

CHAPTER 2

ENGINEERING METHODS AND COGNITIVE TASK ANALYSIS

2.1 Introduction

In the previous Chapter we developed a generic framework for human-machine interaction and we devised the requirements and specifications for generating models, simulations and data of human and machine behaviour. The Information Processing System metaphor, developed over the last 20 years, was chosen as the initial reference paradigm of human behaviour.

Particular focus has been placed onto the cognitive functions of the IPS paradigm. These functions characterise the human as being an active actor rather than a passive one in the control process, by generating and developing diagnoses, plans and decisions, through cognitive processes, as well as by perceiving, through sensory receptors, and by reacting manually to external stimuli.

It is, therefore, very important to compare and define areas of application of some of the most advanced theories and simulations of human behaviour, and programming languages and environments particularly suited for representing cognitive and manual processes.

However, before such comparison may be carried out, it is important to position this body of methods and techniques within a general process of technological evolution, which has demanded, to the human operator, firstly the performance of manual tasks and then, progressively, more and more supervisory control activity. This means that, initially, direct sensorial contact with phenomena were particularly relevant and, now, remote indirect control and planning goals are the locus of human behaviour analyses.

Therefore, in this Chapter, we will review firstly the most important engineering and mathematical techniques that have sustained human behaviour modelling during the first years of modern human-machine interaction studies, in order to arrive at presently applied techniques, based on Artificial Intelligence approaches, higher order logic and qualitative-physics theory.

In particular, we will start by analysing the engineering techniques dedicated to the simulation of perception, interpretation and execution, as they have been developed first.

We will then concentrate on approaches suited for simulating more cognitive processes, typical of supervisory control tasks, like planning and decision making.

Thereafter, we will focus on Cognitive Task Analysis, which is a method based on field studies, largely developed and applied for the evaluation and formal description of human tasks, goals and working conditions, particularly centred on the mental aspects of human behaviour. Cognitive Task Analysis lies at the basis of any simulation development, as it allows the representation of boundary and environmental conditions and of the actual manner in which tasks and jobs are carried out.

These engineering methods and cognitive task analysis are the basic components of any quantitative and numerical simulation. Therefore, their contribution to human behaviour modelling will not be limited to the discussion in this chapter, but they will be recalled later on, in the following chapters, when they will serve as mathematical and numerical formalisms to transform theoretical models of cognition into computational simulations.

In conclusion of this chapter, a first example of implementation of these techniques will be shown, by applying qualitative-physics and artificial intelligence techniques for the simulation of an operator behaviour in controlling a steam generator, during normal operations and emergency management.

2.2 Engineering Methods for Detection, Diagnosis and Action

The objective of this section is to give an overview, with respect to time and evolution of technology, of techniques and methods that have been devised to support the inclusion of the human element in the process of systems design and safety assessment.

While the study and analysis of the fundamental role of humans in their living context is an endeavour as old as mankind and it is the cardinal subject of philosophical debate, the consideration of humans in working environments from an engineering perspective is a much more recent effort, much more limited in scope and objective and certainly manageable by means of numerical and analytical expressions.

Moreover, as we are concerned mainly with the evolutionary process that has led to the formulation of the Information Processing System metaphor, we can identify the origin of modelling and simulation of human behaviour with the formulation of the cybernetic paradigm by Wiener (1948), who developed the analogy between the human operator and the servomechanism. This analogy has increased awareness of the closed-loop nature of human-machine interaction, with the specific goal-oriented behaviour of the human controller.

About the same years, a number of other theories and techniques were developed, or begun to be applied for human behaviour modelling, and in particular: a) the mathematical theory of communication (Shannon and Weaver 1949); b) the theory of signal detection (Peterson, Birdsall and Fox, 1954; Van Meter and Middleton, 1954); and c) the linear control theory (McRuer and Krendel, 1957).

The control role of the operator was then mainly manual or supported by simple computerised machines. Therefore, the principal human functions of those days can be classified as sensory-motor activities. As a consequence of that, the most relevant theories and applications were dedicated to human engineering and to modelling *detection*, by monitoring information, and direct control *action*, rather than planning and problem solving (Pew and colleagues, 1977).

However, as technological progress introduced more and more automation, and the role of operator became of supervisor rather than actor on the control system, the need to model cognitive processes, such as *interpretation* and *planning*, became crucially important for the analysis of human-machine interaction. Therefore, new theories were developed, with the precise objective to represent reasoning and cognitive processes. These new developments were also fostered by the advent of new programming algorithms, languages and environments and by new generation computers. In particular, the latter begun to be conceived as electronic devices dedicated to represent logical processes rather than being simple "number crunching" machines.

Control theory approaches evolved into optimal control theory (Stassen, Johansen, and Moray, 1990) and fuzzy set theory (Dubois and Prade, 1980). Qualitative-physics theory (Bobrow, 1984) became very popular and widely applied for studying human-machine interaction. A substantial push and support to these new developments came from the advent of Artificial Intelligence (Barr, Cohen and Feigenbaum, 1989) and its computerised architectures and languages, which enabled a wider use of such theories and, more in general, of logic reasoning by natural language.

These new methods allowed to numerically represent human cognitive and behavioural processes, by multiple mathematical and computerised architectures. The supporting semi-empirical approach able to capture the complex process of goals setting and tasks distribution and to represent them in a structured and organised fashion is Cognitive Task Analysis (Bonar and colleagues, 1986). Cognitive Task Analysis allows to describe how activities of people are designed and actually carried out, in consideration of environment and working context. In this sense cognitive task analysis can be considered the initial step of development of a human behaviour simulation.

All the above mentioned theories and techniques will now be briefly reviewed, describing their mathematical formalisms and the scope of their application. A sample application of qualitative-physics integrated into an artificial intelligence architecture will follow, as an example of model development.

2.2.1 Theory of Communication

The *Theory of Communication* has been popular and widely applied for the definition of human reactions to signals, in terms of time required to make certain discrete (arm) movement.

The models developed by Hick (1952), or by Fitts (1954), or later by Senders (1964) are all based on the "classic" Shannon communication channel assumptions (Shannon and Weaver 1949), which can be summarised as follows:

- The operator is a "single channel" observer, and his/her eyes can only focus on one area at the time.
- The main task is to reconstruct signals from samples.
- Monitoring process reflects statistical properties of displays.

In this case, considering a random signal i with probability of occurrence p_i, it is possible to evaluate the time required to make a choice of a movement T_c, related to that signal, by the following correlation:

$$H_c = \sum_i p_i \log_2(1/p_i)$$
$$T_c = \alpha + \beta H_c \tag{2.1}$$

where H_c is the average information rate in bits, α and β are constants depending on the task conditions.

Fitts model considered a similar method for defining the time T_m needed for actually performing a movement related to the amount of information and the experimental or working conditions:

$$H_m = \log_2(2A/B)$$
$$T_m = \gamma + \delta H_m \tag{2.2}$$

where A is the distance moved, B is the tolerance within which the move must be made and γ and δ are constants, as above.

The first-order implication was made, by which, if a person has to make a choice first and then move, the resulting model for the total time required T_t, is:

$$T_t = T_c + T_m \tag{2.3}$$

In more recent years, as human tasks and goals have quite substantially changed, the application of this theoretical approach has been criticised and drastically restricted to very simple cases. The main reason for such criticism rests in the fact that these types of model reflect the properties of the context and thus hold only as far as the human-machine system lies in a limited functional domain and "sensory-motor" environment, i.e., when the operator accomplishes only routine tasks nearly unconsciously.

In modern supervisory controlled, highly automated systems, this is seldom the case, and consequently, the application of communication theory based models is not appropriate and would lead to incorrect predictions of human behaviour.

2.2.2 Signal Detection and Other Monitoring Theories

Signal Detection Theory is another successful engineering technique that has given rise to a number of human performance models.

In signal detection theory the observer receives a stimulus related to a continuous decision variable and has to make a binary decision about the stimulus, i.e., either that it consists of a signal-plus-noise or a noise-alone (Green and Swets, 1966). Obviously, it is assumed that the stimulus occurs only as signal-plus-noise or noise-alone.

The important assumption of signal detection theory is that observation is treated as an event in probability theory and that there exist probability densities for each possible observation, given both "signal" or "noise-alone".

Signals and observations are organised in a stimulus-response matrix and a related curve. The matrix contains the response alternative versus the stimulus alternative. Calling the stimulus either s for the signal or n for the noise and the response either S or N the matrix contains four possible alternatives: the probability of selecting correctly a signal, $P(S|s)$, also called "hit"; the probability of identifying a noise only, $P(N|n)$, also called "correct rejection"; the probability of omitting a signal, $P(N|s)$, also called "miss"; and finally the probability of selecting as signal a noise stimulus, $P(S|n)$, also called "false alarm" or "false-detection". Given these definitions, the following relations hold:

$$\begin{aligned} &P(S|s) + P(N|s) = 1 \\ &P(S|n) + P(N|n) = 1 \end{aligned} \tag{2.4}$$

The curve derived from the response matrix, called "Receiver-Operating-Characteristic" (ROC) graph (Figure 2.1), shows the estimation of the probability that the observer performs the correct detection of a signal, versus the probability of a false alarm, e.g., the probability of an observer to respond S, either in presence of a signal-plus-noise *or* noise alone.

The information contained in a ROC graph is very relevant. For example, the area under the ROC curve shows the percentage of correct detections in a two-alternative selection task. As the area approaches 100%, the ROC curve follows the extreme left and top co-ordinates of the graph. This means a "false-detection" probability of almost zero and a hit probability of nearly 100%. As the area nears 50% the ROC curve falls near the diagonal, with the false-detection and hit probability nearly equal.

By the application of these graphs, detection theory provides a means of predicting the percentage of correct detection in forced choices.

Following the signal detection theory, Sheridan and Ferrell (1974) developed an *open-loop action theory,* that is an extension of the original theory. In the *open-loop action theory* the decision maker does not have to make the binary selection between signal-plus-noise or noise-alone, but makes a response along a continuum. Moreover, it is assumed that the stimulus can be a combination of signal-plus-noise and noise-alone, whereas in straight signal detection theory only one of the two stimuli is possible.

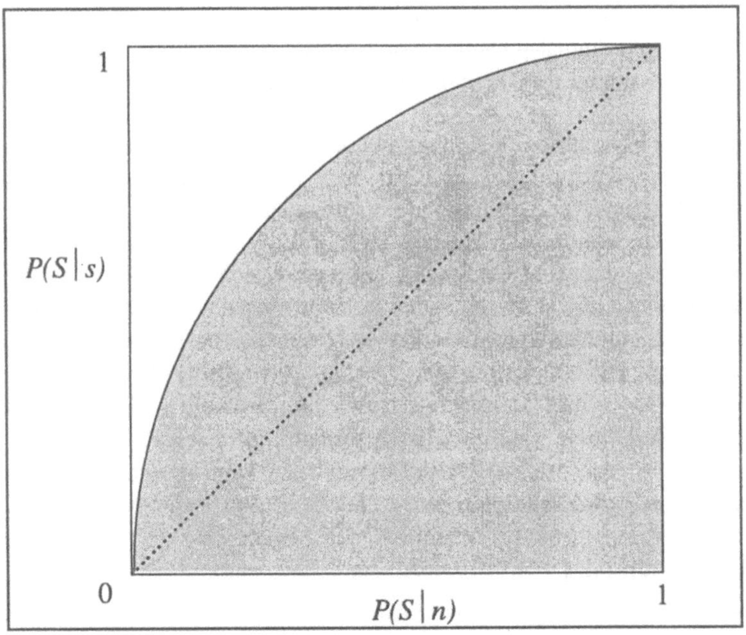

Figure 2.1. Example of "Receiver-Operating-Characteristic" (ROC) graph.

Another theory, developed in connection with monitoring and detection, is the *Queueing Theory* which considers the "cost" associated with choosing in a context of a multiple tasks human-machine system (Rouse, 1980). These concepts were introduced firstly by Carbonell (1966), who studied the "cost" associated with scanning instrument panels and choosing a certain instrument and the probability that the instrument reading exceeded a given limit. The aim of the model is to minimise the cost of instrument reading. The interesting aspect of this theory is that it combines time since last reading and risk-benefit analysis of performing a new reading.

In this domain, an interesting model of detection was developed by Sheridan (1970) for tackling the problem of choosing optimal sampling intervals, in order to maximise the amount of information retrieval from displays. The model combines information content together with costs. In particular, the operator is expected to know: a) the probability distribution of a state function and its rate of change; b) the correspondence between certain values associated with the state function and the related control actions, and a discrete cost of sampling. Then, an optimal sampling strategy, e.g., the best time interval between observation of certain variables, can be derived by maximising the expected gain.

2.2.3 Control Theory

Control theory began to influence human engineering just before the 50ies, when conventional linear control theory was applied to model simple manual control.

A variety of studies followed the appearance of the first models up to the present days, and the progressive complexity of human tasks and supervisory control have been gradually followed by corresponding advances in control theory.

This is the case of the development of model-based controllers, containing model-based state-estimators, or *Kalman filters*, for optimal estimation of process state variables. A particularly important development in this area has been the *Optimal Control Model*, which continues to be quite substantially applied for complex control problems.

Given the relevance of control theory for human behaviour modelling, it is worth following in more detail its historical development from linear control theory to optimal control model.

Linear Control Models

Since the 50ies the analogy was developed of considering humans as if they were servomechanisms, acting as the sole-in-the-loop control element (Figure 2.2).

The initial hypotheses were that the models of operator and machine are independent, so that a human behaviour model could be combined with whatever control process. This was proven to be impractical, and the two upper blocks of the control system were combined in a single loop element.

The idea was that there would be only minor variation from one application to the other of the combined human and process element. This approach was very successful, in particular with the simple *crossover model* developed by McRuer and colleagues (1965).

In this type of approach it is assumed that the human operator plus the controlled process is invariant, as opposed to the invariance of the human operator per-se (McRuer and Jex, 1967).

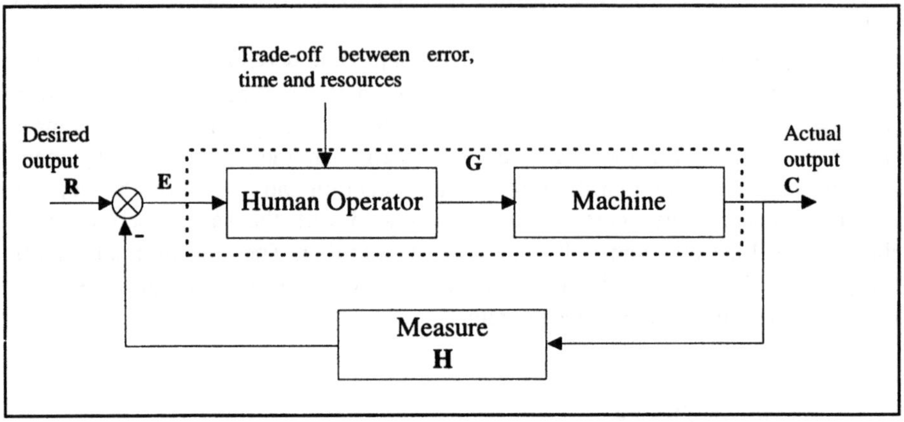

Figure 2.2. Simple manual feedback control system.

In Figure 2.2 the simple manual negative feedback control is depicted, where the *block diagram* notation is utilised to represent the relationship between variables of interest (R and C) and transfer functions (G and H).

Using linear control theory, i.e., assuming that the controlled processes can be represented by linear differential equations and that the operator and control process can be described by a single function (G), the following relations can be established:

$$E = R - HC$$

$$C = EG$$

and, thus:

$$C = G(R - HC)$$

where:
R = desired state
C = actual process output
G = human-machine interaction
H = measure of process state variables
E = difference between measure and desired state (error function).

The following relation between output, desired state, measure and human-machine interaction takes the form:

$$C = R\frac{G}{1 + HG} \tag{2.5}$$

which is the classical expression of the transfer function relating Output to Input in a closed loop, negative feedback, control system.

Optimal Control and Kalman Filter

Expanding further the analogy of humans as servomechanisms in control theory, it can be easily recognised that, in controlling a dynamic process, the most important task becomes the minimisation of the deviations of the process state from the desired trajectory. This process would provide an optimal estimate of the controlled process state variables, and thus the identification of this approach as *Optimal (or modern) Control Theory*.

In real situations moreover, the measured state of process variables are always associated with noise due to many different reasons and thus the "true" state must be estimated using noisy state measures and the best available model. The Kalman filter (Kalman, 1960) is a technique to combine information using both noisy measures and modelling.

In order to outline the task of the human operator in a control theory perspective, we start by representing the dynamic behaviour of the human-machine system with general equations. Using a matrix notation, the control process can be described by:

$$\mathbf{X}(t_{i+1}) = \mathbf{\Phi X}(t_i) + \mathbf{\Psi U}(t_i) + \mathbf{\Gamma W}(t_i) \tag{2.6}$$

$$\mathbf{Y}(t_{i+1}) = \mathbf{HX}(t_{i+1}) + \mathbf{V}(t_{i+1}) \tag{2.7}$$

$$\mathbf{U}(t_i) = \mathbf{C}(t_i)\mathbf{X}(t_i) \tag{2.8}$$

where:

X	Process state variables (**X'** = Estimated Process state variables)
Y	Measures on displays (**Y'** = Estimated Measures on displays)
U	Control actions by the human (inputs to the process)
W	Disturbance (zero-mean Gaussian distribution)
V	Measurement noise (zero-mean Gaussian distribution)
C	Coefficients of the multiple regression problem that minimises certain criteria to optimise control
t	Current time, reaction delay time
$\mathbf{\Phi}, \mathbf{\Psi}, \mathbf{\Gamma}, \mathbf{H}$	Matrices of constants characterising the system

Thus the human observes **Y**, which is a measure affected by noise of certain process variables **X**, and produces an action **U**, which aims to optimise the expected response (Figure 2.3).

In practice, a Kalman filter contains the best available models of the process variables, and of the measurements, as well as the distribution of the disturbance statistics. These models and distributions constitute an *internal model* of the process and are applied in calculating the dynamic behaviour of the process at the time t_{i+1} using the process state variables at the time t_i, which are known. The model output measurement vector (**Y'**) is compared with the actual measure (**Y**) and the residual is used for an iterative correction loop from the prior to the posterior estimate of the process state variables (**X'**).

Under the assumptions that: 1) the controlled process is linear, 2) the internal model matches perfectly the actual process and 3) the noise is random and normally distributed (Gaussian), then this process produces a minimum variance state estimate with white noise residuals.

Figure 2.3 shows the expansion of the simple manual feedback control system to the optimal control model, using the model-based Kalman estimator (Kalman filter).

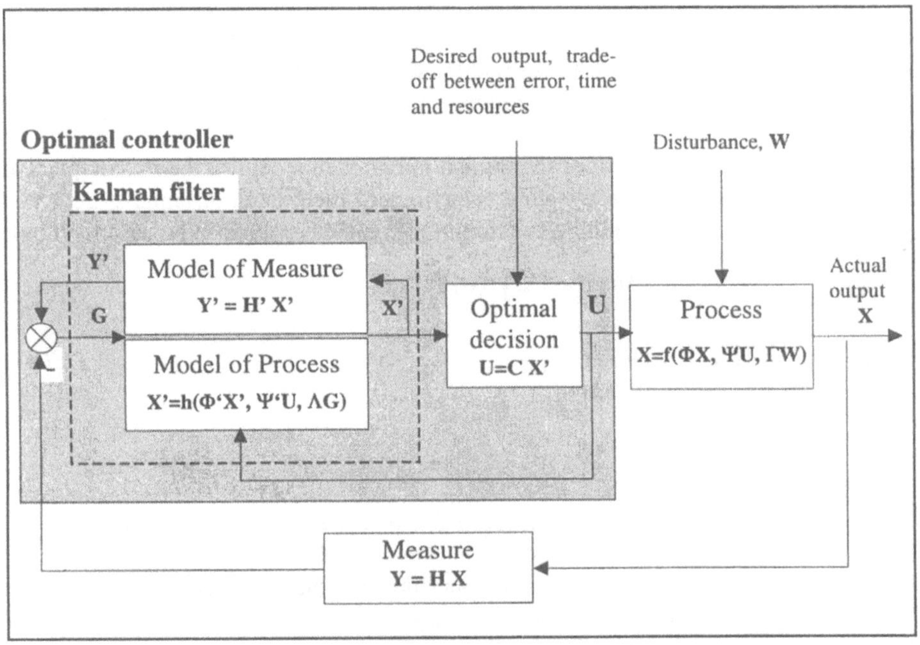

Figure 2.3. Feedback control system by "model based optimal controller", (from Sheridan, 1992, p.31).

As the objective of this section is only to give a short overview of methods and techniques on estimation and control, no further details of these models will be given. However, the reader interested in a more accurate treatment of manual control and optimal filtering techniques can refer to well known textbooks, such as Kelley (1968), Sheridan and Ferrell (1974), Rouse (1980), and Sheridan (1992).

Optimal Control Model

The *Optimal Control Model* (OCM) was firstly developed by Kleinman, Baron and Levison (1970) as an extension of the optimal control paradigm, by adding a predictor element, to compensate for the operator's reaction time delay, and a neuromuscular dynamic component, to simulate human sluggishness.

With reference to Figure 2.4, the human model aims at calculating the action **U**, at time *t*, on the basis of the best estimation of the process behaviour, given the observed display measures, **Y**.

The first step of the model is to give an estimate of the process state, **X'**, in presence of noisy observations, $\mathbf{V_Y}$. As there is a reaction time delay of the human, a prediction is necessary to produce an estimate at the correct time, *t.* The control actions, **U**, are then calculated according to the criteria of the optimal control, by minimising the square deviations of the process output from the desired output.

This output is further altered by the motor noise, $\mathbf{V_U}$, which accounts for the human psycho-physical limitations in producing exactly the control actions **U**, and by the human neuromuscular system which introduces typical idleness in the final actions.

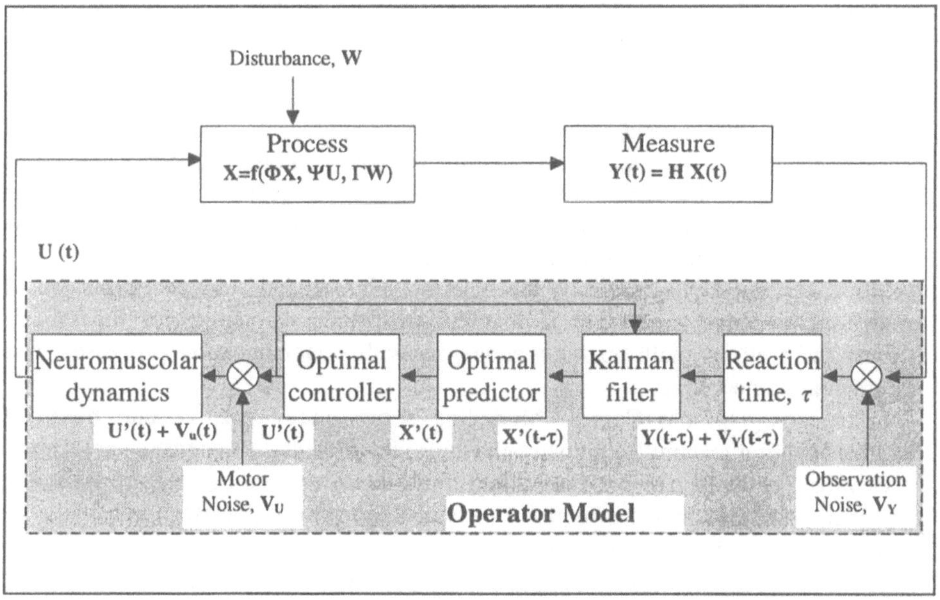

Figure 2.4. The Optimal Control Model, (from Kleinman, Baron, and Levison 1970).

The OCM has been employed to describe several situations with multiple input-output manual control actions. It contains a structured connection with the *human model of reference* developed in the previous Chapter (Figure 1.5) and based on the four cognitive functions, *Perception, Interpretation, Planning,* and *Execution.* OCM is quite able to cover a wide range of supervisory control tasks of modern control systems (Pew and Baron, 1983).

A number of models have been developed on the basis of the OCM, further expanding the control-theoretic approach (Stassen, 1988, 1990). As an example, the OCM was considered for the study of the human control process of super-tankers, which are characterised by quite long time constants, e.g., minutes, to respond to the steering commands of the helmsman. This approach also includes a decision-making element to make new estimates and updates within the observer part of the human supervisor model.

For this reason, the model has been called Observer/Controller/Decision Model (OCDM) (Kok and Stassen, 1980).

Papenhuijzen and Stassen (1987) considered the problem of multiple crew behaviour by extending the previous model to include both navigator and helmsman in large super-tankers.

In further studies, the problem of the navigator's planning and control behaviour was considered from the modelling perspectives of a control theoretic approach and a fuzzy set model (Papenhuijzen, 1994). The objective of this study was to investigate limits and advantages of these two techniques for studying human machine interaction in complex highly automated environments.

A very interesting human behaviour model, based on the OCM paradigm, is PROCRU (Procedure Oriented Crew model) (Baron and colleagues, 1980). This model includes the consideration of reasoning features, not explicitly based on process related variables, and multi-operator modelling, along with communications amongst operators. PROCRU will be discussed in more detail in the next Chapter, when a number of selected models and simulations will be compared with the requirements and specifications for a human-machine interaction simulation.

PROCRU represents, possibly, the most complete and developed expression of control-theory-based models, as it contains algorithms accounting for supervisory activity, as well as multi-operator situations with communication processes, and multiple tasks modelling.

In essence, Optimal Control Modelling shows many positive and powerful features enabling to simulate human behaviour including certain cognitive processes.

However, a number of drawbacks and limitations can also be identified. Firstly, the mathematical formalisms required to describe human supervisory behaviour by means of control theory is very complicated.

Then there exists an intrinsic limitation of the theory to consider cognitive processes not induced by system state variables, such as cyclic reasoning and repeated mental processes.

These two limitations have been well identified since the 80ies, raising some doubts about the possibility of control theory to fully model human behaviour in modern control and plant management activities.

As an example, in his book on engineering models of human machine interaction, Rouse states "... *Although these new formulations* (optimal control formulations) *look promising, one is inclined to wonder if the servomechanism analogy, and hence control theory, may not be approaching the limits of its applicability.*" (Rouse, 1980, p. 59).

Indeed, modelling cognitive types of tasks requires the use of different theories and techniques, which are better suited for simulating heuristic and non-mathematical thinking of humans. Moreover, specific programming environments and computer structures are necessary, which are able sustain knowledge based representations and problem solving strategies.

New theories and techniques have been developed in the 60ies and 70ies, with the specific objective to account for logical, heuristic, imprecise and non-formal reasoning processes typical of people. A typical example of these methods is Artificial Intelligence (AI), which became very popular and was promoted and proposed by many scientists as the revolutionary science for problem solving and simulation in moderns technology environments.

Even if AI has recently shown some limitations and, for certain aspects, has not fulfilled all its proclaimed power and expectations, many interesting and useful techniques and numerical formalisms have sprang from AI related research and development. We will review in the next section some of these theories and techniques, as they are largely applied and represent the building blocks of many cognitive simulations.

2.3 Engineering Methods for Planning and Decision Making

2.3.1 Fuzzy Set Theory

Fuzzy Set Theory has been developed with the purpose to describe human logical problem solving and approximate reasoning about variables or relationships between variables.

In particular, Fuzzy Set Theory allows to account for verbal, qualitative expressions utilised by experts and novices when discussing physical systems in natural language.

Expressions such as "short", "long", "high", "normal", or "low" relative to process variables, are the criteria to decide and perform certain control actions. The situations considered by fuzzy set theory are those that can not be defined by a crisp yes/no answers, but are characterised by a degree of membership to one or more of the possible answers.

Fuzzy set theory and mathematics were first introduced by Zadeh (1965). The central concept of the theory is the *membership function*, $\mu_A(x)$ of an element, which represents the degree of belonging of an element with property x in a fuzzy set A.

Membership functions can take any value between 0 and 1 ($0 \le \mu_A(x) \le 1$). In the special case where $\mu_A(x)$ can only take values 0 or 1, fuzzy set theory reduces to classical set theory.

There exist a variety of operations that can be performed by fuzzy theory, such as fuzzy statistics and fuzzy differential equations. For our purposes, it is sufficient to recall here, firstly, the basic operations of *union* and *intersection*, described by the following expressions:

$$\mu_{A \cup B}(x) = \max\,\left[\mu_A(x);\ \mu_B(x)\right] \tag{2.9}$$

$$\mu_{A \cap B}(x) = \min\,\left[\mu_A(x);\ \mu_B(x)\right] \tag{2.10}$$

These operations represent the degree of membership, of an element with property x, to *either* (Union) or to *both* (Intersection) the fuzzy sets A and B. Both these operations match the classical set theory, which are the special case of fuzzy operations.

The two other relevant operations are: *fuzzy relations* and *fuzzy compositions.* But, they are not as straightforward as union and intersection. These two operations will now be briefly described and discussed using a simple example.

Fuzzy relations

Fuzzy relations are used to describe the degree to which particular values of properties are compatible. Fuzzy relations are built by fuzzy operations according to the logical correlation existing between the variables.

As an example, it is sufficient to consider a maintenance job and the relationship between the "shortness" of the maintenance operation, measured in terms of amount of time needed to perform of the maintenance task, and the "urgency", measured in terms of time available before restarting or returning the plant on the on-line operations (Figure 2.5).

A simple example of car maintenance tasks could be: "substituting a tyre" (short time required for performance) or "changing the gearbox" (long time required for performance) and the need to utilise the car "within few minutes or hours" (short time available to have the car in not-operating state) or "within few days or weeks" (lot of time available to have the car in not-operating state).

The following expression then applies:

$$\mu_{AxB}(s,t) = \min\,\left[\mu_A(s);\ \mu_B(t)\right] \tag{2.11}$$

which denotes the fact that the two membership functions for the values s and t of A and B are combined to produce a new membership function $\mu_{AxB}(s,\ t)$, by a fuzzy intersection. This gives the degree of compatibility of the two properties s and t.

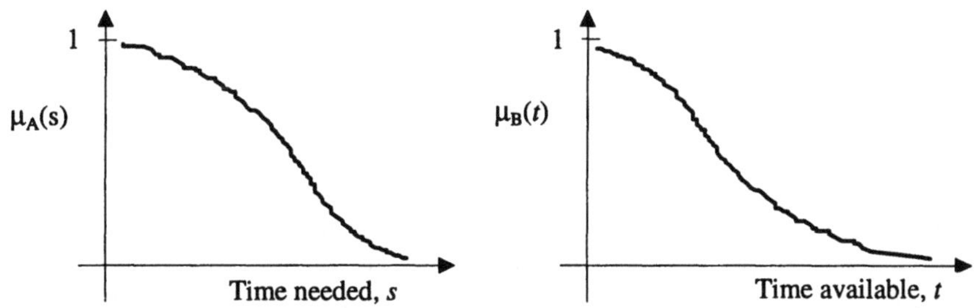

Figure 2.5. Membership functions for fuzzy sets shortness, $\mu_A(s)$, and urgency, $\mu_B(t)$, of a maintenance task.

Logically, a high urgency, i.e., little *time available* and thus $\mu_B(t) \approx 1$, is highly compatible with a rapidly performed maintenance operation, i.e., little *time needed* for the performance and thus $\mu_A(s) \approx 1$. This relation is well captured by the expression (2.11), which leads to high values of the *fuzzy relation,* i.e., $\mu_{AxB}(s,t) \approx 1$: "high" urgency combines well with "high" shortness of performance of a task.

Pursuing further the example of car maintenance tasks, we could say that the need to utilise the car "within few minutes or hours" is highly compatible with the operation of "substituting a tyre".

Moreover, other combinations between shortness and urgency seem to work reasonably well using the fuzzy relation given by the (2.11). In particular, little *time available*, i.e., $\mu_B(t) \approx 1$, and lot of *time needed* for performance of the maintenance operation, i.e., $\mu_A(s) << 1$ combine to give low values of the fuzzy relation $\mu_{AxB}(s,t)$, i.e., $\mu_{AxB}(s, t) << 1$. The same result is obtained when combining lot of *time available*, i.e., $\mu_B(t) << 1$, and little *time needed* for performance of the maintenance operation, i.e., $\mu_A(s) \approx 1$.

In the example of car maintenance tasks, we could say that the need to utilise the car "within few minutes or hours" combines poorly with the operation of "changing the gearbox". A similar result is obtained combining the need to utilise the car "within few days or weeks" with the operation of "substituting a tyre". In both cases, the fuzzy relation (2.11) would lead to $\mu_{AxB}(s,t) << 1$, which may be considered a logical and acceptable result.

It is, however, interesting to point out that the fuzzy relation between "low" shortness of tasks and "low" urgency, i.e., small values of $\mu_A(s)$ and $\mu_B(t)$, produces small values of $\mu_{AxB}(s, t)$, which may not be totally acceptable. Indeed, lengthy (long time needed for the performance) maintenance operations and lack of urgency (long time available) in returning the plant to operation should be compatible.

In the example of car maintenance tasks, the relation between the need to utilise the car "within few days or weeks" is logically compatible with the operation of "changing the gearbox". But this relation is not captured by equation (2.11), which would lead to low values of $\mu_{AxB}(s, t)$, instead of expected high values.

Certainly, the relations that are formulated can not be arbitrary and need to be carefully thought out, in order to avoid obvious non-senses.

Fuzzy Compositions

Fuzzy Compositions are used to represent conditional relationships between fuzzy variables, based of the values of membership. Once again, fuzzy compositions are built by fuzzy operations, i.e., fuzzy intersections or unions, according to the logical correlations between the variables and to the goal of the analysis.

Expanding further the maintenance example, let's assume that the goal of our analysis is to define whether to start or not a certain sequence of actions, given certain values of reference fuzzy variables.

In particular, the decision concerns whether to start or not a maintenance procedure on a specific part of a machinery, which has to be taken out of the on-line operation for a certain time, given certain values of the "duration" of the maintenance operation and the "urgency" to restart the plant:

- The *urgency* of this mission is related to the time available, *t*, to perform the tasks. The time available can be measured by the fuzzy *membership functions* "urgent", "little", "normal", "plenty".
- The *duration* of the mission is related to the time, *d*, needed to perform the operation of maintenance: duration can be measured by the fuzzy membership functions "low", "medium", or "high" (Figure 2.6).

The *rules* that we will assume to define whether or not to start the maintenance procedure can be defined as follows:

- "if the duration of performance is medium and there is little time, or if the time is urgent, then start".
- However, "if the duration of performance is high, and the time available is normal, then the procedure can be delayed".

Applying these two rules, the following expressions are obtained:

$$\mu_{start}(t) = \max\left[\mu_{AxB}(d_{med}, t_{little}); \mu_B(t_{urg.})\right] = \\ = \max\left\{\min\left[\mu_A(d_{med}); \mu_B(t_{little})\right]; \mu_B(t_{urg.})\right\} \tag{2.12}$$

$$\mu_{delay}(t) = \mu_{AxB}(d_{high}, t_{normal}) = \min\left[\mu_A(d_{high}); \mu_B(t_{normal})\right] \tag{2.13}$$

With reference to Figure 2.6 and according to the above expressions for fuzzy rules, Eqs. (2.12) - (2.13), the *membership functions* decision to start or delay the beginning of the procedure are calculated as follows:

$$\mu_{start}(d = x,\ t = y) = \max\{\min\ [0.2;\ 0.9];\ (0.4)\}\ =\ 0.4$$

$$\mu_{delay}(d = x,\ t = y) = \min[0.6;\ 0.1] =\ 0.1$$

This result means that the new membership functions "start the procedure" (μ_{start}) or "delay the procedure", (μ_{delay}), are satisfied at values 0.4 and 0.1 respectively.

A further criterion is then needed to decide whether to actually start or delay the procedure. This criterion can be based either on further fuzzy logical analysis or on assignment of crisp, threshold values.

Therefore, to complete this example, if the following threshold values were assumed:

$$If\ \left(\mu_{start} \geq \mu_{dalay} \cap \mu_{start} \geq 0.6\right)\ Then\ \ (start\ procedure)$$

$$If\ \left(\mu_{delay} \geq \mu_{start} \cup \mu_{start} < 0.6\right)\ Then\ \ (delay\ procedure)$$

then the procedure would be delayed for some time.

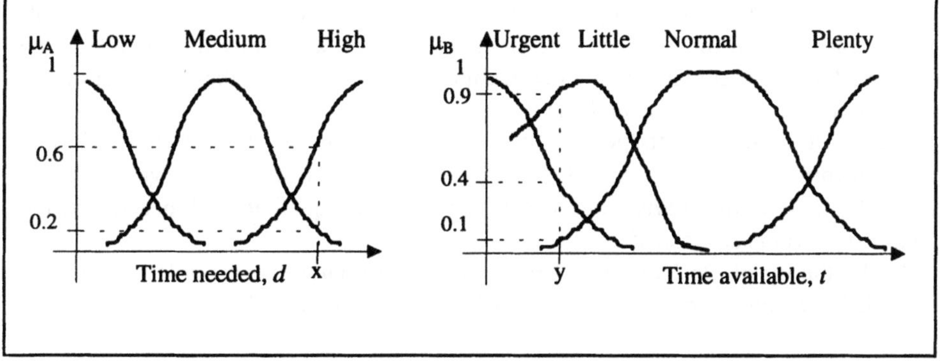

Figure 2.6. Fuzzy membership functions and fuzzy measures for decision making.

Fuzzy theory has been widely applied to represent several kinds of human-machine system interactions. In particular, it has been used for simulating *diagnosis* and *planning*, in real working context, as it captures the slightly imprecise nature of human logic thinking and communication.

This process also includes the technical and highly proceduralised behaviour of skilled operators, such as for examples operators of nuclear power plants, pilots, air traffic controllers or super-tanker navigators, etc.

Fuzzy set theory has been utilised for developing a high number of models of specific diagnostic tasks or decision making processes (Dubois and Prade, 1980, Zadeh, 1973), as well as for studying the ability of humans to develop general, context-free diagnosing faculty (Rouse 1978a, 1978b, 1979).

Other models based on fuzzy theories have been developed, as already mentioned, for simulating the "navigator" of super-tankers during highly demanding activities (Papenhuijzen and Stassen, 1987; Papenhuijzen, 1994)

In conclusion, it is important to stress that fuzzy sets and fuzzy logic are still utilised nowadays as a mathematical technique for larger developments aimed at overall models of reasoning, decision making and simulation of cognitive processes (Kosko, 1991).

These techniques and formal numerical expressions can then be considered, as the previously discussed control theory methods, the building blocks for the implementation of cognitive models in the simulation of human-machine system interaction.

2.3.2 Qualitative-Physics Theory

Qualitative-Physics is another theory for simulating human approximate reasoning about processes, by means of a quasi-numeric representation of control and balance equations (Bobrow, 1984).

A number of theoretical approaches have been developed on this idea. We will review the basic principles of some of them and we will show how they can be combined with Artificial Intelligence techniques, described in the next section, to arrive at a simulation of human behaviour.

The most important standpoint, common to all *Qualitative-Physics* approaches, is the transformation of any variable of a problem, x, in a qualitative variable, Λx, where the Λ is a *qualitative operator,* which gives the qualitative dimension to the quantitative value associated with x.

In the theory of De Kleer and Brown (1984), a qualitative variable can assume three different values, namely *0*, -, or +, which mean respectively:

0 = *normal, reference* or *zero* value;

\- = *negative* or *smaller* value than reference;

\+ = *positive* or *greater* value than reference.

The modelling process then consists in describing, in a heuristic fashion, the relationships between variables.

A mechanisms of inference has to be devised to develop predictions about physical systems and these are transformed by the above analytical description. However, the simplicity of these expressions (*0*, -, +) quickly makes the system unmanageable, in all those cases when correlations between variables are not clearly known.

For example, in combining the effects of two independent variables on a third variable, the overall result will be "+", if one of the independent variables has a positive effect and the other has a non-negative effect. The result would be "-" if one of the independent variables has a negative effect and the other has a non-positive effect; and, finally, the result would be *"0"* if both independent variables are *"0"* themselves. However, the sum of two independent variables of opposite effect (sign) on the dependent variable is undetermined.

This ambiguity can be solved by "branching" the calculations, i.e., considering all possible alternatives, each time such undetermined situation occurs. However, this procedure can quickly lead to a very large number of branches to be analysed, making the process difficult to manage and control.

This method is very useful when the equations describing the system are not known and when a static analysis is required. On the other hand, in dynamic conditions and when the basic equations are known, a more complex approach is needed.

For example, the model of Iwasaki and Simon (1986) or the model of Forbus (1984), allow a more structured representation of physical systems, including an accurate consideration of causal and structural descriptions of processes by referring to the governing equations.

Similarly, the methodology proposed by Kuipers (1984, 1986) is considered particularly appropriate for representing dynamic conditions. In particular, Kuipers' approach couples the three fundamental attributes (*0*, -, +) of a qualitative variable with the qualitative derivative of the variable, which can, thus, be associated with the values *decreasing, steady* or *increasing*. This allows to make a better analytical description of the dynamic evolution of the physical system.

In substance, *qualitative-physics theory* has the goal of simplifying the mathematical formalism and complexity of the equations describing the process state variables. However, one of the major problems that are encountered, when applying the theory, is that the number of qualitative-physics equations (although simple), which have to be considered in order to follow all possible paths and progress of phenomena, quickly diverges. In this way the initial implicitly is lost.

This major difficulty is in many cases solved by the implementation of qualitative-physics into computerised structures and simulations based on artificial intelligence approaches, inference engines and knowledge bases. An application of qualitative-physics implemented by artificial intelligence techniques will be shown in a forthcoming section.

2.3.3 Artificial Intelligence and Expert Systems

The goal of *Artificial Intelligence* (AI) is to capture and reproduce human cognitive processes in a computerised architecture and language, where a human being is seen as a processor of symbols (Cohen and Feigenbaum, 1986; Barr, Cohen and Feigenbaum, 1989)

In this chapter, we have firstly reviewed a number of engineering techniques based on the human-cybernetic analogy, which catch in their mathematical representation the feedback processes of sensory-motor activity. Then we have shown how fuzzy set and qualitative-physics theories can be used to represent, by logical and numerical expressions, the imprecise and approximate reasoning and mental processes typical of people, at any level of expertise.

Artificial Intelligence takes a leap forward, as it tries to account for indirect inferential processes. These are cognitive processes not directly activated by the interaction with the physical process or environment, but steer and resort to the vast amount of knowledge base not immediately linked with the ongoing control activity.

Artificial Intelligence and its major application elaboration, i.e., *Expert Systems* (Hayes-Roth, Waterman and Lenat, 1983), have been developed to support people in performing their tasks, embedding expert knowledge, control heuristics and symbolic representation.

In this sense expert systems can not be utilised for reproducing human behaviour, as they are oriented towards what people "must" know rather than what people "actually" know, and as they are not very well suited for hill-defined tasks.

However, there are many aspects of expert systems that are relevant to human behaviour modelling. In particular, the way to represent and utilise knowledge, i.e., the forms and content of knowledge and the processes of knowledge activation, is an important aspect of expert systems that can be useful also for simulating human behaviour.

The main characteristics and architectures of expert systems technology will now be briefly reviewed.

Expert Systems Characteristics

The techniques utilised in artificial intelligence are strongly dependent upon the way in which the nature of human expertise and knowledge are formalised and represented.

For example, one could consider four main learning processes contributing to build up knowledge and supporting recognition and recall: Rules Induction, Analogy, Abduction, and Knowledge compilation.

Another way to represent knowledge and skill could consider two major stages: a declarative stage, in which facts about the domain are interpreted, and a procedural stage, in which the domain knowledge is embodied in procedures (Anderson, 1983).

This knowledge can be represented in many different ways, which are then utilised during the recall process. As an example, knowledge can be stored in a set of simple production rules between facts, or can be represented by *Frames* (Minsky,

1975), or *Scripts* (Schank and Abelson, 1977), which contain the description of a situation, a matching process and also the necessary or most common control activity associated with it.

Once the knowledge base is filled in by *frames,* or *scripts,* or by any other formats describing knowledge throughout logical expressions, the process of search can take place. In so doing, a large collection of heuristics, or "rules of thumb", are applied, some context-specific, others more formal. *Analogical reasoning* and *symbolic reasoning* are examples of formal heuristics (Vosniadou and Ortony, 1989).

In practice, the basic structure of an Expert System contains a "Knowledge Base" (KB) and "Inference Engine" (IE). The KB contains the basic knowledge of the domain, including facts, beliefs and heuristics own by the expert. The IE incorporates reasoning algorithms, which act on the KB and make use of input data in order to solve the stated problem and to provide an explanation for the proposed solution. In other words, the inference engine fires rules according to the built-in reasoning protocol.

In addition to providing solutions, the IE may modify and expand the knowledge base by adding new rules and knowledge. This process may also be considered as a learning mechanisms.

Expert Systems Generations and Architectures

The so called "first generation" of Expert Systems (Guida and Alty, 1985) is based on the concept of *production rules,* which rely substantially on *predicate calculus*:

< *If* (condition) *Then* (action) >

Two different procedures are used: *Forward Chaining* and *Backward Chaining.* In Forward Chaining, if a stored fact in the KB is verified in a rule, then the actions associated with the rule are performed and the content of the KB is updated.

In Backward Chaining, the right part of the rule ("action") is evaluated with reference to the goals contained in the current database. If it matches some of the goals, then it is checked whether the left part of the rule ("condition") refers to known parts of the current database. In case of positive answer, the matched goal is assumed achieved, and it becomes a new known fact.

Another important issue is the consideration of two different types of knowledge: *shallow* or *deep knowledge.* Shallow knowledge is made of the empirical assets and associations derived from observing a process with little or no knowledge about the physical phenomena and models that are associated with it.

Deep knowledge is made of theoretical background and field experience on phenomena and processes, and can be obtained by merging academic and empirical expertise.

Shallow knowledge is very plant, process or goal dependent, while deep knowledge is independent of any specific goal or functional view.

Deep knowledge is usually found in more recent developments of Expert Systems, where the knowledge bases and inference mechanisms are coupled by means of representations like *Frames* or *Scripts*.

These knowledge descriptions are usually included in computing architectures and languages which are particularly suited for non-mathematical formats and allow the development of very large sets of rules, reasoning processes, chaining of rules, inheritances links etc.

Artificial Intelligence languages, architectures and programming environment are nowadays very much developed and largely used. Some examples are: "Object Oriented Programming" methodology, "Blackboard Architecture", "Smalltalk" programming environment, and the programming languages "C", "C^{++}", "Lisp", etc. (Engelmore and Morgan, 1988; Capper, 1994).

The flexibility of these means of programming and of computer architectures makes them very interesting for any simulation purpose, concerned with describing knowledge and logical processes.

Moreover, the similarity existing between the concepts contained in Expert Systems and the objectives of human behaviour simulation, make them very useful for programming the human component in a human-machine interaction architecture.

2.4 Cognitive Task Analysis

Task analysis, like many of the other theories described earlier, was firstly developed in the 50ies, from the effort of attempting to formally decompose and describe overt human behaviour by a series of simple and elementary components (Skinner, 1957).

The analysis of tasks was, and still is, intended as a prerequisite for the assessment of the overall human-machine interaction, as it offers a way to structure, by formal expressions, the envelope of procedures, actions and contextual facts characterising human behaviour in working environments.

Task analysis theory, originally applied to language learning and consequently to learning performances manifested through verbal expressions, was naturally extended to wider contexts of working processes (Payne and Green, 1986).

However, as we have already mentioned several times, over the last decades the roles of humans in the control of processes have changed, becoming mainly of supervisors rather than active actors.

Similarly, also the analysis of tasks has been adapted to these new duties of the operators: from task analysis the approach became the well known and much debated *Cognitive Task Analysis* (CTA).

In the following sections, we will analyse scope, structures and forms of Cognitive Task Analysis. The outcomes of CTA will be discussed, with particular reference to the Human-Machine Interaction architecture, identifying how and which models and data may be affected by CTA.

2.4.1 Scope of Cognitive Task Analysis

By adding the attribute "cognitive", the focus of task analysis has been placed on mental and internal processes rather than on overt manifestation of behaviour.

Thus, while the objectives of task analysis have not changed, the methods of Cognitive Task Analysis had to be adapted to the new locus of interest of human-machine interaction analysis.

CTA can be defined as follows (Bonar and colleagues 1986):

Cognitive Task Analysis is a method that attempts to specify the interaction of mental procedures, factual knowledge and task objectives in the process of job performance.

The aspects of operator behaviour that are studied by cognitive task analysis can be summarised in:

- *Contents* of technical skills, or procedures, by which the human-machine interaction takes place, and consist mainly in the behavioural elements of task analysis.
- *Contexts*, in which technical skills are applied, which represent the contribution of working conditions, contextual and socio-technical factors affecting cognitive processes and overt behaviour.
- *Mental models*, which are the intermediary element between overt behaviour and cognitive process.

From this general definition, the importance and wide scope of Cognitive Task Analysis in the process of studying and simulating human-machine interaction is immediately clear. In particular, CTA can affect the whole study of human-machine interaction by establishing correlations, dependencies and links that are maintained and exploited throughout the analysis.

As an example, by CTA it is possible to perform the preliminary analysis of the working environment and working practices. Moreover, CTA can be applied in defining structures of cognitive plans, goal setting and links between tasks and goals.

CTA can be utilised for analysing people's level of expertise in order to identify training strategies and methods. Furthermore, CTA can be applied for the design of interfaces and control procedures, or for retrospective analysis of events in defining the context of human erroneous behaviour and for guiding the study and search for root causes.

This broad spectrum of application explains the important role associated with CTA both for *prospective* and *retrospective* types of analyses.

Different attempts have been made to categorise CTA methods, which can be grouped in many different ways. For example, Grant and Mayes (1991) have characterised existing approaches according to their objectives, namely: a) techniques that provide a task decomposition formalism matching the human's cognitive structure; b) techniques which make reference to a theory and model of human cognition; and c) techniques based on the observation of salient features of human cognition, not related to logical analysis of tasks, nor to models of cognition.

2.4.2 Structures and Forms of Cognitive Task Analysis

We will now discuss Cognitive Task Analysis from three different perspectives, which represent the main lines of development to be followed in order to reach the aims of CTA:

1. The first perspective is the *theoretical standpoint*, which is necessary to define what is contained in a CTA and how to obtain it.
2. The second perspective concerns the *formalisms* by which it is possible to represent, in a structured fashion, human planning and action sequences.
3. The third perspective relates to methods and empirical approaches to be applied in order to collect information and data in the field and working environments.

Theoretical Standpoint for Cognitive Task Analysis

Following the definition of Cognitive Task Analysis and considering that a task is always described with reference to a goal, it results essential to define how goals and tasks correlate in the process of job performance.

Goals, as the name implies, are the objectives to be reached by a procedure and a plan to be implemented. Goals are usually expressed in terms of state of the plant or physical process to be achieved by implementing a certain sequence of actions.

Tasks are aggregations of elementary actions or procedures (sub-tasks) to be carried out in a certain sequence or combination in order to reach the goal. Sub-tasks are made of their own specific goals and task sequences.

There is a "recursiveness" between goals and tasks by which the principal goal ("top goal") of a procedure may be reached. This "recursiveness" has been properly identified and discussed by Hollnagel, and can be referred as Goals-Means Task Analysis (GMTA) (Hollnagel, 1993a) (Figure 2.7).

Another important relationship between tasks and goals is the representation of tasks contributing to the attainment of a common goal. In particular, following the methodology MAD (Méthode Analytique de Description) (Scapin and Pierret-Golbreich, 1990; Sebillotte and Scapin, 1994), the following modes of tasks organisation can be envisaged: sequential tasks (*Seq*), parallel tasks (*Par*) alternative tasks (*Alt*), iterative tasks (*Loop*), and optional tasks (*Op*).

The meaning of these different modes of organisation are intuitive, as "sequential tasks" are tasks which must be performed in a pre-defined sequential order to reach the goal; "parallel tasks" are tasks whose order is not relevant for the attainment of the goal; "alternative tasks" are different ways to reach the goal and thus are performed in alternative to each other; "iterative tasks" are tasks to be carried several times, in the same form, to reach the final goal; and, finally, "optional tasks" are tasks not strictly necessary, but add certain characteristics that improve the final result, without changing its substantial effect.

Apart from *goals* and *tasks* it is important to define other basic components of a cognitive task analysis, such as: *elementary actions, pre-conditions,* and *post-conditions.*

Each task, in addition to its own goal, elementary actions and sub-tasks, is associated with a number of *pre-conditions* and *post-conditions*. Pre-conditions correspond to success conditions of previous tasks, plant state and environmental events that must be identified and permit the start-up of the task.

Post-conditions are the results of the performance of a task and affect the plant state, related tasks, context and working environment.

From a more formal viewpoint, the combination of interconnected goals, tasks, and elementary actions leads to a "tree like" structure which is well suited to graphically represent the development of a procedure (Figure 2.8).

However, if each goal and task is associated with pre- and post-conditions the tree structure may become too cumbersome to develop and would need to be supported by a computerised tool and a specific programming language for managing inheritance and combination of attributes and dependencies.

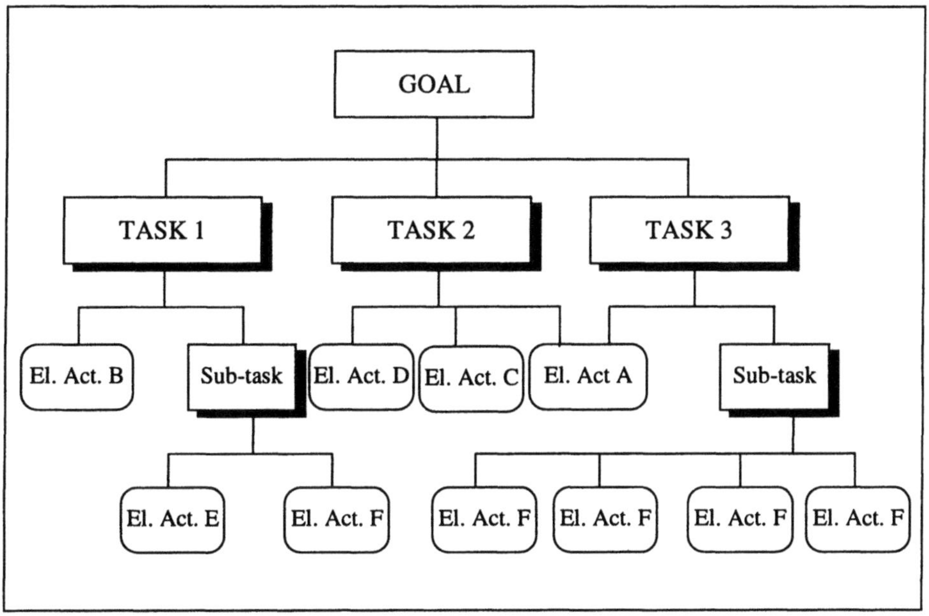

Figure 2.7. Structure of a goals-means task analysis.

Formalism for Representing Cognitive Task Analysis

Following the formalism of MAD, the combination of tasks can be organised hierarchically and can be represented graphically.

This formalism supports the analytical study of human planning processes and action sequences (Mo, 1997) (Figure 2.8).

Moreover, the use of object-base structure facilitates the development of a computerised method, which becomes a necessity when the tasks to be analysed are complex and contain many sub-tasks and elementary actions.

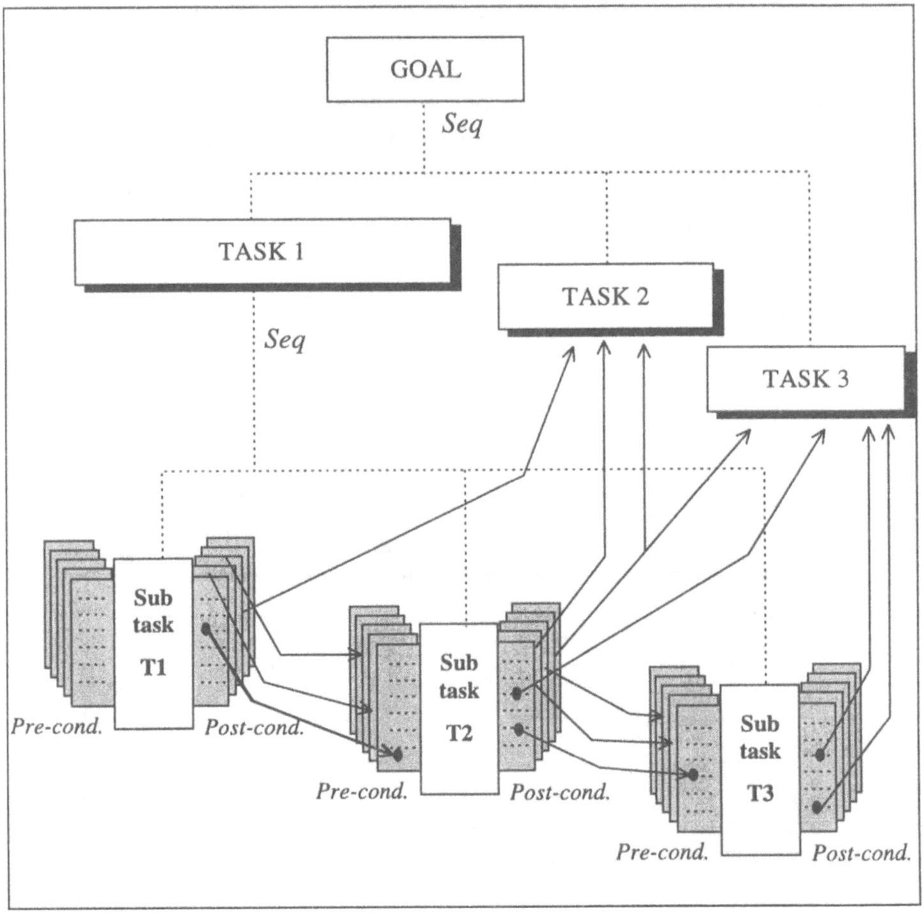

Figure 2.8. Basic structure of Plans and Tasks.

The structure described here for developing task analysis and building a tree-like graph containing the representation of conditions and modes of performance of tasks has to be utilised in combination with approaches for understanding and studying practices and behaviour in real working environments.

MAD contains, as part of the methodology, a well developed approach of data gathering (Sebillotte, 1995). Two specific techniques are proposed, namely interviews and trace analysis.

However, several other techniques exist for analysing field and working environment and for collecting data for task analysis. Some of these techniques will now be briefly described.

Data Collection Approaches

A number of approaches can be exploited to develop Cognitive Task Analysis, and to determine data relative to cognitive structures and processes underlying job performance.

These methods and techniques cover mainly field assessment of working environment and modelling techniques of logical and behavioural performances (Redding and Seamster, 1994).

In brief, the mostly utilised approaches are:

- *Cognitive interviewing* techniques, based on semi-structured or structured interviewing methods designed to elicit information about mental processes.
- *Protocol analysis*, by which operators are involved in reviewing, either concurrently or retrospectively, the actual performance of a task, carried out either by themselves ("auto-confrontation") or by others (Ericsson and Simon, 1984).
- *Psychological scaling* methods, which are applied for determining knowledge structures not readily derivable by more direct methods. Operators are involved in rating, scaling and ranking task-relevant knowledge. Questionnaires and statistical methods are used for these methods.
- *Cognitive and performance modelling*, by which tasks performance, planning and goal setting are analysed with respect to a theoretical paradigm of cognitive behaviour. Such analysis will then support and establish the boundary of application of a simulation of cognition.
- *Error analysis*, which requires the differentiation and consideration of error types and modes, and cognitive processes. Error analysis aims at establishing a correlation between 1) internal causes, related to functional characteristics of human cognition ("genotypes"), 2) external manifestations of overt behaviour ("phenotypes"), and 3) processes of detection-interpretation-planning and action typical of cognitive modelling (Hollnagel, 1991).

2.4.3 Outcomes of Cognitive Task Analysis

The effort of applying the theories of Cognitive Task Analysis in a formal development, coupled with field study, as described in the previous section, demands an extensive amount of work of preparation and, if the working environment and the tasks to be studied are complex, the application becomes very time consuming.

Fortunately, this type of methodological and analytical work does not need to be fully carried out in all studies of human-machine interaction.

Sometimes, scope and objective of the analysis are limited and the task analysis can be performed by simple observations or by theoretical description of existing procedures. In other words, the outcome of the task analysis varies according to the objectives of its application.

In our case, we are interested in the impact that task analysis has on the development of a human-machine simulation framework.

Therefore, if we consider the reference Data and Human-Machine Interaction Architecture of Figure 1.13, we can argue that the cognitive task analysis contributes: 1) to the development of formal connections between Human-Machine Models with Taxonomies and Data, and 2) to generating and structuring the Human Model itself.

As shown in Figure 2.9, CTA leads to the definition of three major elements of the Human-Machine Interaction Architecture:

1. The definition of the way in which procedures are theoretically and actually carried out (*performance of procedures*);
2. The determination of internal/external *factors affecting behaviour*, i.e., the environmental contextual and socio-technical elements influencing cognitive and behavioural performance;
3. The identification of *forms of erroneous behaviour*.

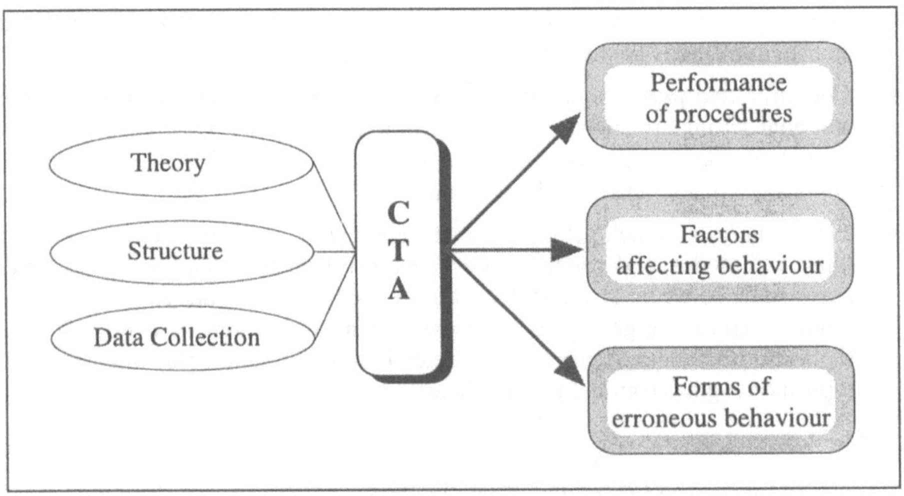

Figure 2.9. Outcome of Cognitive Task Analysis.

Performance of Procedures

Operating and emergency procedures are usually designed and written during the early phases of plant design and development.

The human behaviour model can be developed on the basis of the procedure alone. In this case, tasks and goals can be organised in a hierarchical tree-like structure, able to include cognitive and behavioural aspects, but without heuristics or real environmental aspects.

This can be considered a *normative* model of human behaviour, embodying only the required performances, as planned by the designer.

However, this type of modelling does not capture actual performances, as they result from their application in a certain working context and, in many cases, in environmental conditions that change over time. Moreover, these performances represent essential contributors to the development of a model and simulation of human behaviour.

The need to include in a simulation either actual human performances, or *normative* representations of procedures depend on the objectives and uses of the human simulation.

Cognitive Task Analysis is able to consider both types of behaviour, by performing more or less detailed field and data analysis.

For example, the evaluation of the proficiency of procedures to fulfil their goals can be carried out by studying the human-machine interaction with a human behaviour model based on a *normative* simulation.

On the other hand, the study of a special accident configuration for safety analysis, demands the application of a human behaviour model based on the actual performance of tasks.

In any case, formal theoretical representation of procedures and field studies of real environments of human-machine interactions, are essential contributors to the development of sound cognitive task analyses and human behaviour models.

Factors Affecting Behaviour

Contextual and environmental factors affecting behaviour can be identified by field studies and assessment approaches.

These factors are not necessarily important for the development of a model of behaviour, as far as the simulation of cognitive functions are concerned. However, they become essential in the definition of the dynamic aspects of the human-machine interaction.

During the simulation of plant and human behaviour, a number of conditions are generated and evolve in time:

- External events, component failures and random facts may modify the context and plant performance, causing inappropriate human behaviour.
- Socio-technical conditions and internal factors, such as training, culture, workload, and stress, may also engender inappropriate behaviour and human error.

One of the aims of CTA is the identification of these factors.

It is important to emphasise, here, that these factors are studied, independently of a specific erroneous behaviour to which they may be associated with, but they are rather evaluated as a sort of "catalysts" of behaviour.

The following principle, that is obvious in reality, is crucial, but is not easy to respect in developing simulation:

> **Internal/external factors affect human behaviour in a dynamic fashion, i.e., they intervene and modify the cognitive function or process which is in progress at time of their occurrence.**

Consequently, if one of such factors appears during a *perception*, then the perception function is modified by, for example, altering the perceived values or quantities or by short-cutting the function as a whole. If the same factor had appeared during a *planning*, the effects would have been different, e.g., it would have caused an erroneous plan or the selection of an inappropriate task.

Both these types of errors would then actually appear as erroneous performance of actions (overt behaviour), even if the real errors were made at different cognitive levels.

The internal/external factors affecting behaviour are part of the variables and parameters that contribute to coupling the architectures of data and human-machine simulation.

They are generated during the interaction between the machine and the human simulations and may also appear as consequence of random events.

Internal/external factors are then fed from the Human-Machine Interaction Architecture to the Data Architecture, in order to search in the database for consequences and related events. These are then returned to the Human-Machine Interaction Architecture in terms of *modes and types* of human *errors* and plant *failures* (Figure 2.10).

Forms of Erroneous Behaviour

The result of an inappropriate behaviour, as the consequence of either an external event, or a random or internal occurrence, must be defined in numerical or formal expressions, in order to allow the development of the simulation of human-machine interaction.

The need to define a "reasonable" spectrum of inappropriate behaviour derives from the fact that it is impossible to consider all "rational" forms of behaviour which differ from the expected one, in presence of external or internal factors.

These forms of inappropriate human behaviour and malfunctions of components are the connection of taxonomies of erroneous behaviour and plant failures with the human and machine simulations (Figure 2.10).

In other words, once the fact that an erroneous behaviour or a malfunction might occur has been ascertained, either as a consequence of an external, internal or random factor, the problem becomes to identify *how* and *which form* the error or the failure could take.

By the application of Cognitive Task Analysis, it is possible to determine a spectrum of manifestations of erroneous behaviour and of matching modelling characteristics that that should be included in the model and simulation human behaviour and that are sufficient for the study being carried out.

In any case, also this third step of CTA is essential for defining and limiting the scope and range of cognitive processes described by the Human Model.

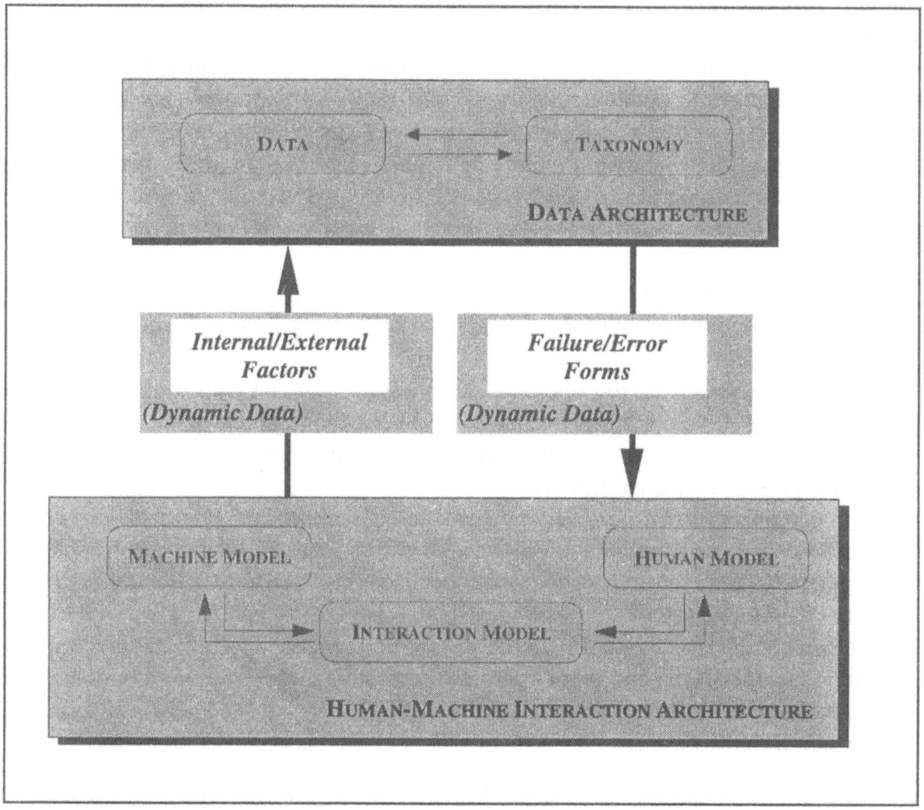

Figure 2.10. *Internal/External Factors* and *Forms of Errors and Failures*, derived from CTA for human -machine system simulation.

2.5 Qualitative-Physics Model for the Control of a Steam Generator

We will now show how *Artificial Intelligence* approaches and *Qualitative-Physics* can be combined for the modelling and simulation of an operator reasoning and action performance in controlling a Steam Generator (SG).

2.5.1 Structure of Models, Simulations and Data

In this case study a first example is shown of the Human-Machine Interaction Architecture discussed in Chapter 1.

The objective of the study is the development of a model and simulation of an operator's reaction to sudden steam generator faults or malfunctions, demanding a rapid formulation of a strategic plan and a series of appropriate control actions.

It is assumed that the steam generator is controlled automatically during normal operations and the operator's only functions are to supervise the control process and to intervene in case of dynamic operational or emergency conditions (Cacciabue, Guida and Pace, 1987).

The main standpoints of our work is that the cognitive processes underlying operator's behaviour can be of three different types, according to the actual situation facing the operator:

1. In *normal situations*, the operator can observe the plant evolution in automatic control and develops *deep reasoning* and knowledge about plant structures and functions. To develop deep knowledge the operator relies on qualitative-physics principles of SG processes.
2. In *dynamic operational situations* the operator can resort to *empirical reasoning*, based on shallow knowledge which directly relates plant behaviour to possible interventions. The operator relies on a "qualitative" model of control and regulation system.
3. In *emergency situations*, the operator must perform the appropriate interventions in a very short time and often operates under stress. The operator exploits a kind of *immediate reasoning*, mostly based on remembering ready-for-use solutions, acquired in the past.

The use of an *AI* approach coupled with *Qualitative-Physics* has been assumed as a fundamental design choice for the representation of the three levels of mental processes considered, namely (Chandrasekaran and Milne, 1985; Guida, 1985):

- *qualitative modelling* for *deep reasoning*;
- *qualitative production rules* for *empirical reasoning*; and
- *simple transition rules* for *immediate reasoning*.

With respect to the structure of reference Human-Machine Interaction Architecture (Figure 1.13) and to the Human and Machine Models (Figures 1.5, 1.6), a number of observations can be made.

Human Model

- *Memory/ Knowledge Base*

Normal situations are utilised to build up the *Knowledge Base* (KB) with the necessary rules, heuristics and frames. These are then activated in transient or emergency conditions. Such rules, heuristics and frames are developed by observing the behaviour of physical state variables and control system in normal operations and by applying a qualitative-physics model of the plant and physical processes.

- *Allocation of Resources*

The process of *Allocation of Resources* is simply based on the "gravity" of the initiating event and related transient, and determines whether the operator will use shallow knowledge and empirical reasoning or will select the actions to be carried out directly from the stored knowledge in the KB.

- *Cognitive Functions*

As far as *Cognitive Functions* are concerned, this model presumes perfect perception of all available information and correct diagnosis of the ongoing transient. Only *Planing* and *Action Performance* are simulated in more detail. Planning is the main function, represented by the qualitative-physics model, while action performance is a straightforward consequence of planning.

Machine Model

- *Plant*

All the physical quantities and state variables governing the heat transfer in the steam generator are assumed to be known.

In particular, the power exchanged at all times between the primary and secondary circuits is represented by an analytical, time dependent, expression. All steam generator components, such as valves, indicators, regulators etc., are assumed to operate according to design specifications.

A quantitative model of the steam generator has been developed with the objective to generate the necessary body of data and rules for filling the knowledge base. The same model and simulation has been applied for simulating the interaction of the operator with the steam generator.

- *Context* and *Environment*

Context and *Environment* are considered only for what concerns the effect that the seriousness of the transient has on the level of reasoning of the operator, i.e., on whether the operator reacts to the context and external stimuli by using empirical reasoning or immediate knowledge.

Interaction Model

The *Interaction Model* has been kept very simple, as the objective of this study is not related to the analysis of system failures or human errors and the Data Architecture has also been kept very simple.

Consequently, the interaction model allows transfer of all necessary and available information to and from human and machine behaviour model.

Data and Taxonomy

No analysis nor *taxonomy* of human error was developed for this case study and the *data* necessary for the study consist only in input and initial conditions of the plant simulation and in the context characteristics that define the gravity of the situation.

No extended *Cognitive Task Analysis* has been performed of the likely expected and actual performances of the operator, as the sole objective of this case study is to show how to develop of a model based on qualitative-physics and artificial intelligence techniques and not the study of a real working context.

For our purposes, the only control task of the operator is to maintain the level of water in the steam generator within acceptable ranges. In so doing the operator regulates the inlet mass flows.

2.5.2 Quantitative and Qualitative-Physics Models of Physical Processes

The basic functions of a Steam Generator (SG) and the most important process state variables are simply represented in Figure 2.11. This simplification is used for describing *quantitative* and *qualitative-physics* behaviour of the *physical processes* starting from the basic conservation principles and balance equations.

Moreover, as the elaboration of shallow knowledge requires also reasoning about the behaviour of the regulation and control system of SG, an appropriate model needs to be developed of the *regulator* of SG.

We will describe first the quantitative and qualitative-physics models of physical processes and then the quantitative and qualitative-physics models of the regulator. These four models will lead to the overall quantitative and qualitative-physics model of the human machine system.

The *qualitative-physics* models will serve for representing operator's behaviour and will generate the control actions, while the *quantitative* models will describe the machine response and dynamic physical behaviour.

Quantitative Model of Physical Processes

The *quantitative model* of the *physical processes* of a Steam Generator is devoted to simulate the main plant state variables and uses traditional system modelling techniques.

A simplified representation of a SG is shown in Figure 2.11.

In a SG, heat is transferred from a primary (hot) fluid to a secondary (cold) fluid which is vaporised. Primary and secondary fluids flow in separate circuits and do not intermix.

Two simplifying assumptions are made:

1. The exchanged power between primary and secondary circuit is known and, thus, only phenomena occurring in the secondary side of the steam generator can be considered.
2. Complete mixture exists at each time instant between vapour and liquid inside the steam generator.

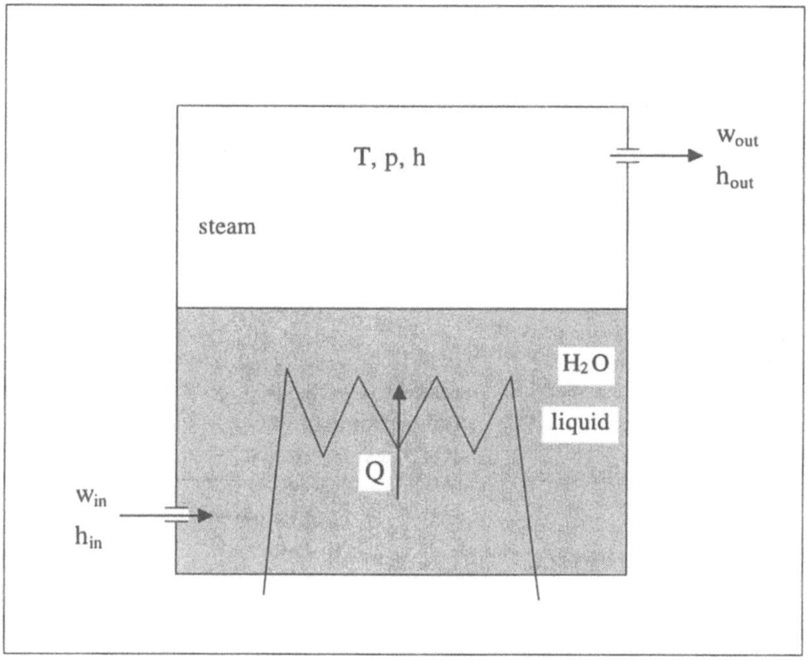

Figure 2.11. Simplified representation of a Steam Generator for qualitative-physics modelling.

The overall mass and energy conservation laws can be applied and the following equations are obtained:

$$\frac{dM}{dt} = w_{in} - w_{out} \tag{2.14}$$

$$\frac{d(Mh)}{dt} = V_{tot}\frac{dp}{dt} + Q + w_{in}h_{in} - w_{out}h_{out} \tag{2.15}$$

After few steps of development and simplification, the following equations can be obtained:

$$\frac{dh}{dt} = \frac{1}{V_{tot}(\frac{\partial v}{\partial p} + v\frac{\partial v}{\partial h})}\left[v\frac{\partial v}{\partial p}Q_{eff} + v^3(w_{out} - w_{in}) \right] \tag{2.16}$$

$$\frac{dp}{dt} = \frac{1}{V_{tot}(\frac{\partial v}{\partial p} + v\frac{\partial v}{\partial h})}\left[-v\frac{\partial v}{\partial h}Q_{eff} + v^2(w_{out} - w_{in}) \right] \tag{2.17}$$

where:

h	enthalpy	t	time	p	pressure
v	specific volume	w	mass flow	x	mass vapour quality
M	total mass of water	T	temperature	Q	exchanged power
V	volume	z	water level	c	specific heat

and

$$Q_{eff} = Q + w_{in}(h_{in} - h) - w_{out}(h_{out} - h) \tag{2.18}$$

and the following definitions apply:

$$x = \frac{h - h_{liq}}{h_{vap} - h_{liq}} \tag{2.19}$$

$$v = v_{liq} + x(v_{vap} - v_{liq}) \tag{2.20}$$

$$\frac{\partial v}{\partial p} = \frac{v_{vap} - v_{liq}}{h_{vap} - h_{liq}} \tag{2.21}$$

$$\frac{\partial v}{\partial p} = \frac{v[T(p)] - v[T(p - 0.5)]}{0.5} \tag{2.22}$$

The equations for the state variables enthalpy and specific volume are evaluated for the liquid and vapour conditions, under the assumption of equilibrium temperature and pressure.

The following relations are utilised:

h_{liq}, h_{vap}, v_{vap}, $T(p)$ *from tables or polynomials (Meyer - Pittroff)*

or

$$
\begin{aligned}
h_{liq} &= c_p T \\
h_{vap} &= c_p T + \lambda \\
v_{vap} &= v_{vap}(T) \\
T &= T(p) = \textit{equilibrium liquid - vapour}
\end{aligned}
\tag{2.23a}
$$

$$
\begin{aligned}
h_{in}(t) &= \textit{constant} \;\; \textit{or} \;\; = h_{in}(t - \Delta t) \\
h_{out}(t) &= h_{out}(t - \Delta t) \approx h(t - \Delta t) \\
w_{out}(t) &= f\left[p(t - \Delta t)\right] \;\; \textit{or} = \; w_{out}(t - \Delta t) \\
w_{in}(t) &= w_{in} \;\; \textit{from operator model}
\end{aligned}
\tag{2.23b}
$$

A computer simulation of the steam generator has been developed by the numerical solution of equations (2.16) and (2.17) using the simple explicit Euler method.

By this simulation, the values of pressures and enthalpies are calculated over time intervals, in consideration also of the control and regulation actions, which in essence affect inlet mass flows, w_{in}.

Qualitative-Physics Model of Physical Processes

The *qualitative-physics model* of the *physical processes* of a Steam Generator is based on the above quantitative equations (Eqs. 2.16 - 2.23).

Further knowledge about the SG performance is also exploited by analysing the actual dynamic variations of the involved variables during normal steam generator operations.

Using the quantitative simulator it is indeed possible to evaluate the qualitative behaviour of the SG in a very large number of situations, and to discover, by statistical analysis, new qualitative properties of system behaviour.

This information can then be combined with the arithmetic of integer numbers in order to solve the major problem encountered in traditional qualitative-physics simulation and thus control or limit the exponential generation of alternative trajectories of physical state variables in dynamic conditions. For this purpose, it is necessary, however, to modify the definition of a qualitative-physics variable, given in the previous sections.

Following the definitions of *qualitative-physics,* if a is a generic quantity, Λa denotes the corresponding qualitative-physics variable. But now, the qualitative-physics operator Λ, may be *any* integer number, and it represents a qualitative measure of the corresponding quantity a (Pace, 1987).

For example, if Λp=0 is the steady-state value of the pressure, or the normal value of pressure level in the steam generator, Λp= -1 represents a value that is generically lower than the landmark, while Λp= +1 denotes a generically higher value. Values of Λp= -2 or Λp= +2 represent differences from the landmark somewhat lower (Λp= -2) or greater (Λp= +2) than the previous ones. These values are, to a large extent, arbitrary and can be chosen according to the characteristics of the system under study.

Following this approach, the equations of SG, (2.16) and (2.17), are first transformed into qualitative-physics expressions, keeping the sign of partial derivatives, and then, using the information and insight derived from the study of the quantitative simulation, the actual contribution of each individual derivative is evaluated. These contributions are represented in the qualitative-physics equations by appropriate integer values associated with the qualitative-physics operator Λ.

In particular, it can be observed that in Eqs. (2.16) and (2.17), the dependent variables, p and h, are related to v, V_{tot}, Q_{eff}, w_{in}, and w_{out}. Moreover, from Eqs. (2.18) - (2.23) it can be shown that:

$$v = v(p,h)$$

$$V_{tot} = constant$$

$$Q_{eff} = Q_{eff}(Q, h_{in}, w_{in}, w_{out})$$

As pressure and enthalpy do not depend qualitatively from the values of h_{out}, the equations (2.16) and (2.17) can take the following qualitative-physics form:

$$\Lambda \frac{dh}{dt} = \mu'_Q \Lambda Q + \mu'_{h_{in}} \Lambda h_{in} + \mu'_{w_{in}} \Lambda w_{in} + \mu'_{w_{out}} \Lambda w_{out} \tag{2.24}$$

$$\Lambda \frac{dp}{dt} = \mu''_Q \Lambda Q + \mu''_{h_{in}} \Lambda h_{in} + \mu''_{w_{in}} \Lambda w_{in} + \mu''_{w_{out}} \Lambda w_{out} \tag{2.25}$$

where:

μ'_i, μ''_i are weighting coefficients (integer numbers) for the qualitative-physics variable i, as derived from experience (in our case from a number of simulation runs performed with the quantitative simulator).

The model represented by these two equations is banal and unrealistic, as it simply shows that time dependent behaviour of pressure and enthalpy are exactly similar, apart from the weighting coefficients. Whereas it is known that one variable undergoes much faster dynamic transients than the other!

A refinement of the model is thus necessary, and can be obtained by considering more carefully the qualitative-physics relation between dependent and independent variables.

In particular, considering the dependence of the outlet mass flow, w_{out} , from steam generator pressure p and outlet pressure, the following equation applies:

$$w_{out} = k_{out}\sqrt{p - p_{out}} \tag{2.26}$$

where, k_{out} and p_{out} are constants.

In this way the variable w_{out} becomes a calculated variable, instead of an independent variable, which depends only on p.

Moreover, it can be shown that the qualitative-physics significance of w_{out} on the derivative of the enthalpy ($\Lambda dh/dt$) is irrelevant, while it continues to affect the derivative of the pressure ($\Lambda dp/dt$) through the value of p itself, as shown in Eq. (2.26).

Implementing the above considerations into equations (2.24) and (2.25), the following final qualitative-physics model is obtained:

$$\Lambda\frac{dh}{dt} = \mu'_Q\,\Lambda Q + \mu'_{h_{in}}\,\Lambda h_{in} + \mu'_{w_{in}}\,\Lambda w_{in} \tag{2.27}$$

$$\Lambda\frac{dp}{dt} = \mu''_Q\,\Lambda Q + \mu''_{h_{in}}\,\Lambda h_{in} + \mu''_{w_{in}}\,\Lambda w_{in} + \mu''_p\,\Lambda p \tag{2.28}$$

As already mentioned, the values of the weighting integer coefficients μ'_i and μ''_i are evaluated by experience. In our case, they have been obtained by observing the response of the steam generator in several transients, calculated using the quantitative simulator. The following values have been found:

$$\begin{aligned} &\mu'_Q = \mu'_{h_{in}} = 1 \\ &\mu'_{w_{in}} = \mu''_{w_{in}} = \mu''_p = -1 \\ &\mu''_Q = \mu''_{h_{in}} = +2 \end{aligned} \tag{2.29}$$

The dynamic qualitative prediction of the system behaviour is obtained by integrating equations (2.27) and (2.28), applying the coefficients defined in (2.29) and using Euler integration method.

In particular, as the qualitative-physics reasoning concepts must be applied also to numerical and simplifying algorithms, it is assumed that the qualitative measure of the *time step* is equal to the unity of reference, i.e., $\Lambda\Delta t=+1$. Therefore, the explicit Euler integration method implies that:

$$\Lambda h(t+1) = \Lambda h(t) + \Lambda \frac{dh(t)}{dt} \tag{2.30}$$

$$\Lambda p(t+1) = \Lambda p(t) + \Lambda \frac{dp(t)}{dt} \tag{2.31}$$

The system of equations (2.27) - (2.31) represent the final qualitative-physics model of a steam generator and shows the following main characteristics:

- The model is qualitative and causal but not ambiguous.
- The model is much simpler than both quantitative and traditional qualitative-physics approaches.
- The model is transparent to the operator and appropriate to support deep reasoning.

The application of the qualitative-physics model of state variables, *p* and *h*, for explaining normal steam generator behaviour simulates the *deep reasoning* process and contributes to generate rules and control responses ready for use.

These rules build up the knowledge base of the operator and represent a collection of qualitative transition rules concerning changes in the state of the steam generator obtained through application of appropriate inputs.

As an example, rules of the knowledge base have the following structure:

< *current-state, inputs* => *future-state* >

This means that, if the steam generator is presently in the *current-state*, and certain *inputs* are applied, it will reach the *future-state* in the next time step.

It can be assumed that this knowledge has been mostly gathered from experience, and is used for "*immediate*" reasoning about the plant every time a transient or emergency situation requires rapid understanding from the operator.

2.5.3 Quantitative and Qualitative-Physics Models of the Regulator

The analysis of a steam generator, oriented to the description of realistic behaviour, can not be carried out without evaluating the control system.

It is thus essential to include in the simulation some design principles of the *regulator*, which automatically controls the behaviour of the steam generator during normal operations.

The major concern of the control tasks is to keep the water level inside the steam generator at a steady level, by an appropriate regulation of inlet flow. To this purpose, three different actions can be considered with respect to the deviation of the level in the steam generator from its normal value: *Proportional*, *Integral*, and *Derivative* (*PID*) regulation.

The control system examined in this case study is shown in Figure 2.12. The system considers level and flow transducers (*LT*, *FT*) and a *PID* regulator, which evaluates the inlet mass flow (w_{in}) proportionally to:

- Measured deviations in levels and mass flows.
- Time derivatives of measured deviations in levels and mass flows.
- Integrals over time of measured deviations in levels and mass flows.

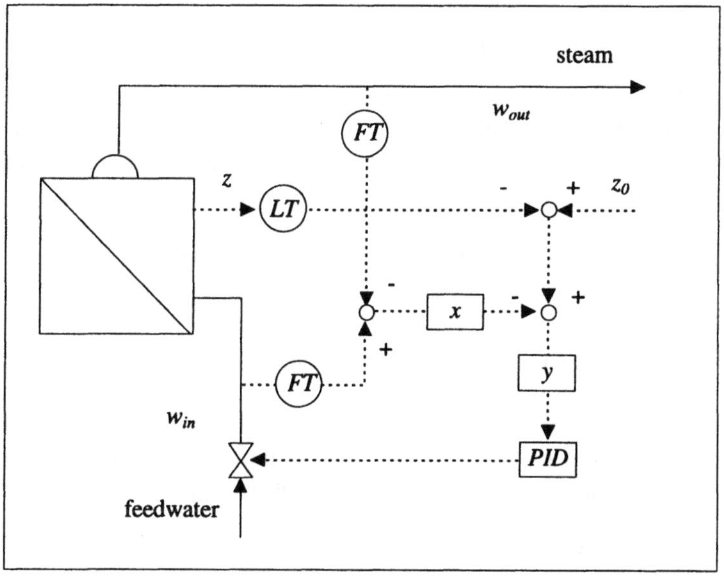

Figure 2.12. The automatic level controller of the Steam Generator.

Quantitative Model of Regulator

The measure of deviation (error) in level in the Steam Generator, considers the difference between the existing level (z) and the desired one (z_0), while the error in mass flows accounts for the difference between inlet and outlet mass flows. In steady conditions, these are equal. However, if they are different, a condition of initial draining or filling of water could occur and the level of the SG would gradually start decreasing or increasing, accordingly.

With reference to Figure 2.12 the following equations apply:

$$\varepsilon = y\left[(z_0 - z) - x(w_{in} - w_{out})\right] \tag{2.32}$$

$$w_{in} = k_p \varepsilon(t) + k_i \int_{t_0}^{t} \varepsilon(\tau)\, d\tau + k_d \frac{d\varepsilon(t)}{dt} \tag{2.33}$$

where:

ε is the error;

x, y are weighting factors, e.g., $y=3$, $x=0.6$;

k_p k_i k_d are determined utilising empirical methods well know in control theory.

Qualitative-Physics Model of Regulator

In order to develop a *qualitative-physics model* for the regulator, a method similar to the one developed for the qualitative-physics model of physical variables is used.

By applying the qualitative-physics operator Λ, the above control equations (2.32) and (2.33) take the following qualitative-physics expressions:

$$\Lambda w_{in} = \alpha_w \Lambda w_{in,proportional} + \beta_w \Lambda w_{in,derivative} + \gamma_w \Lambda w_{in,\ integral} \tag{2.34}$$

where:

α, β, γ are coefficients determined by experience.

In order to simplify the development of the qualitative-physics equations, we will limit the dependence of w_{in} only to the dynamic evolution of water level, z. Therefore, the following qualitative-physics equations apply:

$$\begin{aligned}
\Lambda w_{in,proportional} &= -\Lambda z(\Lambda t) \\
\Lambda w_{in,derivative} &= -\Lambda z(\Lambda t) + \Lambda z(\Lambda t - 1) \\
\Lambda w_{in,integral} &= -\Lambda z_{integral}(\Lambda t) = -\sum_{\Lambda\tau=\Lambda t_0}^{\Lambda t} \Lambda z(\Lambda \tau)
\end{aligned} \tag{2.35}$$

The negative correlations between the qualitative-physics variables Λw_{in} and Λz derives from the main control goal of maintaining constant the level inside the steam generator. Therefore, an increase in water level implies a decrease of inflow and vice-versa.

The final qualitative-physics model of the regulator can be represented by the following equation:

$$\Lambda w_{in} = \alpha\Lambda z(\Lambda t) + \beta\Lambda z(\Lambda t - 1) + \gamma\Lambda z_{integral}(\Lambda t) \quad (2.36)$$

As in the case of the qualitative-physics model of the physical processes, the coefficients α, β, γ, are determined by the study of the results of several runs of the quantitative simulator, and take the following values:

$\alpha = -2/3;$

$\beta = 1/3;$

$\gamma = -1/9.$

The intuitive meaning of equation (2.36) is straightforward. When the operator is controlling the steam generator, he/she can read the measure of the actual water level on the instrumentation panel. According to this observation the decision can be taken to increase, decrease, or keep constant the inlet flow.

The three effects outlined above can influence such decision: the first effect (*proportional*) appears whenever the level is not at its normal value and must be corrected; the second effect (*derivative*) predominates when the level is rapidly moving in one direction; and the third effect (*integral*) prevails when the level differs from its normal value for a long time, not-withstanding previous interventions.

This model is utilised to simulate the *empirical reasoning* and it is processed by a traditional rule interpreter: an inference engine, using an appropriate search strategy, concatenates rules together and develops, in this way, the deduction chains needed for solving the control problem at hand.

The qualitative-physics model of the regulator is used for generating more rules ready-for-use: when empirical reasoning is performed during a transient situation, a number of novel rules may be developed. These new rules can be fed to the knowledge base in order to further improve the possibility to generate, by *immediate reasoning*, frames and plans of action readily available for use.

2.5.4 Qualitative-Physics Simulation of Human Machine Interaction

The general configuration of the Human-Machine Interaction for the Steam Generator case is illustrated in Figure 2.13. It includes two parts connected by an interface.

The *Interface* plays the role of the *Interaction Model* of the reference Human-Machine Interaction Architecture (Figure 1.13) and, as already discussed, in this case, it is a simple model, that allows transfer of all required and calculated information between the *Human Model* and the *Machine Model*.

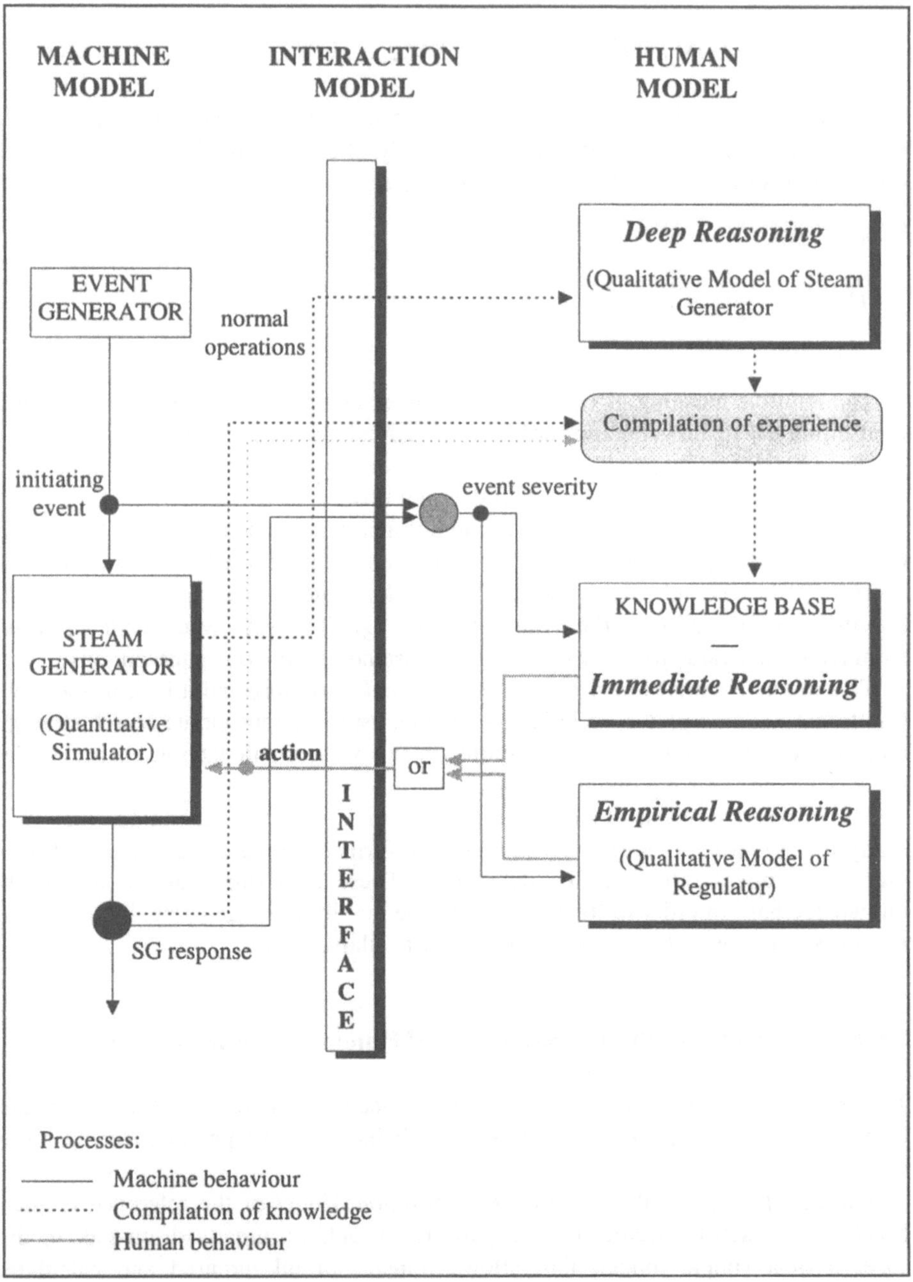

Figure 2.13. Overall architecture of HMI system based on qualitative-physics modelling.

The Machine Model is made of the SG simulator (quantitative simulation of physical processes, and regulator) and an Event Generator, that produces initiating and boundary conditions of transient sequences. All models and simulations are based on quantitative "classical" techniques.

The Human Model is characterised by the three different levels of operator's behaviour previously discussed: *deep*, *empirical* and *immediate reasoning*. The interface allows conversion from quantitative to qualitative-physics representations and vice versa.

The paths and successive iterations that are considered in the final qualitative-physics simulation are described in Figure 2.13. In particular, the following comments can be made:

- In normal operations the behaviour of the steam generator, controlled by the automatic regulator, is interpreted through the qualitative-physics model. *Deep reasoning* can be developed and the experience is compiled into rules for the *Knowledge Base* (dotted lines in Figure 2.13).

  Moreover, the SG responses are utilised for further compilation of empirical rules that are also fed to the *Knowledge Base*.

- During transient conditions, initiated by the *Event Generator*, the severity of the situations determines whether the control actions are produced by *empirical reasoning*, throughout the qualitative-physics model of the regulator, or are selected from the vast repository of built-in rules contained in the *Knowledge Base*. In this second case, we can outline the operator's behaviour as based on *immediate reasoning*.

  As an example, in the case of an emergency, a similarity matching algorithm looks in the *Knowledge Base* for the rule whose current-state best fits the present state of the steam generator and whose future-state is the desired normal value. If a suitable rule is found, the suggested inputs are considered as an appropriate intervention for facing the transient situation at hand.

  If no rules are found, then *empirical reasoning* is utilised for defining the most appropriate control action.

- The response of the steam generator to the control actions generates the successive set of decisions and actions and the dynamic evolution of the human-machine interaction.

  All control actions carried out by the operator and all responses of the steam generator are utilised to develop and improve the experience, in that they are fed back to the *Knowledge Base*, so as to expand the amount of rules available for controlling successive transients and events.

2.6 Summary

In this Chapter we have reviewed a number of engineering methods and cognitive task analysis approaches based on:

- *Numerical and computational means*, by which models may be transformed into working simulations.
- *Data and parameters*, derived from working contexts, which are necessary to sustain human-machine simulations.

This Chapter has been structured into four Sections: Engineering Methods for Detection, Diagnosis and Action; Engineering Methods for Planning and Decision Making; Cognitive Task Analysis; and an example of Qualitative-Physics Model for the Control of a Steam Generator.

Engineering Methods for Detection, Diagnosis and Action

Firstly, we have discussed a number of engineering and mathematical methods developed in the past and widely applied for simulating mainly sensory and physical activity of humans in direct manual control of plants.

In particular, we have examined modelling techniques based on Theory of Communication, Signal Detection Theory, and Control Theory.

Special attention has been dedicated to Control Theory, as this method has been substantially developed over the years and, in more recent formulations, based on optimal control and Kalman filtering techniques, has been utilised to describe more cognitive mechanisms and functions, as decision making and planning of operators in managing and supervising complex systems.

Engineering Methods for Planning and Decision Making

Subsequently, we have reviewed three analytical and numerical methods developed and applied for simulating approximate reasoning, planning and decision making, namely: Fuzzy Set Theory, Qualitative-Physics and Artificial Intelligence.

The most important rules and algorithms of Fuzzy Set Theory, Qualitative-Physics and Artificial Intelligence have been presented and discussed, in combination with simple examples of application to trivial cases, in order to demonstrate how they can be applied in practice.

The large diffusion of Fuzzy Set and Artificial Intelligence techniques lies in the fact that their programming languages and computational means are very much developed from the formal viewpoint and offer a very wide range of solutions to specific problems. Moreover, their algorithms are ready for application and are easily applied for transforming theories into numerical simulations.

Cognitive Task Analysis

The third section of the Chapter focused on Cognitive Task Analysis, which is a generic technique used for the definition and identification, in real working contexts, of data, parameters and factors, that support the development of models and simulations.

Cognitive Task Analysis offers the theoretical means and formalisms by which:

- Field studies can be designed and carried out with the aim of defining parameters and variables.
- Control procedures and tasks can be formalised and structured for detail examination and testing.

Cognitive Task Analysis is therefore a complementary technique for engineering methods, as the latter develop analytical and numerical algorithms for computerised simulations, while Cognitive Task Analysis supports the structuring and formalisation of tasks and procedures, and produces data, parameters and coefficients, from real working contexts.

Qualitative-Physics Model for the Control of a Steam Generator

The last section of this Chapter, showed how a model and simulation of human behaviour controlling a technical plant, i.e., the operator of a steam generator, could be developed by using one of the previously described engineering theory, i.e., Qualitative-Physics, coupled with an Artificial Intelligence architecture.

The simulations of the operator and steam generator have been developed according to the requirements of the reference Human-Machine Interaction Architecture described in Chapter 1.

The models of the steam generator and control system have been developed by applying standard balance equations and conservation principles. The quantitative simulators have been based on simplifying assumptions and classical numerical solution techniques. The qualitative-physics models of the physical processes and of the controller have been developed by applying the principles of *Qualitative-Physics*. They have then been applied to simulate operator reasoning and to compile rules for the knowledge base, so as to build a set of heuristics ready for use in the case of need in emergency and time demanding situations.

This first sample case contains a complete development of a simulation of human-machine interaction and gives the flavour of the complexity of this endeavour. Moreover, it prepares the basis for the case studies to be developed in the forthcoming Chapters.

CHAPTER 3
MODELS AND SIMULATIONS OF COGNITION

3.1 Introduction

This Chapter is the natural continuation of the previous one, where we have reviewed engineering methods, analytical means and experimental techniques to quantify human behaviour and to derive, from field observation, the data necessary for such quantification.

In this Chapter we aim at studying and comparing some of the most advanced theoretical models and at discussing a number of simulations of human behaviour, developed over the last 20 years.

In this way, we will complete the review of theories, simulation approaches and computational environments, which need to be merged for the development of a simulation of a human-machine interaction system. We will provide guidelines for the identification and development of the most appropriate modelling structure of human behaviour.

To reach these goals, we will firstly consider a number of theoretical models, containing the basic structure of *cognitive functions*, *memory/knowledge base*, and *allocation of resources*, described in Chapter 1. Slight differences between models will appear due to their specific theories, aims and domains of application.

These models will be compared with the *requirements* of the reference architecture of human-machine interaction, discussed in Chapter 1, and will also be compared with each other.

A number of *simulations* based on these theoretical models, utilising the engineering methods and the techniques of cognitive task analysis, discussed in Chapter 2, will be reviewed.

Guidelines will be discussed consisting of detailed procedural steps for developing a simulation of an operator engaged in supervisory and manual control.

Finally, as an example of application, a simulation described in the preceding sections will be discussed in more detail, showing how it can be developed, how it works and what type of results are produced in a sample case.

3.2 Review of Models of Cognition

The aim of this section is to review and compare some of the well known and commonly applied theoretical models developed in the last two decades.

We will disregard the computational and numerical aspects involved in transforming such models into simulations, as this will be considered in the following section, where we will discuss different computerised cognitive simulations.

In performing the survey of existing theoretical models of human behaviour, we will focus on the *requirements* previously developed for the formulation of a *Reference Model of Cognition* (RMC) in the human-machine interaction architecture. We will consider the ability of models to account for the *cognitive functions* and *cognitive processes*, that need to be included in a simulation approach to describe human behaviour. These are:

- Cognitive functions: *Perception, Interpretation, Planning* and *Execution (PIPE)*
- Cognitive processes: *Memory/Knowledge Base (KB)* and *Allocation of Resources (AoR)*

In addition to these basic constituents of the human model we will also consider:

- Types of connection that are active within the human model and combine cognitive functions and cognitive processes.
- Mechanisms that couple the human model with the machine and socio-technical context.

These connections and mechanisms describe and govern the dynamic behaviour of the model, by regulating transfer of data *within* the cognitive model, from one cognitive function to another and from cognitive processes to cognitive functions and vice-versa, but also *between* the cognitive model, machine model and data architecture, i.e., between the Human-Machine Interaction Architecture and the Data Architecture.

In particular, in surveying different cognitive models, three important aspects of these mechanisms will be considered, namely:

1. Whether a model is governed by sequential or cyclic connections between *cognitive functions*, *cognitive processes* and interactions with machine and context; and, in particular, if cognitive mechanisms and processes are only resulting from a machine and context stimuli or can be generated by pure cognitive activity.
2. Whether a model simply gives general indications or describes in detail the interconnections existing between cognitive functions and cognitive processes, by giving parameters, variables and basic principles by which they are generated and can be formalised.
3. Whether a model gives a shallow and generic description or develops a complete theoretical representation of effects and links between the human model and the context and machine.

In substance, as we are concerned with the analysis of theoretical models, we will mainly focus on whether a model offers a generic or a detailed account and description of the theoretical architecture and of the connections between its various constituents.

In the following review, we will describe shortly the principal peculiarities of a number of models and will examine forms and means proposed to represent cognitive functions, cognitive processes, types of connections and mechanisms of data transfer. Each model will be compared with the requirements of the *Reference Model of Cognition*, described in Chapter 1 (Figure 1.5).

At the end of the section, a comparison of these models will be carried out focusing on their original objectives and scope and on their compliance with the requirements. This comparison will aim at creating a general overview of the range of applicability for each model, as well as at offering, in a compact format, a catalogue of theoretical layouts and formalisms.

The format that will be adopted for comparison is shown in Table 3.1.

It contains major requirements and, for each requirement, a number of detailed formalisms and modelling peculiarities. In practice, the major requirements are the *cognitive functions* (*Perception/Interpretation/Planning/Execution*) and *cognitive processes* (*Allocation of Resources* and *Memory/Knowledge Base*) and the detailed formalisms are the connections and mechanisms governing them.

Six models will be described. They are typical archetypes of models developed from 1972 to 1993, and will be presented in chronological order, so that the progresses made by cognitive and human sciences over these two decades can be seen reflected in the advances shown by each model. The six models are: SHEL (Edwards, 1972); Human Problem Solving (Rouse, 1983); Step Ladder/Skill, Rule, Knowledge model (SL/SRK) (Rasmussen, 1986); Model of Fallible Machine (Reason, 1990); Basic Supervisory Control Paradigm (Sheridan, 1992); and Contextual Control Model (COCOM) (Hollnagel, 1993a).

Table 3.1 Requirements for models of cognition.

1.	***Perception/Interpretation/Planning/Execution (PIPE)***
1.1	*Types of Connection within model (Sequential/Cyclic)*
1.2	*Effects of Context/Machine on PIPE*
2.	***Allocation of Resources (AoR)***
2.1	*Connections of AoR and PIPE*
2.2	*Connections of AoR and Memory/KB*
2.3	*Effects of Context/Machine on AoR*
3.	***Memory/Knowledge Base (KB)***
3.1	*Connections of Memory/KB and PIPE*

3.2.1 SHEL Model

The SHEL model has been developed with the idea to describe the relationship between people and their activities in a Human Factors and Ergonomics perspective (Edwards, 1972, 1988).

Theoretical Description

The schematic representation of the SHEL model is shown in Figure 3.1. The original objective of SHEL is mainly the development of a framework to describe the connections existing between humans, called *Liveware* (L), and other elements of the working environments.

These other elements consist of:

- Physical sources, such as equipment, materials, interfaces, and machines, which may be termed *Hardware* (H).
- Rules, regulations, laws, procedures, customs, practices and habits governing the way in which a plant or a machine are operated. These may be called *Software* (S), as to characterise the difference between the other class of elements, i.e., Hardware, that are made of more visible physical characteristics.

As L-H and -S operate in real contexts, a further element is considered to complete the basic structure of the SHEL model: the *Environment* (E).

The Environment covers social and technical aspects of working contexts and, therefore, it considers all those factors over which a preliminary control can not be devised, but may only be evaluated and studied. In other words, the Environment consists of physical, economic, political and social factors that may indirectly influence human performance.

A basic "unit" of the SHEL structure consists of three elements, a *Liveware,* a *Hardware* and a *Software*, interacting with each other, and, each of them, interacting with the *Environment.* In reality, a system of humans and machines comprises several "units" operating in a certain environment. Thus, an additional dimension is considered, which accounts for the interaction within a group of people, and machines and this generates a modified SHEL diagram, as shown in Figure 3.1.

An important characteristic of the SHEL model is that it can be applied for studying any interaction occurring in working contexts.

All possible combinations between the four constituents of SHEL can be accounted for, bearing in mind that only one working Environment is considered. Therefore, the following inter-relations can be envisaged: L-H, L-S, L-E, L-L, H-H, H-S, H-E, S-E.

Naturally, the human interactions are of particular interest. In this sense, the SHEL architecture offers a complete framework and generic taxonomy to be used by designers and analysts in order to describe formally working environments and human-machine interactions.

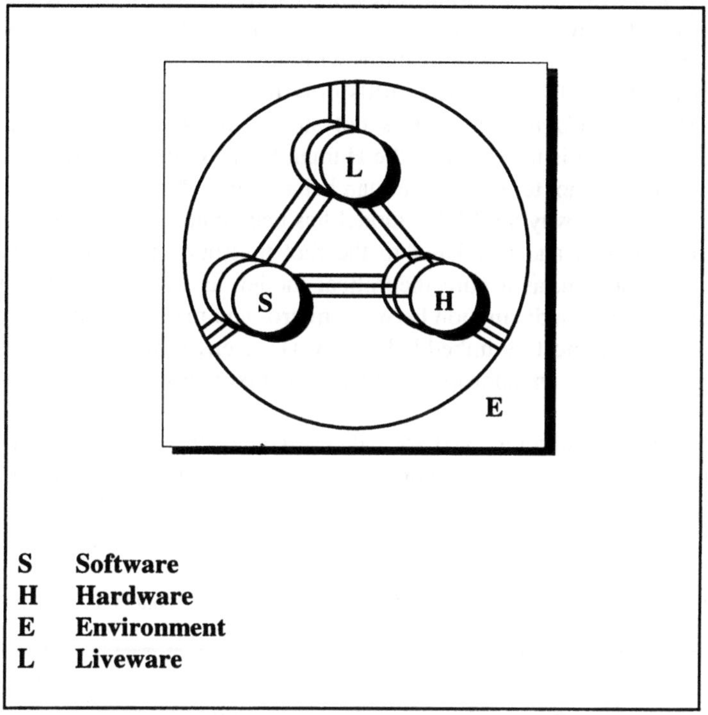

Figure 3.1 The SHEL Model (from Edwards, 1988).

Compliance with Requirements

SHEL has not been developed as a model of human behaviour for simulating dynamic human-machine interactions, but can be considered a paradigm of how to structure and decompose working environments with special attention to human aspects and conditions.

SHEL assumes the Information Processing System metaphor as reference paradigm and, consequently, it treats in an implicit way the four basic functions of human behaviour, i.e., Perception, Interpretation, Planning, and Execution (*PIPE*). However, only minor consideration is given to the connections between cognitive functions (Table 3.2).

As SHEL is not concerned with modelling dynamic interactions of humans and machines, it does not contain a model for active memory and knowledge base. This means that in SHEL there is no account for a model that can be applied to represent how the content of *KB* and the Memory participates to the development of plans and actions. A first consequence of this is that SHEL does not need to consider a mechanism for allocating resources to different human functions and activities.

Only human interactions with machine and working environments are considered in the model, through the connections L-H, L-S. This justifies a minimal correspondence of these modelling features with the requirements.

In summary, it results that SHEL fulfils only marginally the requirements of the *Reference Model of Cognition*. This is not surprising when we consider that the objective of the model is to represent the Human Factors and Ergonomics aspects of working contexts and not to describe human behaviour in dynamic conditions.

One could then ask why the SHEL model has been considered in this review. The reason is very simple, and is related to the fact that human modelling can not be developed in a vacuum, as a pure theoretical model, as we discussed in the initial chapter when we made a distinction between micro- and macro-cognition.

Human modelling must be imbedded in a working environment and technological system and, therefore, an appropriate reference taxonomy and data structure are needed.

SHEL offers such structure and formalism, which remains, more than 20 years after its original development, a complete reference paradigm for data organisation and classification.

Table 3.2 The SHEL model and the requirements for models of cognition.

REQUIREMENTS †		SHEL
1.	***Cognitive functions PIPE***	★
1.1	*Connection within PIPE*	★ - **Seq.**
1.2	*Effects of Context/Machine on PIPE*	★★
2.	***Allocation of Resources (AoR)***	★
2.1	*Connections of AoR and PIPE*	–
2.2	*Connections of AoR and Memory/KB*	–
2.3	*Effects of Context/Machine on AoR*	★★
3.	***Memory/Knowledge Base (KB)***	–
3.1	*Connections of Memory/KB and PIPE*	–

† The following measures apply:

–	Not treated	★★★	Modelled in detail
★	Implicit consideration	**Seq.**	Sequential
★★	Modelled at minimal level	**Cyc.**	Cyclic

3.2.2 Model of Human Problem Solving

The model of Human Problem Solving (HPS) of Rouse (1983) represents a simple, but complete, description of human cognition and behavioural processes. The model defines a number of activities and considers a simple mechanism that is recursively applied for all levels of problem solving activities.

Theoretical Description

The model developed by Rouse reflects the vast repertoire of research and studies carried out by this author in the domain of human-machine interaction (Rouse, 1978a, 1978b, 1979, 1980).

The model contains the notion of Human Problem Solving as being a generic modelling framework (paradigm), that considers **three levels** of human behaviour:

1. *Recognition and classification*;
2. *Planning*; and
3. *Execution and monitoring*.

The first level, *recognition and classification*, implies detection of an event or system state and its categorisation according to certain known scheme or classification. *Planning*, consists in selecting or devising a solution to the problem. *Execution and monitoring* is the actual implementation of the control actions.

This generic modelling framework incorporates the specific aspects of human behaviour typical of system failure management, namely: *detection, diagnosis,* and *compensation*. *Detection* is defined as the process whereby the human operator decides that an event or failure has occurred. *Diagnosis* is the process by which the causes and types of events are identified. *Compensation* is the activity by which a system is kept in operating condition, in presence of a fault or an ab-normal situation.

The second fundamental concept of the Human Problem Solving model consists in the formulation of a *single mechanism* which is applied to simulate human behaviour for each level of problem solving. This mechanism is based on the assumption that humans tend to use pattern recognition in context-specific conditions (as problem solving mechanism) rather than other forms of reasoning, such as for example heuristics, analytical or analogycal thinking etc. Hence, the mechanism considers two different ways of handling information: *state information* and *structural information* (Figure 3.2).

State information represents the attitude of people towards the search for familiar and recognisable "states" or configurations of the plant or system under control. This may be, in many circumstances, only a shallow representation of the true condition of the system. This approach to problem solving is always attempted first in order to find an immediate solution.

The next step is to decide whether such a solution, derived from state information, is acceptable or if a more complex problem solving approach has to be carried out.

In the first case, a "state oriented response" is applied. Otherwise, a process is started which requires the search for "structural information", based on deeper knowledge of functional system behaviour and properties as well as on generic knowledge of physical properties and logical chaining rules.

The results of such a process lead to a "structural oriented response" which is expected to be more accurate, but much more difficult to carry out than the previous one.

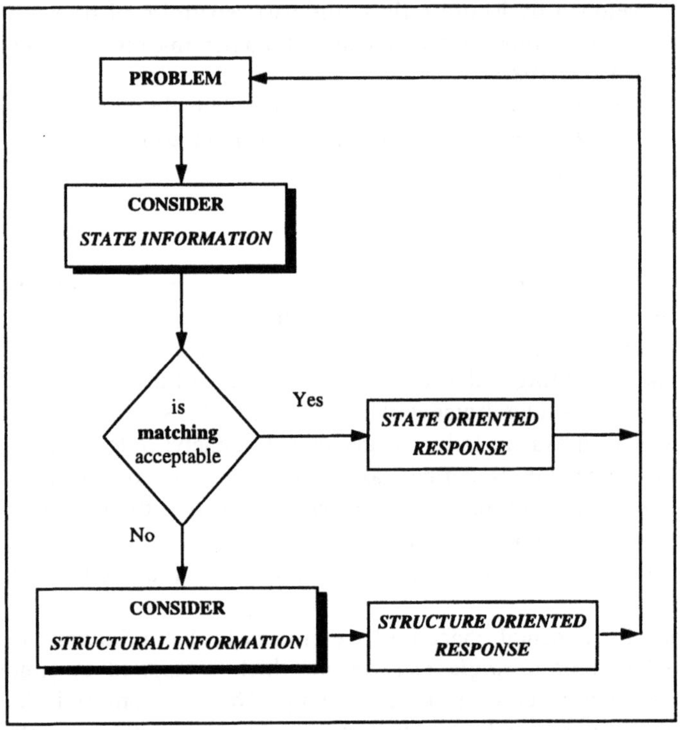

Figure 3.2. Basic Mechanism of HPS model (from Rouse 1983).

This basic mechanism applies to all three levels of problem solving.

The correspondence between mechanism and levels is shown in figure 3.3. In particular, at *recognition and classification* level, human operators are inclined firstly to search in their knowledge base for a "frame" (Minsky, 1975) that matches the perceived system state (State-oriented response). Only if this does not occur, the use of analogy and/or basic principles is proposed as process for recognition and classification (Structure-oriented response).

At *planning* level, the use of *state information* leads to the identification of "scripts" (Schank and Abelson, 1977) or standardised procedures. If these are not available, then a laborious and complex planning process must take place which relates to structural information and deep knowledge.

At *execution and monitoring* level, two types of rules are applied according to whether a "familiar" pattern is accessible for implementing the plan or the script devised at the previous level. In the first case, a context-specific Symptomatic Rule (*S-Rule*) can be applied, otherwise structural information could lead to the use of Topographic Rules (*T-Rule*) for tackling the problem in a more theoretical form.

Examples of *S-Rules* and *T-Rules* are: "If the alarm on a Steam Generator is activated, then check pressures and verify reactor shut-down" (*S-Rule*); "If a component input-data are good and its outputs are bad, then the component is failed or the interface is malfunctioning" (*T-Rule*).

In summary, the Human Information Processing model of Rouse represents a very generic framework by which it is possible to describe many aspects of human behaviour in a human-machine interaction architecture.

	PROCESSES		
LEVELS	***Matching***	***State-oriented response***	***Structure-oriented response***
1 ***Recognition and Classification***	Familiar Frame Available	Select Frame	Apply Analogy and/or Basic Principles
2 ***Planning***	Familiar Script Available	Select Script	Generate Plan
3 ***Execution and Monitoring***	Familiar Pattern Available	Apply S-Rule	Apply T-Rule

Figure 3.3. Basic Mechanism and Levels of HPS model (from Rouse 1983).

Compliance with Requirements

Rouse's model of Human Problem Solving offers a very broad matching with the requirements of the *Reference Model of Cognition*, in the sense that a number of cognitive functions and processes are well identified and correlated.

The cognitive functions *PIPE* and their connections are well represented by means of the three levels of information processing, by the use of "frames" and "scripts", and the *S-* or *T-Rules* (Table 3.3).

The basic mechanism, that governs all levels of the Human Problem Solving model, matches well the process of *Allocation of Resources* and its connections with the cognitive functions. It is not clear if a cyclic process is possible between different levels of the problem solving process. It seems that a pre-defined route is identified starting from some stimuli and the recognition of a plant situation. This leads to the execution of control actions and monitoring plant response.

As an example, when a frame that matches the current situation is found and its associated script is invoked, an *S-Rule* can be applied at execution and monitoring level. Therefore, if a state-oriented response is selected at the first level of the problem solving process, this approach is followed all the way through to the execution level. The same is valid for a structure-oriented approach.

As far as Working Contexts, Machine and Memory/Knowledge Base, it can be noticed that the HPS model gives only a generic and implicit modelling description. In particular, no detailed account is given of how they affect cognitive processes and levels of activities.

Table 3.3 The HPS Model and the requirements for models of cognition.

REQUIREMENTS †		HPS
1.	***Cognitive functions PIPE***	★★★
1.1	*Connection within PIPE*	**★★★-Seq.**
1.2	*Effects of Context/Machine on PIPE*	–
2.	***Allocation of Resources (AoR)***	★★★
2.1	*Connections of AoR and PIPE*	★★★
2.2	*Connections of AoR and Memory/KB*	–
2.3	*Effects of Context/Machine on AoR*	–
3.	***Memory/Knowledge Base (KB)***	★
3.1	*Connections of Memory/KB and PIPE*	★

† The following measures apply:

–	Not treated	★★★	Modelled in detail
★	Implicit consideration	**Seq.**	Sequential
★★	Modelled at minimal level	**Cyc.**	Cyclic

3.2.3 Step-Ladder, Skill-Rule-Knowledge Model

The most known and widely applied approach that implements the IPS metaphor was developed by Rasmussen in the 70ies and 80ies (1974, 1976, 1983, 1986) with the Step-Ladder (SL) model, which includes the notion of the three levels of cognitive behaviour, known as Skill, Rule and Knowledge Based (SRK) (Figure 3.4).

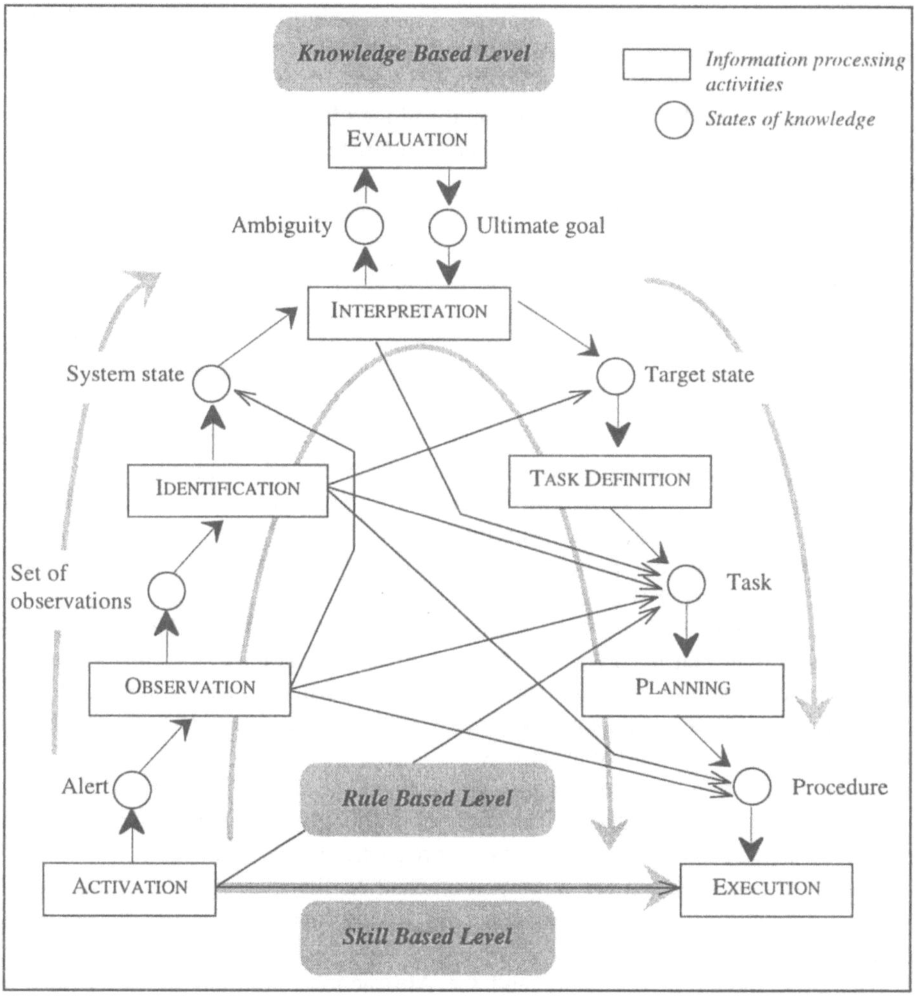

Figure 3.4. The "Step/Ladder" model of decision making (from Rasmussen, 1986). The *information processing activities* and the *states of knowledge* are linked by the SRK model.

Theoretical Description

The SL/SRK model considers human behaviour and control decisions of operators in complex physical systems. The model indicates a set of *states of knowledge* in the decision making process, and links these states of knowledge through a set of *information processing activities.*

The *information processing activities* and the *states of knowledge* are logically connected for the activation of the decision making process up to actions taken by the operator.

In a basic sequence, all information processing activities of the operator defined in the model are exploited and the flow of information follows known patters of rules and well known procedures (*Rule Based Behaviour*).

However, any of the activities may be abandoned by making shortcuts and/or by-passing intermediate stages, on the basis of heuristics or immediate activation of frames for action.

A typical situation of shortcut occurs at *Skill Based Behaviour* level, when the performance of an action is released and guided by the sensory perception of stimuli linked to well experienced and established behavioural modes.

When an unfamiliar and unexpected situation is encountered, more sophisticated reasoning processes may be demanded, which go beyond the acquired rules and frames. These situations require the use of knowledge of systemic behaviour and causal functional reasoning (*Knowledge Based Behaviour*).

Another important feature of the model is the structured representation of the physical system, as it appears to the human supervisor and controller. The model considers a means-ends *Abstraction Hierarchy* for the identification of the functional properties of a technical system. In particular, 5 levels of abstraction are considered: *Physical Form; Physical Functions; Generalised Functions; Abstract Functions;* and *Functional Purpose* (Figure 3.5).

By means of the 5 levels of the Abstraction Hierarchy it is possible to define the type of reasoning that is carried out by the human operator when facing a decision making process with respect to plant behaviour. In particular, according to the level of abstraction at which reasoning takes place, different plant aspects and components are accounted for in cognitive processes.

In the model, moving from lower to higher levels of abstraction, general functions are added to the knowledge of the system, while physical appearances, process components and standardised functions are progressively disregarded.

The Abstraction Hierarchy is very important in the management of accidents, when different gaols and requirements may be conflicting, specially during early stages of the accidents.

The SRK modes of behaviour and the Abstraction Hierarchy are well defined theoretical approaches for describing the cognitive processes and mechanisms within the Step Ladder representation.

In the SL/SRK model very little attention is devoted to modelling memory and knowledge base or to recognising of the effects of context and physical plant behaviour on the development of human information processing.

As a consequence of this, no detailed discussion takes place in the model on the effects of the SRK/Abstraction Hierarchy on the memory and knowledge base and vice-versa.

Finally, it is important to note that, the processes involved in the SL/SRK model are always sequential. The connections between information processing activities and states of knowledge always tend to develop from *activation* towards *execution* and the other activities, without the possibility to return and cycle within the "ladder".

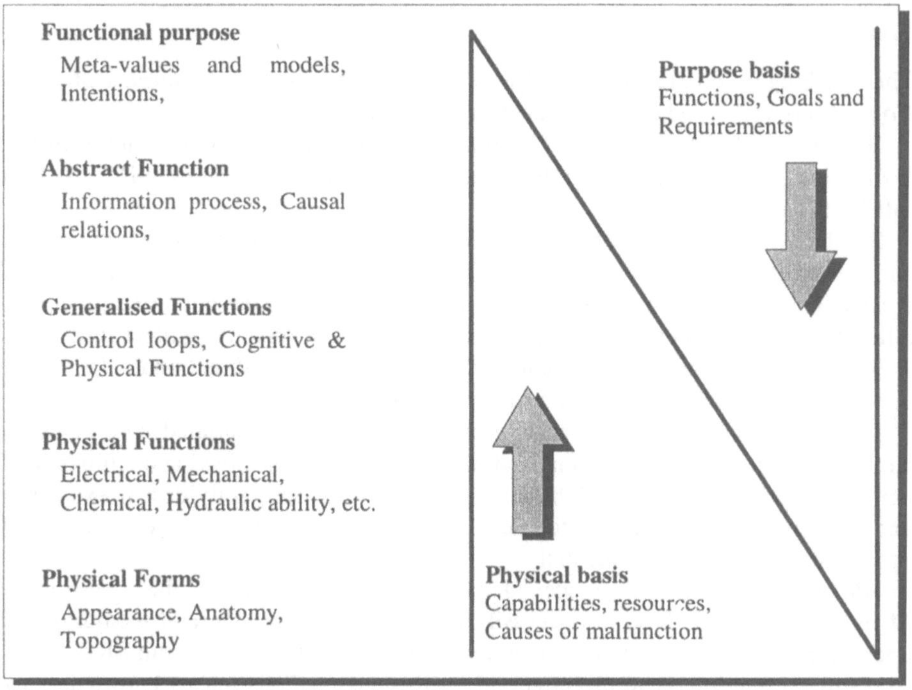

Figure 3.5 Means-Ends *Abstraction Hierarchy* for the functional properties of a technical system (from Rasmussen, 1986).

The SL/SRK model was not developed by Rasmussen into a computerised program. This may be one of the main reasons why the model does not discuss memory and knowledge base aspects in more detail.

The SL/SRK model, is certainly the most known paradigm of human behaviour as information processing device, and many researchers have applied the principles of the model for the evaluation and prediction of operators behaviour in a computational form.

Therefore, many simulations can be found in the literature containing the SRK based behaviour principles and referring to the SRK as their reference model (Fujii, Ogino, and Tanaka, 1988; Goodstein, Andersen, and Olsen 1988; Yoshikawa, 1989; Stassen, Johannsen, and Moray 1990; Johannsen, 1992).

Many other examples of implementation of the SL/SRK model can be found in several papers discussed at conferences like the triennial IFAC Conference on Design, Analysis and Simulation of Man Machine Systems, or the biennial Conference on Cognitive Science Approaches to Process Control, or the Annual Conference of Manual Control.

Compliance with Requirements

The matching of the SL/SRK model with the requirements of the *Reference Model of Cognition* is very good in some aspects and poor in others.

By means of the concepts of *information processing activities* and *states of knowledge*, the matching with the reference cognitive functions *PIPE* is fully satisfied (Table 3.4).

Similarly, the levels of behaviour defined by the Skill-Rule-Knowledge paradigm and the Abstraction Hierarchy are perfectly adequate to represent: a) the internal connections between the four cognitive functions *PIPE*; b) the cognitive process of Allocation of Resources (*AoR*); and c) the effects of *AoR* on the cognitive functions.

The SL/SRK model is substantially sequential and unidirectional. Very little consideration is given to the cyclic nature of cognition.

In practice, the flow of information processing is always initiated by an external stimulus coming from the machine and the sole iteration conceived at cognitive level can take place between the highest information processing activities of evaluation and interpretation, i.e., at the top level of cognition, while all other information processing activities are sequential.

Although the SL/SRK model dedicates little attention to the question of *Memory/Knowledge Base,* the necessity to represent knowledge and memory in an appropriate form and to exploit such knowledge through the paradigm SRK, or the levels of abstraction, is well recognised by Rasmussen. This makes knowledge and memory implicit constituents in a modelling architecture.

The principal reason for not having developed in detail the question of memory and knowledge base is that the major interest of the of the SL/SRK model lies with the generic representation of a paradigm of behaviour that captures the information processing nature of cognition.

Indeed, this specific focus of the model has also reduced the interest in responding to issues raised by the need to develop the theory into an analytical or numerical simulation, able to predict behaviour and mental processes in a mathematical and computational form.

The absence of, or minor consideration for, the representation of Memory and Knowledge Base leads to the lack of consideration for the effects of the Allocation of Resources on the Memory/KB processes and functions.

For similar reasons, the effects of the context and physical plant behaviour are only marginally considered, at least in the paradigm of behaviour described in the original reference (Rasmussen, 1986).

In later research development Rasmussen, has revisited this area of modelling, and a detailed and complete discussion on the subject can be found in more recent publications (Rasmussen, Pejtersen, Schmidt, 1990), expanding on taxonomies and classifications of human behaviour with respect to contexts and socio-technical working environments.

Any further discussion on this matter is outside the scope and purposes of the present review, and thus we can only refer interested readers to the above references for a more complete discussion on this subject.

It is important to note here that, when developing a numerical simulation of human behaviour based on the SL/SRK model, also this last aspect of the modelling architecture should be considered, in order to avoid making unnecessary assumptions and simplifications, which may not match those of the originator of the model.

Table 3.4 The SL/SRK model and the requirements for models of cognition.

REQUIREMENTS †		SL/SRK
1.	***Cognitive functions PIPE***	★★★
1.1	*Connection within PIPE*	★★★ - Seq.
1.2	*Effects of Context/Machine on PIPE*	–
2.	***Allocation of Resources (AoR)***	★★★
2.1	*Connections of AoR and PIPE*	★★★
2.2	*Connections of AoR and Memory/KB*	–
2.3	*Effects of Context/Machine on AoR*	★
3.	***Memory/Knowledge Base (KB)***	–
3.1	*Connections of Memory/KB and PIPE*	–

† The following measures apply:

–	Not treated	★★★	Modelled in detail
★	Implicit consideration	**Seq.**	Sequential
★★	Modelled at minimal level	**Cyc.**	Cyclic

3.2.4 Model of "Fallible Machine"

The paradigm of human behaviour called model of a *fallible machine* has been developed by Reason (1990).

The acronym reflects the two main features of this model, namely: the erroneous (fallible) behaviour of humans, that is the central theme of Reason's research in cognitive psychology, and the idea to express the model in a potentially computable form, as it is usual when dealing with "machines".

In this sense the theory is borrowing terminology and structures from engineering and computer sciences, such as Artificial Intelligence, and its application and computational implementation is expected to be more manageable.

Theoretical Description

The theory contained in the model of human behaviour of Reason can be derived from a number of publications, the most important one being a reference book on human error (1990). In this book, an entire chapter has been dedicated to the description of the model, which will now be reviewed focusing on the principal characteristics of the theory.

The model of a *fallible machine* can be described by starting with the "machine" part, i.e., the structure and framework by which human cognition is described (Figure 3.6).

The model has two principal constituents: *Knowledge Base* (*KB*) and *Working Memory* (*WM*). *Working Memory* is further subdivided into *Focal Working Memory* (*FWM*) and *Peripheral Woking Memory* (*PWM*).

This subdivision preserves the distinction between the two areas: the area of vast capacity, that is *PWM*, which receives information *directly* from the *KB* and the outside world and makes selection; and the area of limited capacity, that is *FWM*, which continuously receives *filtered* information, through the *PWM*.

The filtering mechanisms which support the prioritisation of the information that reaches *FWM* are based on a number of sensory and cognitive principles, namely: "visual dominance", "change detection", "coherence principles", "activation principle", and "dedicated processors" (Reason, 1990, p. 127-128).

Knowledge Base is a vast repository made up of two types knowledge:

1. Procedural knowledge made of *Frames* or *Knowledge Structures.*
2. Unstructured knowledge made of *Units* of declarative knowledge.

Either procedural *Frames* or knowledge *Units* are activated during the interaction with the working context throughout *sensory inputs*.

Frames are built of descriptions of plant, context conditions and configurations that characterise known abnormal or accidental situations. These are associated with known and pre-constructed plans and procedures to be implemented in order to recover or counteract the specific situation. This description of Frames shows many similarities with the analogous concepts of Frames of Minsky (1975) combined with the Scripts of Schank and Abelson (1977).

Units are more basic "chunks" of knowledge that need to be combined with other units to make a plan in less familiar or unclear situations.

The most original idea of this model lies in the retrieval mechanisms that are applied to: a) compare perceived information and *KB* content; and b) to select between alternative solutions.

These mechanisms are the "primitives of cognition" of *Similarity Matching (SM)* and *Frequency Gambling (FG)* and the inferential process of *Direct Inference* or *Search* (*DI* or *DS*).

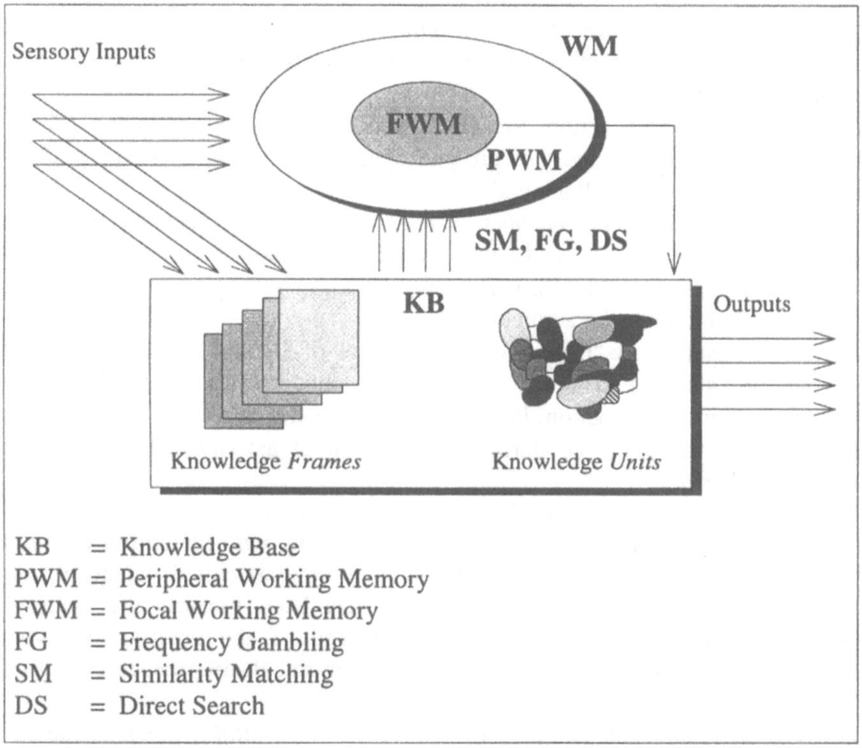

Figure 3.6 The model of "fallible machine" of Reason (1990).

Similarity Matching is applied for searching within the *KB* for knowledge *frames* that better match the information perceived from the external world, or "cues" of the world. This search usually selects several frames that satisfy the matching conditions of *SM*.

The Frequency Gambling primitive resolves the selection problem by favouring the candidate frame that is more familiar or has been most frequently encountered in the past.

Sometimes, no matching process takes place and the *FG* mechanism becomes even more perverse and symptomatic of inappropriate behaviour.

In this theoretical framework, *SM* and *FG* can always provide an answer. However, this answer may be very inappropriate for the current situation and might be induced as consequence of insufficient time or resources to perform deeper reasoning.

On the other hand, when *SM* and *FG* primitives have difficulty to deliver a "definite" appropriate plan of action, and the situation allows accurate reasoning and planning, then the Direct Search mechanism can take place.

Direct Search requires frequent and complex iteration between *KB* and Memory in order to develop a sequence of actions, built utilising several knowledge *Units* and combining them in a "novel" sequence, not previously exploited, and thus not yet structured into a *Frame*.

In other words, these actions are already individually present in the *KB* in the form of knowledge *Units*, and they are put together to provide a new plan of action able to deal with the unfamiliar situation.

This is the main difference between *Frames* and *Units*, i.e., the latter are blocks of individual or simple sequences of actions which are combined to produce plans and complex sequences, while the former are already known plans of interventions, fully built in the *KB* and ready for application.

In Reason's theory, the "primitives of cognition" of *SM* and *FG*, that guide human behaviour, are also the prime generators of human errors. The *fallible machine* theory expands in this direction, looking firstly at modes and types of human errors that can be directly observed during performance of activities. These are the so called "active" errors.

The model then develops the issue of dynamic interactions within the organisations at different levels of the decision making, with the aim of tracing and highlighting the hidden root causes of an accident: these are called "latent" errors.

In other words, the focus of this theory has been towards the analysis of organisations and socio-technical contexts instead of the analysis of dynamic interactions of operators with the plants under control, during the unfolding of an accident.

In essence, the concepts and principles adopted in the *fallible machine* model can be applied either to study how latent errors creep through an organisation affecting human performance at the "sharp" end of the control chain (operators and controllers of plants and systems) or to handle situations in which a changing context and working environment influences the behaviour of humans in direct contact and control of plants (Reason, 1992).

An example of application of this theory to study the case of human-machine interaction in accident management will be shown in a forthcoming section of this chapter.

The model of *fallible machine* has been computerised (Marsden, 1987) for the study of the way in which people, with different background knowledge, respond to a specific question: the request of identifying a President of the United States.

This application, although complete and accurate with respect to the theoretical model, does not really tackle the problem of dynamic interaction of people with a changing environment. Hence, it can be considered a static implementation of the model. It shows however that the *fallible machine* model can be quite straightforwardly computerised and applied.

Compliance with Requirements

As for the model SL/SRK, the *fallible machine* model complies very well with some of the requirements of the *Reference Model of Cognition*, but is less satisfactory with respect to others (Table 3.5).

In particular the *Memory* and *Knowledge Base,* as designed by Reason, correspond perfectly to the same constituents of the *Reference Model of Cognition.*

Similarly, the cognitive primitives of *SM*, *FG* and the inferential process *DI* outline very well plausible and logical mechanisms to simulate the cognitive process of *Allocation of Resources*. Indeed, by *SM*, *FG* or *DI* it is possible to express very well "when" and "which" different *cognitive functions* operate and interact with each other.

Moreover, the theory develops further the concepts, especially of *SM* and *FG*, and presents a set of prioritising principles that link sensory inputs to Knowledge Base and Working Memory. These principles aim at explaining how external cues are perceived by the cognitive model.

However, the *fallible machine* model fails in matching two aspects of the *Reference Model of Cognition*: the representation of a clear distinction between cognitive functions and the consideration for environmental and contextual factors.

The question of cognitive functions is not deeply developed in the *fallible machine* model because the model focuses on the way in which the primitives of cognition affect the error making, rather than the actual process of decision making and performance of actions.

The *cognitive functions* are thus only implicitly considered in the model, and further assumptions needs to be made by the developer of a simulation in order to properly account for them.

The environmental and contextual factors affecting human behaviour are only marginally considered in the model, except for the process by which errors are matured and fostered through the socio-technical and organisational structure of a company or institution. But, these are only one set of the causes that may contribute to manifestations of erroneous behaviour.

Other aspects need be considered, such as local and plant effects, to create the overall family of "external" and "internal" factors affecting human behaviour. In the *fallible machine* model this need is recognised, but only developed as far as the socio-technical and organisational aspects are concerned.

Table 3.5 The *Fallible Machine* model and the requirements for models of cognition.

REQUIREMENTS †		*Fallible Machine*
1.	***Cognitive functions PIPE***	★
1.1	*Connection within PIPE*	★
1.2	*Effects of Context/Machine on PIPE*	–
2.	***Allocation of Resources (AoR)***	★★★
2.1	*Connections of AoR and PIPE*	★
2.2	*Connections of AoR and Memory/KB*	★★★
2.3	*Effects of Context/Machine on AoR*	★★
3.	***Memory/Knowledge Base (KB)***	★★★
3.1	*Connections of Memory/KB and PIPE*	★

† The following measures apply:

–	Not treated	★★★	Modelled in detail
★	Implicit consideration	**Seq.**	Sequential
★★	Modelled at minimal level	**Cyc.**	Cyclic

3.2.5 Basic Supervisory Control Paradigm

The Basic Supervisory Control Paradigm was developed by Sheridan (1992) as a generic model for covering the fundamental functions of humans in their role of supervisory controllers of "modern" plants.

Theoretical Description

The key concept which sustains the Supervisory Control Paradigm is the role assigned to operators in human-machine interaction systems: the human is expected to control, at a distance, the development of process and machine performance, which are carried out by several dedicated computers, that perform "remotely" specific tasks.

To carry out such duty, the operators are supported by a specific "local" computer that governs the subordinated remote computers and represents the interface between the plant and the operator. This specific computer is called "human interactive computer", while the remote computers are called "task interactive computers" (Figure 3.7).

Figure 3.7 represents very clearly the existing situation of process control in many industrial and technologically developed systems, such as: nuclear power plants, chemical plants, tankers, oil platforms etc.

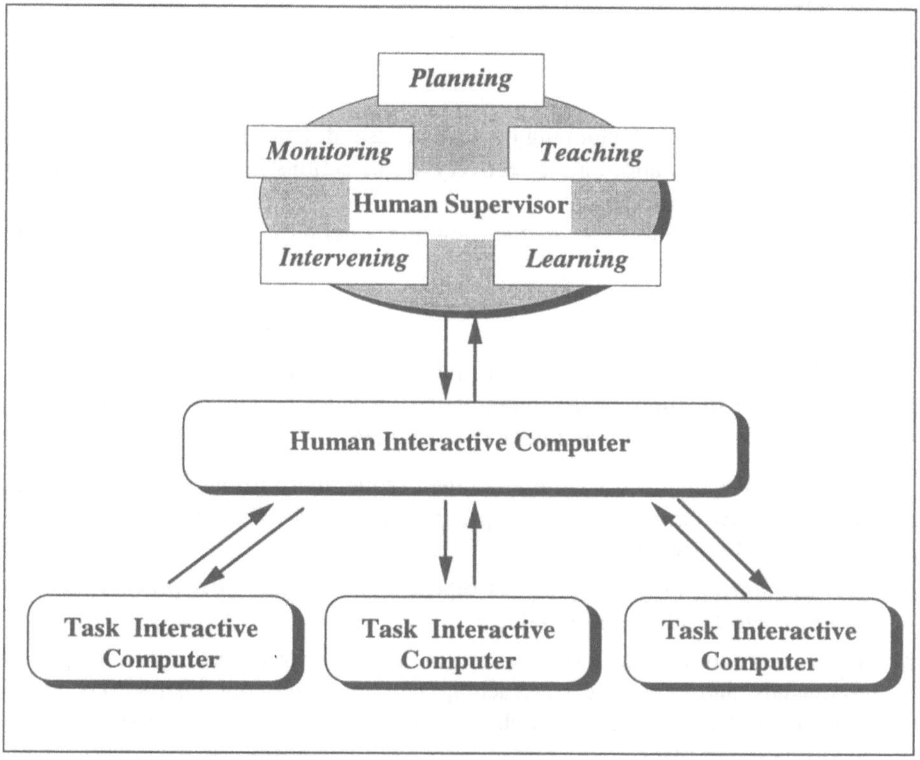

Figure 3.7. Basic Supervisory Control Paradigm and Control Computers, (from Sheridan, 1992).

Five main functions are carried out by operators in performing their supervisory control task: *planing, teaching, monitoring, intervening*, and *learning*.

- *Planning* implies:
  a) Gaining experience and understanding the whole process under control;
  b) Setting goals and objectives; and
  c) Formulating a general strategy.

- *Teaching* means:
  a) Selecting specific control actions related to the strategic plan; and
  b) Performing the related commands that are carried out through the human-interactive computer.
- *Monitoring* means observing the performance of tasks by the automation. This demands that the supervisor is able to:
  a) Acquire, calibrate and combine" measures of process state;
  b) Estimate states from current measures and past control actions; and
  c) Detect and diagnose discrepancies or accordance between actual and expected system response.
- *Intervening* implies:
  a) Aborting and/or abandoning the task in case of major fault; or
  b) Developing an appropriate recovery set of actions in case of occurrence of rectifiable conflict; or
  c) Initiating a new task if the ongoing operation has been successfully completed.
- *Learning* means:
  a) Recording experience; and
  b) Updating models of system performance for future use.
  This last function clearly requires the presence of memory storage usually supported by the human interactive computer.

The 5 supervisory functions are combined sequentially from planing to learning and cyclically by three nested loops, which characterise different cognitive processes, namely (Figure 3.8):

1. The monitoring process, which is repeated to evaluate new process states in order to detect and diagnose failures.
2. The intervention and modification of task performance, by issuing new commands and therefore by teaching new programming instructions to the human interface computer, which are necessary in case of recovery actions or modification of plans.
3. The feedback, which implies learning from the results of one task and utilising this knowledge for the performance of the following ones.

The supervisory control paradigm has been compared by Sheridan himself (1992) with a number of other models and engineering formulations, such as the SL/SRK model or the Optimal Control Model (OCM).

The supervisory control model has been shown to represent a valid generic paradigm able to accommodate many other modelling and simulation approaches.

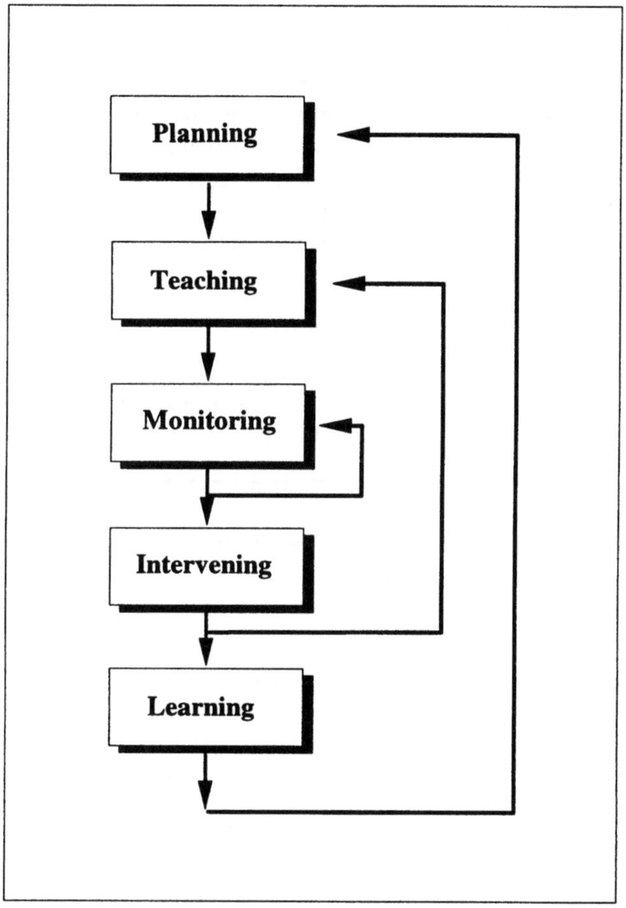

Figure 3.8. Nested loops of supervisory functions, (from Sheridan, 1992, p.16).

Compliance with Requirements

The supervisory control paradigm, like all generic frameworks describing human behaviour, captures quite well and expands the general features of human information processing mechanisms.

In general, this model matches very well the requirements of the *Reference Model of Cognition* .

In particular, in this model the functions *PIPE* are well recognised and their interconnections are represented by sequential and cyclic mechanisms, according to the nature and type of human activity (Table 3.6).

It is also interesting to note how the 5 supervisory functions are organised: planning comes first, and learning last. This implies that, in this paradigm, human cognitive activity is somehow independent from the machine and plant activity, or at least it is not subordinate to the behaviour of the plant. Hence, cognitive processes can start and be engendered by other cognitive processes and do not require to be initialised by plant or environment stimuli.

This way of representing nested cognitive activities differentiates cyclic from sequential types of models.

Table 3.6 The *Supervisory Control Paradigm* and the requirements for models of cognition.

REQUIREMENTS †		***Supervisory Control***
1.	***Cognitive functions PIPE***	★★★
1.1	*Connection within PIPE*	★★★ - Cyc.
1.2	*Effects of Context/Machine on PIPE*	–
2.	***Allocation of Resources (AoR)***	★
2.1	*Connections of AoR and PIPE*	★
2.2	*Connections of AoR and Memory/KB*	–
2.3	*Effects of Context/Machine on AoR*	–
3.	***Memory/Knowledge Base (KB)***	★
3.1	*Connections of Memory/KB and PIPE*	★

† The following measures apply:

–	Not treated	★★★	Modelled in detail
★	Implicit consideration	**Seq.**	Sequential
★★	Modelled at minimal level	**Cyc.**	Cyclic

The theoretical description of the model does not consider activities referring to the cognitive processes of *Allocation of Resources* and *Memory/Knowledge Base*. These processes are only lightly treated by some implicit considerations. In particular, in connection with the *learning* function, certain mechanisms for knowledge development and memory activity are foreseen.

Similar theoretical limitations can be found for what concerns the effects of context and working environment. These are simply defined in relation to the supervisory role of operators, thus focusing only on the human-computer interaction architecture.

All other aspects affecting human performance, like physical working conditions, socio-technical and organisational factors are not considered.

It is important to note here that the supervisory control paradigm has been developed with the aim of representing the specific role of operators and controllers of automated plants, in an engineering and design framework. In this perspective, the model offers a valid and sound paradigm of reference for designers.

3.2.6 Contextual Control Model

The Contextual Control Model (COCOM) of cognition (Hollnagel, 1993a) has been developed mainly as a theoretical framework for representing the behaviour of humans through two fundamental characteristics:

1. The *cyclical nature of cognitive mechanisms* and
2. The *inherent dependence of human actions on context and environment.*

Theoretical Description

The model considers human behaviour as characterised by two main elements: the *model of competence* and the *model of control.*

<u>The competence model</u>

The *competence model* is substantially characterised by the basic functions of cognition and by the combined repository of knowledge base and memory retrieval process.

This part model is considered, by Hollnagel, "very classical" and well described by literature and previous works. For this reason, the competence model has been formalised in a *Very Simple Model of Cognition* (V*SM*oC), which offers, in any case, a complete structure able to describe the most relevant functions of cognition, their inter-connections, and their links with memory and knowledge base (Figure 3.9). The correspondence of the *VSMoC* model and the *Reference Model of Cognition* is obvious and no further discussion is therefore needed.

The important feature of this modelling framework is that the connections envisaged in *VSMoC* are able to handle a cyclic model of cognition, and, therefore, to simulate more complex and complete processes of human cognition and interaction with the machine and context.

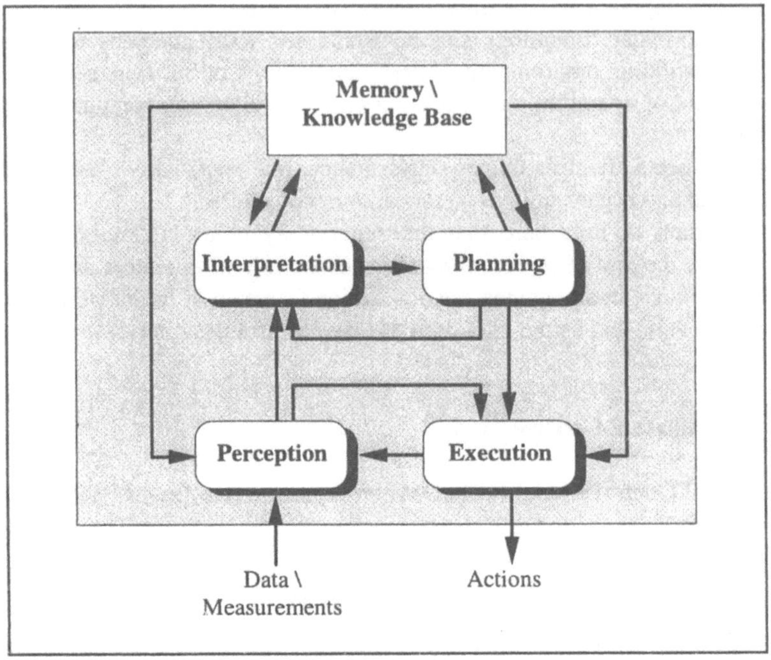

Figure 3.9. The Very Simple Model of Cognition (from Hollnagel, 1993a).

The control model

The *control model* operates in direct combination with the *competence model* and constitutes a sort of model of meta-cognition.

The control model represents the most innovative feature of COCOM and contains the theoretical development and description of the mechanisms (and *parameters*) that govern the sequence and operability of the cognitive functions contained in the *competence model*.

Four different control modes are envisaged: *Scrambled, Opportunistic, Tactical* and *Strategic*.

They represent attitudes of the operator versus the use of his/her competence and determine ultimately the sequence of actions carried out.

The environment strongly affects the control mode, by causing dynamic changes of control mode, and, ultimately, affects performances through change over in the control mechanism (Figure 3.10).

The model enters in more detailed analysis of the parameters and factors that affect the control mode, by defining a number of *main control parameters* and *additional dimensions* of control.

In particular two main parameters are identified, as a sort of primitives of meta-cognition or regulators of control mode:

1. The *outcome of the previous task* or action.
2. The *subjective estimation of available time*.

Additional dimensions defined by Hollnagel as parameters of control mode are: *number of simultaneous goals*; *availability of plans*; *event horizon*; and *mode of execution*.

These parameters represent the connection between control mode and context and allow to describe, dynamically, during the human-machine interaction process, the selection of a certain control mode as well as the transition from one control mode to another.

Although these main control parameters and additional dimensions may not be an exhaustive set of variables, as Hollnagel points out, they show how the theoretical background of COCOM has been very accurately developed, in consideration of all the different aspects for modelling cognition and human interactions.

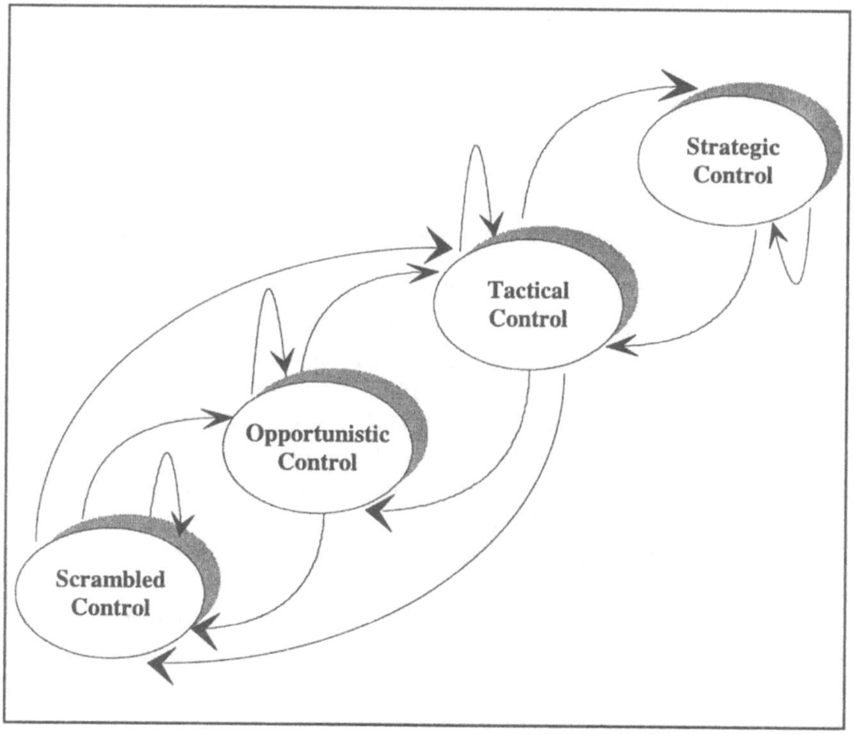

Figure 3.10. Control modes and transitions (from Hollnagel, 1993a).

The computation implementation of COCOM has not been fully carried out. However, the author has quite extensively described all the necessary requirements and also the possible specifications to follow in order to develop a computerised version of COCOM.

A first simplified version of COCOM, where the model of competence has been included, but the contextual control mechanisms have only superficially been considered, has been developed in a computer simulation program (Hollnagel and Cacciabue, 1991) and it has been applied to the analysis of transient conditions in civil aviation systems (Cacciabue, Cojazzi, Hollnagel, and Mancini, 1992). Another simulation has recently been attempted using COCOM in the nuclear environment (Yoshida and colleagues, 1997).

COCOM is a relatively recent model, and a certain time is needed before it becomes properly known and spread in the scientific community.

Certainly, thanks to its completeness and accuracy, COCOM will be largely applied and many simulations based on its modelling features will appear in the future.

Compliance with Requirements

COCOM is the most recent of the models that have been reviewed and has been devised as a model of cognition to represent a guide for further development in a computer simulation.

In particular, many educated suggestions have been made by Hollnagel (1993a), which make the actual transformation of COCOM in an analytical and numerical simulation readily feasible. This means that one does not need to make many assumptions when developing a simulation based on COCOM. This is a very relevant peculiarity of the model.

The fact that Hollnagel has been working for many years in the area of cognitive psychology and human-machine interaction explains why COCOM seems to represent the essence of human modelling paradigms, containing all necessary features to transform such a model into a working simulation.

It is, therefore, not surprising that COCOM, together with the V*SM*oC model, show the highest degree of compatibility with the requirements of the *Reference Model of Cognition*. In particular, all the requirements are covered and most of them are satisfied at detailed level (Table 3.7).

The cognitive functions *PIPE* are completely matched by the theoretical framework of V*SM*oC.

The interactions and links within the *PIPE* functions are discussed focusing on cyclic processes, which include the sequential development occurring in certain conditions.

The effects of dynamic behaviour of context and plant are also well covered, mainly for what concerns socio-technical working environments.

The meta-cognition processes and mechanisms that govern the Control part of COCOM correspond completely and in detail to the Allocation of Resources.

The only part of the model which is not fully detailed is the interaction with memory and knowledge base. However, this area of development has been one of the specific fields of research of Hollnagel for a number of years, especially in the domain of definition and classification of human errors (Hollnagel, 1991).

The step of combining the architectures of human-machine simulation and data with a correlated classification of human erroneous behaviour is part of the most recent work carried out by this author, in combination with the development of a working model utilising COCOM as theoretical background, both for retrospective and prospective type of studies (Hollnagel and Marsden, 1996; Hollnagel, 1998).

Table 3.7 *COCOM* and *VSMoC* and the requirements for models of cognition.

REQUIREMENTS †		*COCOM + VSMoC*
1.	***Cognitive functions PIPE***	★★★
1.1	*Connection within PIPE*	★★★ - Cyc.
1.2	*Effects of Context/Machine on PIPE*	★★★
2.	***Allocation of Resources (AoR)***	★★★
2.1	*Connections of AoR and PIPE*	★★★
2.2	*Connections of AoR and Memory/KB*	★★
2.3	*Effects of Context/Machine on AoR*	★★★
3.	***Memory/Knowledge Base (KB)***	★★
3.1	*Connections of Memory/KB and PIPE*	★★

† The following measures apply:

–	Not treated	★★★	Modelled in detail
★	Implicit consideration	**Seq.**	Sequential
★★	Modelled at minimal level	**Cyc.**	Cyclic

3.2.7 Comparison of Cognitive Models

Objective of Comparison of Models

The objective of the comparison of the six theoretical models described in the previous sections is dual:

- The rank ordering of the six models with respect to different areas of application, types of analysis, and simulation that may be required.
- The identification of the accuracy or level of matching of each model with the requirements of the *Reference Model of Cognition.*

We have already defined, in Chapter 1, four main types of application, i.e., *Design*, *Safety Assessment*, *Training*, and *Accident Investigation*, two types of analysis, i.e., *Prospective* and *Retrospective*, and two types of simulation, *Qualitative* and *Quantitative* (Figure 1.3), for which a model of human-machine interaction is necessary.

The six models described above share a common basis and starting point in the metaphor of human behaviour as Information Processing System. However, while some of them have made full reference to it, others have used it as a starting point to make progress and develop more advanced and comprehensive theories of human cognition. The latter are very appropriate to represent the role of humans in managing complex systems of present technology.

The most relevant differences identified in these models derive from two major circumstances:

1. They have been elaborated over the last two decades, at different stages of development of human sciences and technology.
2. They tackle different aspects (types of application, analysis and simulation) of human-machine interaction.

This variance, instead of a drawback, offers the analyst a wider option and possibility of choosing the most appropriate model for the problem at hand.

In particular, even if the most comprehensive model will, in general, be able to cover all types of application, analysis and simulation, its use for a specific type of study may require quite an extensive effort of adaptation and simplification.

On the other hand, a less sophisticated, but more focused model, may be perfectly adequate and readily available for application.

The forthcoming comparison aims at this main objective.

Comparison of Models and Areas of Application

With the improvement of theoretical models and simulation means, a number of specific aspects have become progressively more important for the completeness of the analysis of human-machine interaction.

While the consideration of *cognitive functions* and *cognitive processes* have now been widely accepted as necessary constituents of a human behaviour modelling architecture, the innovative factors deemed essential to complete a model are:

- The assessment of *environment and contexts*;
- The consideration of *cyclic* and *dynamic* interaction "within" the human model and "between" the human model and machine model (plant and socio-technical working contexts).

In particular, theoretical models accounting for cyclic and dynamic interactions are being considered as extensions of the "classic" sequential and static approaches. However, the work of modelling involved in such expansion is quite important, and requires major changes in the human-machine architecture, as well as extended work of classification and task analysis for the collection of appropriate data. This leads to the development of more complex models.

In substance, a greater diversification exists between complete, but complex, models and simple, but more focused, models. This difference should be enhanced by the following comparison, together with the definition of the area of application of each model.

Table 3.8 shows the overall result of the matching of each of the six models with the requirements of the *Reference Model of Cognition* derived in Chapter 1. The areas of application and the types of analysis and simulation for which each model is best suited can be inferred from their description.

SHEL

The SHEL model is the oldest of the six models considered and presents the least compliance with the requirements. Only the basic cognitive functions are considered implicitly or at minimal level. These are combined with external and internal factors affecting human behaviour. Internal factors are the most important aspects for the SHEL model, as the ergonomics of the working conditions is fully covered by the model, including complicated interactions, such as those concerning teamwork and automation.

This feature makes the model particularly suited for classifying human factors and working environments and for developing data frameworks that can be utilised in combination with and in support to Cognitive Task Analysis.

The same classification structure can also be used for accident investigation and, therefore, for retrospective analysis.

Similarly, SHEL could be applied as paradigm of reference in training. In this case, though, further detail analysis and evaluation of parameters and variables would need to be associated with the various levels of the SHEL structure, such as, for example, the definition of factors affecting L-L (team work) or L-H and L-S (situation awareness) interactions.

The use of SHEL in dynamic prospective studies of human-machine interaction is less appropriate, as the user would have to make many assumptions, mainly for the development of a method representing the meta-cognition mechanisms of *Allocation of Resources* and *Memory/Knowledge Base*.

Table 3.8. Overall matching of the six models of cognition with the requirements of the *Reference Model of Cognition.*

REQUIREMENTS[†]	**Edwards** *SHEL* (1973)	**Rouse** *HPS* (1983)	**Rasmussen** *SL/SRK* (1986)	**Reason** *Fallible Machine* (1990)	**Sheridan** *Supervisory Control* (1992)	**Hollnagel** *COCOM* (1993a)
1. *Perc./Interp. /Plan./Exec. (PIPE)*	★	★★★	★★★	★	★★★	★★★
1.1 *Connections within PIPE*	★ Seq.	★★★ Seq.	★★★ Seq.	★	★★★ Cyc.	★★★ Cyc.
1.2 *Effects of Cont./Mach. on PIPE*	★★	–	–	–	–	★★★
2. *Allocation of Resources (AoR)*	★	★★★	★★★	★★★	★	★★★
2.1 *Connections of AoR and PIPE*	–	★★★	★★★	★	★	★★★
2.2 *Connections of AoR and Memory/KB*	–	–	–	★★★	–	★★
2.3 *Effects of Cont./Mach. on AoR*	★★	–	★	★★	–	★★★
3. *Memory/KB (KB)*	–	★	–	★★★	★	★★
3.1 *Connections Memory/KB and PIPE*	–	★	–	★	★	★★

[†] The following measures apply:

–	Not treated	★★★	Modelled in detail
★	Implicit consideration	**Seq.**	Sequential
★★	Modelled at minimal level	**Cyc.**	Cyclic

SL/SRK and HPS

The SL/SRK model and Rouse's model of Human Problem Solver (HPS) were developed approximately during the same period, i.e., the late 70ies and early 80ies, and show many similarities.

Both models exhibit the advances made by cognitive sciences, with respect to the SHEL model, as far as the formalisation of the Information Processing metaphor is concerned. The cognitive functions and the mechanisms for meta-cognition are accounted for in detail and the theoretical formalisms are fully described, even if little consideration is given to cyclic and dynamic interactions of humans with machines and working environments.

Both SL/SRK and HPS models make little reference, or treat in an implicit fashion, the effects of context and the Memory and Knowledge Base.

The SL/SRK model is much more known and applied than the HPS, even if, at first, they can be considered substantially similar from the mere application viewpoint and for the scope of their theory. The main reason for the popularity of SL/SRK derives from the fact that this model has been much more extensively described and elaborated in a number of publications. Therefore, its implementation in a computerised simulation form, useful to designers and safety analysts has been made much easier. As already mentioned, many application of the SL/SRK model have been attempted, as we will see in the next section, when reviewing analytical and numerical simulations.

In substance, the SL/SRK and HPS models are very valid paradigms and give a thorough set of parameters and variables for modelling cognitive human behaviour, even if only sequential and static interaction are considered.

When developing a simulation of human-machine interaction based on these models, a number of assumptions are left to the developer, specially for the Memory and Knowledge Base and for the effects of context and environment.

Moreover, the low consideration for working environments and contexts makes these models less applicable for classification and retrospective studies, which require a smaller focus on cognition and wider scope of factors affecting human behaviour.

Fallible Machine model

The *Fallible Machine* model of Reason is more focused on cognition than on engineering aspects.

The primitives of cognition and the attention given to Memory and Knowledge Base make the model particularly suited to represent the related cognitive processes and the Allocation of Resources. The basic cognitive functions are considered as implicit and, therefore, no theoretical formalism is given for their representation in a numerical computational form.

One of the objectives of the modelling architecture is to consider the "pathological" development of errors made at different levels of a company management. For this reason, the context and socio-technical environment are considered in the model mostly from the social and management contribution.

The theory is useful for classification of errors, as far as they relate to human aspects, but less applicable for physical and external factors in general. The theoretical development of Reason, associated with the model of *fallible machine,* has given rise to a model of error chain and generation which has been widely applied for retrospective analysis and definition of root causes of real accidents, such as the accidents of Zeebrugge, Chernobyl, Bophal and the Challenger.

As far as simulation is concerned, the *fallible machine* model is very useful because it offers, as the SL/SRK and HPS models, a complete variety of parameters and variables which allow the development of a numerical and computational application. The user still needs to make some assumptions, mainly for what concerns the basic cognitive functions and the mechanisms of interaction of these functions with machine and environment stimuli. In particular, the questions of cyclic and sequential processing of cognitive activity and dynamic interactions is left to the developer of the simulation.

In a forthcoming section of this chapter, a simulation derived from the *fallible machine* model will be presented. The specific assumptions necessary for transforming the theory into a working structure will be discussed in detail.

Supervisory Control Paradigm

The *Supervisory Control Paradigm* of Sheridan shows a low degree of matching with the requirements.

As in the case of SHEL, the objectives of the paradigm are not really aimed at simulating human behaviour in a human-machine interaction architecture, but are more focused on offering an engineering architecture, acting as reference for designing human-centred automated systems.

For this reason, the interaction between humans and environment are considered mainly for what concerns the cognitive process of a human supervisor, controlling the plant performance through the Human-Interactive Computer. Cognitive functions are well defined and described.

Relevant peculiarities of this model are that, in the cognitive *loop*, planning precedes monitoring, and that the modelling architecture is based on a *loop* of nested functions. This implies that a cyclic process has been envisaged, as an extension and alternative to the sequential process.

The low consideration of context and environment as well as of memory and knowledge base make this model not very suitable for simulation purposes. Too many assumptions would have to be made and thus the nature of the theory could be strongly affected by the needs of the numerical simulation.

However, the application of the paradigm for design and assessment purposes could be very well considered.

COCOM

The COCOM model, specially in combination with the V*SM*oC model, shows the most extensive degree of matching with the requirements of the *Reference Model of Cognition.*

COCOM is definitely the best theoretical framework and model to which we should refer for developing a numerical simulation of human cognitive behaviour in a human-machine architecture.

In particular, cyclic and dynamic interactions between context and machine are discussed and theoretically justified. The parameters and variables governing competence model and control processes, i.e., the cognitive functions *PIPE* and the process of *AoR*, are described in detail.

Only the question of memory and knowledge base are modelled at minimal level, in the sense that the theoretical formalisms to define data, memory and knowledge base are discussed but not completely integrated in a modelling architecture.

The amount of work and assumptions required to develop a simulation from COCOM and V*SM*oC is certainly minimal.

However, the price paid by the completeness of this model lies in the difficulty to collect and organise the data necessary to support the various parts of the model. Indeed, the idea of Means Goals Task Analysis (GMTA) in support to this question and the more recent work of the author (Hollnagel, 1998) on taxonomies and classification of human performance aim at fulfilling this gap.

The use of COCOM for retrospective studies and other applications, such as design and safety assessment or training, is certainly possible, but a certain amount of reduction and simplification of the overall theory would be needed.

The model is best suited for dynamic prospective simulation of human behaviour in human-machine interaction studies, specially if it can be complemented by a solid architecture and taxonomy of data and by an equally comprehensive model of the machine (including working contexts) interacting and surrounding the human operators.

Comparison of Models and Types of Analysis

Finally, the result of the above discussion, for what concerns the applicability of the six cognitive models of cognition to perform *prospective* or *retrospective* types of analysis can be summarised as follows (where *A* stands for Applicable, and *NA* for not-applicable):

Types of Analysis	**SHEL**	**HPS**	**SL/SRK**	**Fall. Mach.**	**Sup. Cont.**	**COCOM**
Prospective	*NA*	*A*	*A*	*A*	*A*	*A*
Retrospective	*A*	*NA*	*NA*	*A*	*NA*	*A*

3.3 Review of Simulations of Cognition

This section will focus on a number of working simulations, which are able to describe numerically and to quantify human behaviour models in a human-machine interaction architecture.

The main objective of this section is to show how engineers and analysts may combine theories of cognition, computational means and programming environments, in association with task analysis and data collection techniques, to reach a "simulation" model.

In practice, a simulation is made up of three fundamental elements:

1. The *theoretical cognitive model*, which defines conservation principles, criteria, parameters and variables, that allow to describe cognitive and physical behaviour of humans in a conceptual form. The models described in the previous sections of this Chapter will be taken as reference, as they are representative of the most commonly utilised theories.
2. The *numerical algorithms* and the *computational architecture*, by which a theory is implemented in a working computerised form. Data and parameters necessary to "calculate" human behaviour, are identified, but not evaluated at this level. The engineering methods, and the various techniques described in Chapter 2, like control theory, fuzzy set and artificial intelligence approaches, represent a good variety of means that can be utilised to solve the problem of quantification and computation of theoretical formulations.
3. The *task analysis technique*, which is applied to evaluate human tasks and associated working contexts, and to describe procedures and actual human performances in a formal way. By using this technique, data and parameters needed by the simulation are quantified. The formalisms and the methods that can be classified as Cognitive Task Analysis (Chapter 2) will be taken as reference.

These three elements, or levels of analysis, can be considered necessary and sufficient for the development of a simulation of cognition.

However, a simulation can not be developed independently of the specific machine and context in which humans are operating. Each specific working context and machine design requires a particular type of behaviour, and, consequently, also a certain amount of specific simulation.

Moreover, the identification of the most suitable simulation model is usually the result of an iterative process by which a *theory* has to "compromise", and reduce its generality and holistic attitude, in order to comply with the *algorithms* available as well as with the data that can actually be found in real work settings by *cognitive task analysis* (Figure 3.11).

This basic principle has already been encountered in Chapter 1 when discussing the need and principle of modelling the man-plus-the-controlled-element enunciated by McRuer and colleagues (1965).

Therefore, while we have been able to describe a number of well formulated and generically applicable theories of cognition, aiming at covering all possible and foreseeable characteristics of cognitive behaviour, the step of developing a numerical simulation requires a more engineering attitude. This attitude helps in selecting, almost on a case by case basis, the "good" combination of theories, analytical studies, computational and numerical means leading to a simulation output that best suits the case under study.

In the following part of this Chapter, a number of examples of such approaches are reported in generic terms and only one of them will be discussed in detail, also showing what type of results can be expected and obtained.

In this way, by analysing how engineers and analysts have dealt with the problem of combining the above three elements to develop cognitive simulations in certain domains and working environments, the reader may be able to find hints on how to develop his/her own simulation.

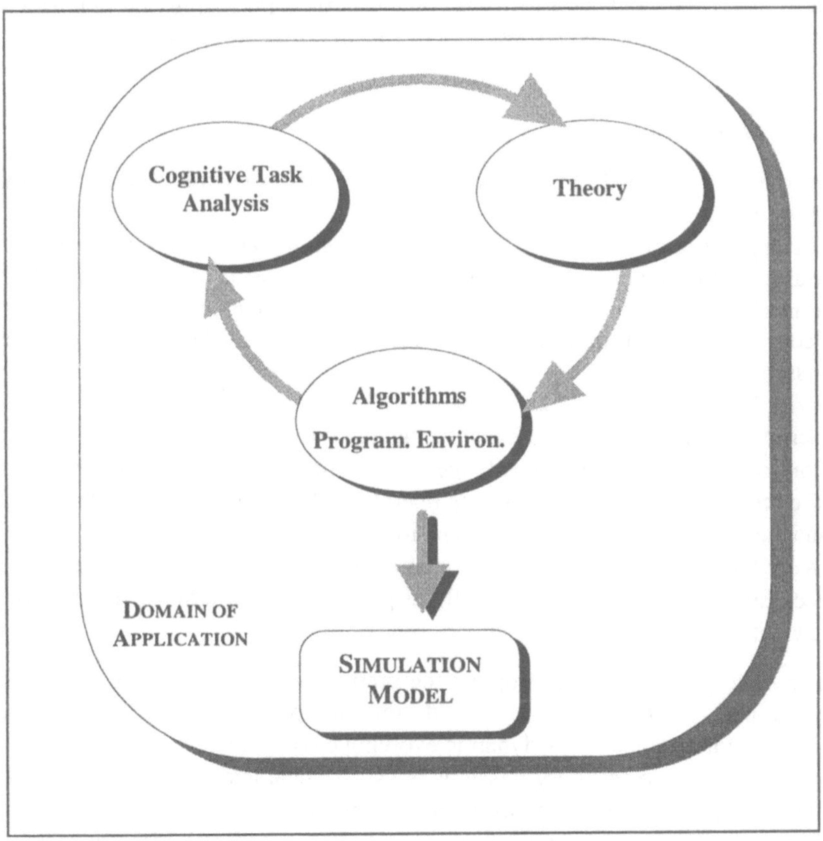

Figure 3.11. Levels and elements of cognitive simulation.

It is very unlikely that any of the simulations described in the forthcoming paragraphs can be applied to a problem and working environment different than the ones for which the simulations have been developed, no matter how similar these may be.

The development of a "generic" simulation and architecture, able to cover many working contexts and situations, would require an enormous effort and it would remain unsatisfactory and too cumbersome to apply in practice.

3.3.1 The Simulation AIDE

The model AIDE (Amalberti and Deblon, 1992; Valot, Amalberti, and Cibiel, 1991) has been developed with the precise aim of representing and simulating the behaviour of military pilots.

The model is based on a vast amount of data collected in the field and during simulated experiments as well as from elicitation of experts (pilots). This is a very important feature of the model and shows the importance of task analysis for the development of a simulation.

Theoretical Description

The overall model is made of two elements: the *model of competence* and the *model of performance*.

The *competence* of the pilots is expressed through the knowledge of goal-oriented procedures, with their associated knowledge of mission completion ("tactical procedures") and system handling ("system management procedures"). The overall knowledge is organised, in accordance to the three levels of the SRK base model of Rasmussen and contains:

- Script level, built of chunks or frames of highly proceduralised knowledge (the "scripts"), which contain the basic factual knowledge of pilots.
- Control level, containing knowledge needed for monitoring and adapting scripts to circumstances. These two levels operate with the same abstraction condition and support skill and rule based behaviour.
- Deep knowledge level, which contains mission independent functional and declarative knowledge, based on flight principles, instructions, flight laws, and navigation principles.

The *model of performance* supports the preparation of the mission and executes the plans and procedures developed during the pilot-aeroplane interaction, by adapting the competence to the dynamic evolving environment.

The performance of military missions is usually characterised by high speed and low altitude flight. Pilots are confronted with two broad temporal spaces. A short-term horizon, which implies navigation, routine checks, and activities connected with the handling of the ongoing flight. A medium and long-term horizon develops the plan for the whole mission, establishes the links with other missions and monitors and adapts the ongoing mission with respect to possible future dangers and incidents or to an ideal plan of maximum efficiency.

In order to cover the differences between short-term and medium/long term activities, two levels of cognitive processes are considered: a *local level* which handles short-terms activities, and a *global level* covering the medium and long term activities (Figure 3.12).

The knowledge obtained from empirical experiments and task analysis of missions, as will be described in the forthcoming parts of this sub-section, is coupled with specific short or long term activities, such as controlling, planning, time sharing, performing procedures etc.

Computational Implementation

AIDE has been developed using an AI formalism, namely the *actor* modelling and language approach (Hewitt, Bishop, and Steiger, 1973; Hewitt, 1977).

This type of computerisation allows the simulation of parallel (synchronous) and asynchronous communications between *actors*, which are essential constituents of pilot simulation.

Actors regulate the whole flow of information and action performance. *Actors* communicate with each other by means of messages which generate the respective behaviour, namely:

- When sending synchronous messages, actors can continue operating after the message independently;
- When sending asynchronous messages, actors are bounded to wait for the answers; and
- When sending "rendezvous" messages, actors continue to operate up to a predetermined state and then have to wait for the answers.

The formal "actor language" allows to represent the two key features of pilot's behaviour, that is the short-term/long-term parallel processing and the direct communication between knowledge poles.

Task Analysis and Work Context

The development of the simulation AIDE has been supported by a very extensive work of analysis of the tasks and working contexts of military pilots.

Three types of studies have been carried out in order to outline the main characteristics of pilots' behaviour and influence of the specific working context and activity.

In particular, a "classical" job analysis was firstly carried out, covering the study of official documents and the decomposition of a mission in a sequence of tasks, sub-tasks, goals, sub-goals and connections between them.

Then a number of field observations of the behaviour expert pilots were performed on a number of missions. These operations were also videotaped.

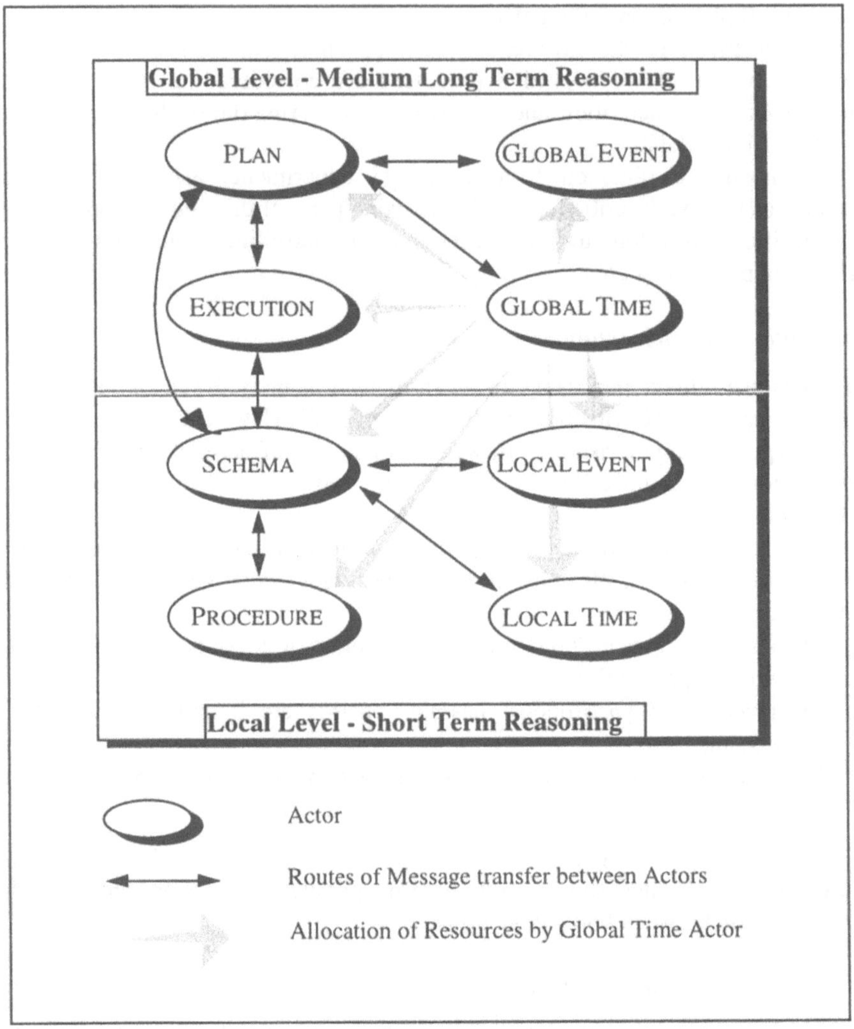

Figure 3.12 Architecture of the cognitive simulation AIDE, (from Amalberti and Deblon, 1992).

Finally, a series of semi-structured interviews took place, in order to get more insight on human behaviour, including the self-evaluation of the pilots on analysing their own performance from the videos (auto-confrontation).

This vast amount of field study and evaluation of the pilot's tasks, covered a substantial amount of the work of development of the model and has been very important for the definition of the boundaries and the relevant parameters and variables of the simulation.

In conclusion, it is remarkable to note how the development of AIDE has been focused on the basic elements of cognitive simulation, namely, a theoretical model of reference, a numerical and computational formalisms for the implementation of the quantitative part of the simulation and a strong analysis of the working contexts and tasks.

AIDE has utilised a well developed computerised approach, i.e., the *actor* formalism, and well established formal methods for the analysis of the task, i.e., job analysis, auto-confrontation and semi-structured interviews. The reference to a theoretical cognitive model has not been so explicit, as the model SRK has been utilised only in part for the development of the "competence model". The remaining theoretical basis of AIDE are original and due to studies of its developers (Amalberti, Valot, and Menu, 1987; Valot and colleagues, 1989).

3.3.2 The Simulation CAMEO

The simulation CAMEO (Cognitive and Action Modelling of an Erring Operator) was inspired by the information processing paradigm of human cognition and it is aimed at simulating appropriate and erroneous behaviour of a supervisory operator in complex working environments, such as nuclear power plants (Fujita and colleagues, 1993).

Theoretical Description

The background theory contained in CAMEO does not refer to any of the general theoretical models described earlier, even if it could be argued that it fits perfectly Rouse's Human Problem Solving model, for what concerns the levels of human behaviour, and Rasmussen's Abstraction Hierarchy, for the ways of operation and application of knowledge (Figure 3.13).

In CAMEO, three modules, sequentially connected, describe three basic cognitive processes. These modules are governed by a controller and exploit knowledge through the mechanisms of *Long-Term* and *Working Memory.*

The *Long-Term Memory* is the repository of knowledge generated and utilised by the decision making and the action modules.

The *Perception and Recognition* module considers auditory and visual abilities, to perform the basic functions of perception and, to a certain extent, interpretation.

The *Decision Making* module takes into account the tasks associated with vigilance, diagnosis, response selection and confirmation. This module represents the real core of CAMEO, as it concentrates the information gathering and elaboration for the evolution of the entire cognitive process.

The *Action* module selects and performs the sequence of actions to be carried out, from a predefined database of procedures, in agreement with the Decision Making module.

The cognitive processes are controlled by the *Attention Resource Controller* module, which supplies resources to the other modules, and by which prioritisation and ordering of cognitive activities is carried out. This mechanism is also very interesting for error generation during human-machine interaction sequences.

The *Working Memory* module is a sort of commonly shared notebook for communication amongst the other modules, also for recording information on the ongoing control process.

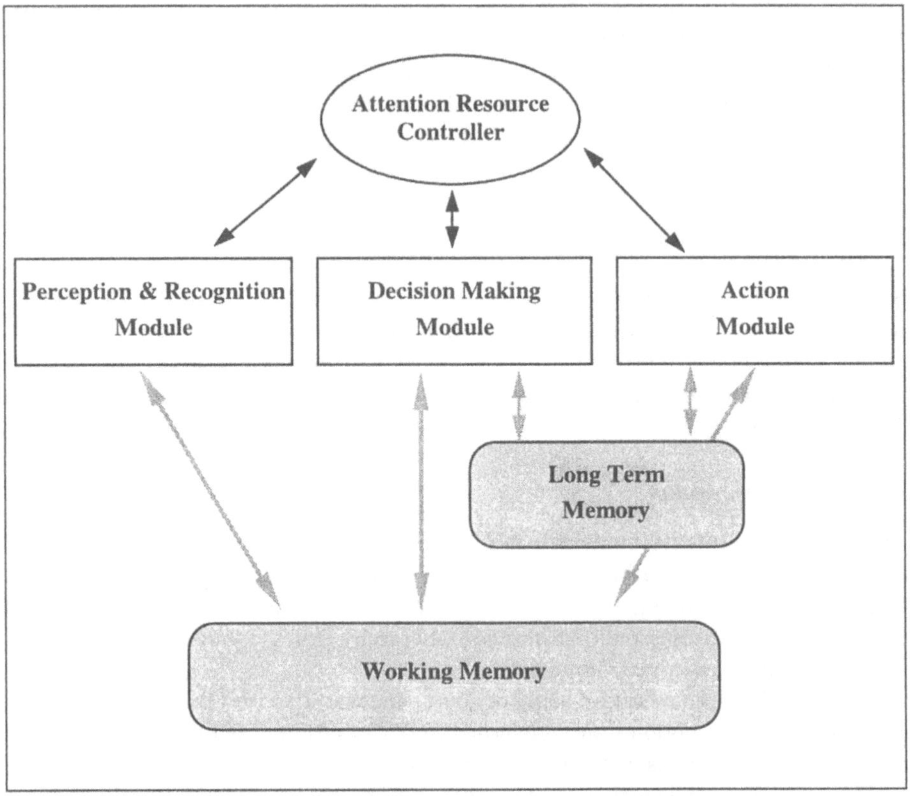

Figure 3.13. Functional representation of CAMEO (from Fujita and colleagues, 1993).

Computational Implementation

CAMEO has been fully developed into a computational tool. It applies the basic concept of Artificial Intelligence related to logical reasoning and rule based shells.

In particular, it has been developed in G2 language (Barr, Cohen, and Feigenbaum, 1989).

This language is a "classical" object-oriented programming shell that allows maximum flexibility and enables the user to develop easily new objects and simulation algorithms, specific to the case at hand. The representation of particular types of erroneous behaviour is expected to be simplified by the use of an object-oriented programming language like G2.

Task Analysis and Work Context

CAMEO has been designed to evaluate procedures and interfaces rather that to simulate configurations outside the design basis.

In particular, CAMEO has been adapted to the analysis of nuclear power plant control systems, and to their effectiveness in supporting operators' diagnosis and decision making (Fujita, 1989; Fujita and colleagues, 1991).

In this sense the model contains a built-in database of hierarchically combined control mechanisms, tasks, rules, and control mechanisms. These structures are a sort of computerised requirements and specifications of the design process, rather than the result of the analysis of the real working context and behaviour of operators. Therefore, it can be argued that, the task analysis is not a standpoint on which CAMEO relies.

This peculiarity of the model can be exploited in two alternative ways. That is, CAMEO can be utilised either for designing or for evaluating the adequacy of procedures, tasks, and interfaces with respect to workload, skill and, simply, basic knowledge of operators.

CAMEO can be seen as an important tool in support of designers, as it allows to evaluate the level of cognitive difficulties associated with the performance of certain tasks, and the effects of erroneous behaviour.

In a simplified version and coupled with a database of error probabilities, CAMEO could be utilised to perform Human Reliability Assessments in the framework of Probabilistic Risk Assessment studies (Fujita, 1992).

3.3.3 The Simulation CES

The Cognitive Environmental Simulation (CES) model (Woods, Roth, and Pople, 1987) generates predictions concerning operator actions as the result of a combined interaction of the simulations of the plant behaviour and human response.

CES was developed as a research endeavour, sponsored by the US Nuclear Regulatory Commission, aimed at modelling cognitive behaviour of nuclear power plant operators during emergency conditions.

Theoretical Description

The CES model is imbedded in a human-machine interaction architecture, where the plant dynamic evolution affects human behaviour through the virtual display board of potentially observable plant behaviours, which affects the human intention formation (Figure 3.14).

Three basic kinds of cognitive activities (processing mechanisms) are represented in CES:

- *Monitoring* and analysing the plant behaviour in order to decide if the plant is in an expected or unexpected evolution.
- *Building explanation* for unexpected conditions.
- M*anaging the response* by correcting the abnormalities through automatic or manual interaction, by monitoring the plant response and by adapting pre-planned responses to unusual situations.

This representation of human cognitive behaviour is similar to the models of Rouse or Sheridan described earlier. In this case, the modelling of allocation of resources and of knowledge base are considered in much more detail, as they are necessary for the simulation of interactions of the cognitive model with the plant performance and working environments.

The basic psychological principle of CES is that people have limited resources, particularly in high workload situations, and therefore not all the available knowledge can be exploited by the operator in an optimal way. The way in which knowledge is activated at any given time in CES depends on three different types of interactions:

- The interaction between *knowledge-driven* and *data-driven* processing. While a knowledge-driven process is governed by cognition and mental work, a data-driven process is generated by the plant and perceived by external sensory receptors.
- The interaction between *resources and workload.* This mechanism represents the onset of an activity which may be bounded by the amount of necessary resource and the amount already allocated to the ongoing process.
- The processing of the most *evident and relevant information.* This interaction follows several layers of criteria by which the importance of a process may be defined with respect to the ongoing one.

These three criteria are original and characterise the theoretical structure of CES. Moreover, they allow the consideration of those cyclical cognitive processes which are particularly important for the theory described, for example, in the COCOM model.

The performance of CES in different workload and environmental conditions is governed by Performance Adjustment Factors (PAFs) by which the analyst can explore variability in human behaviour. The PAFs can modify the knowledge and the processing mechanisms of CES.

The CES model, although originally developed for simulating behaviour of nuclear power plant operators, shows a great flexibility, thanks to wide scope of its theoretical basis. Therefore, it can be applied to many domains, for different types of applications and analyses, such as dynamic perspective studies for design and safety assessment as well as retrospective evaluation of root causes of accidents or explanation of specific sequences of human-machine interactions.

This is a peculiarity of the CES simulation that has to be applied with carefulness by the user, in relation to the application being carried out, so as to avoid the omission or the unnecessary enhancement of aspects which may, or may not, be very important for the specific domain and plant under study (Woods and Roth, 1995).

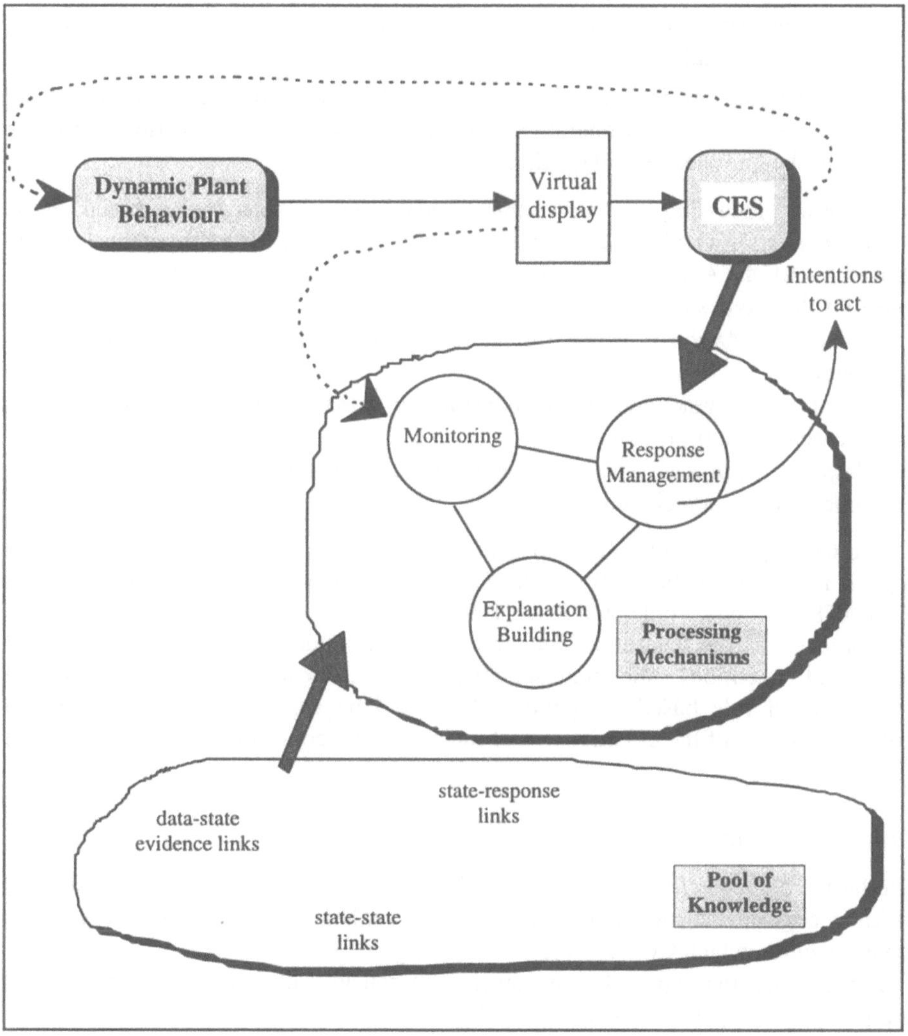

Figure 3.14. CES mechanism and cognitive processes (from Woods, Roth, and Pople, 1987).

Computational Implementation

CES is formally described as an Artificial Intelligence based computer program that follows the formalisms of the EAGOL/CADUCEUS AI problem solving system (Pople, 1977, 1985).

As an instance of EAGOL, CES contains two major elements, which are typical of expert systems and AI structures:

1. A *knowledge base*, that represents the know-how of operator(s) in regard to the plant and its behaviour, including physical inter-relationships between processes, physical variables, goals and functions of the control systems, meaning of signals, alarms and corrective plans of actions.
2. An *inference engine*, which is formulated in the form of the processing mechanisms, by which operators elaborate the available information and make use of the knowledge contained in the Knowledge Base.

The basic processing mechanism that simulates human behaviour is based on the use of "*analysts*", which are AI operators equivalent to the actors or agents, already utilised by AIDE. Analysts are spawn when certain triggering criteria are met. CES considers three types of *analysts*:

- *Behaviour Analysts* which monitors the plant behaviour for unexpected conditions.
- *Situation Analysts* which are responsible for explaining the perceived situations in case of unexpected findings.
- *Response Plan Analysts*, which select the control actions or develop a plan to cope with the perceived abnormal conditions.

The CES model is fully developed in a computer program and has already been applied for probabilistic safety in the Cognitive Reliability Assessment Technique (CREATE) (Woods, Pople, and Roth, 1990).

Moreover, as the theoretical foundations of the model allow the formulation of goals and intentions, the representation of sensory-motor activities and the application of rule based behaviour, CES could be applied for the development and/or validation of the effectiveness of interfaces and control procedures in the case of emergencies.

Task Analysis and Work Context

The basic *competencies* contained in CES are generated and fed into the basis pool of knowledge of the model from a number of studies and analysis of the working environment in which CES operates.

In particular, two types of competencies are considered: theoretical competencies and practical competencies (Figure 3.15).

The theoretical knowledge of structures and functions of the plant under control are formalised to build the first set of competencies.

A set of empirical studies are conducted to observe and evaluate human performance in presence of system faults and emergencies, during simulator training sessions.

Retrospective studies of operators' behaviour during real accidents can also be conducted to generate and formalise more practical competencies.

The process of formalisation of the competencies from the empirical studies into computable forms is quite complex and time consuming. It follows the rules and mechanisms of EAGOL, and produces knowledge structures that can be made readily available for the processing mechanisms of CES during the simulation of human behaviour.

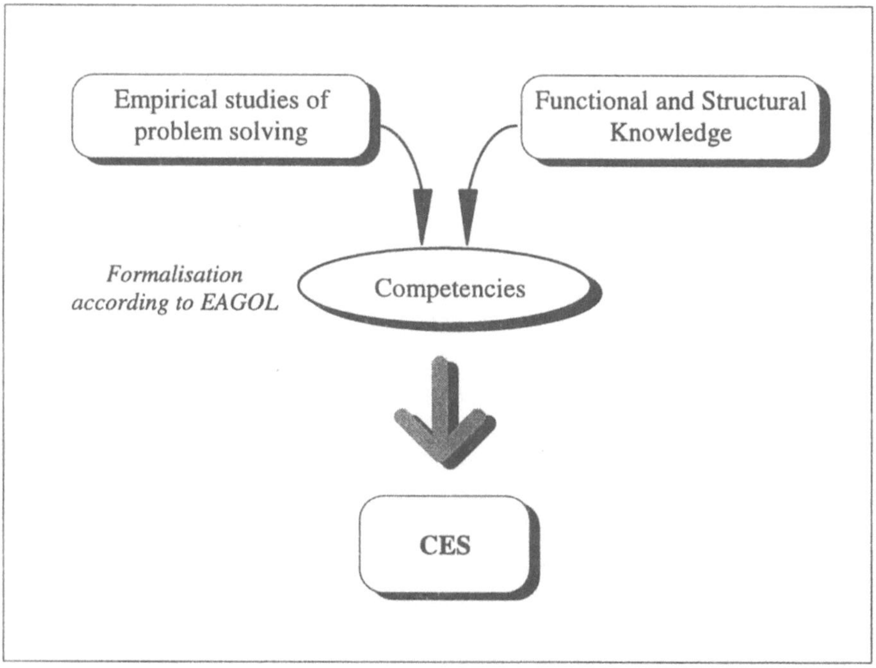

Figure 3.15. Building Competencies for CES (from Woods, Roth, and Pople, 1987).

3.3.4 The Simulation COSIMO

The COgnitive SImulation MOdel COSIMO (Cacciabue and colleagues, 1992a) was developed with the objective to describe and predict quantitatively human behaviour during dynamic human-machine interactions, mainly in highly automated working contexts like the control rooms of nuclear power plants and air-traffic control rooms.

Theoretical Description

The COSIMO model is based on the cognitive under-specification theory developed by Reason (1990). In particular, it formalises the cognitive primitives of *Similarity Matching* (*SM*) and of *Frequency Gambling* (*FG*) (Figure 3.16).

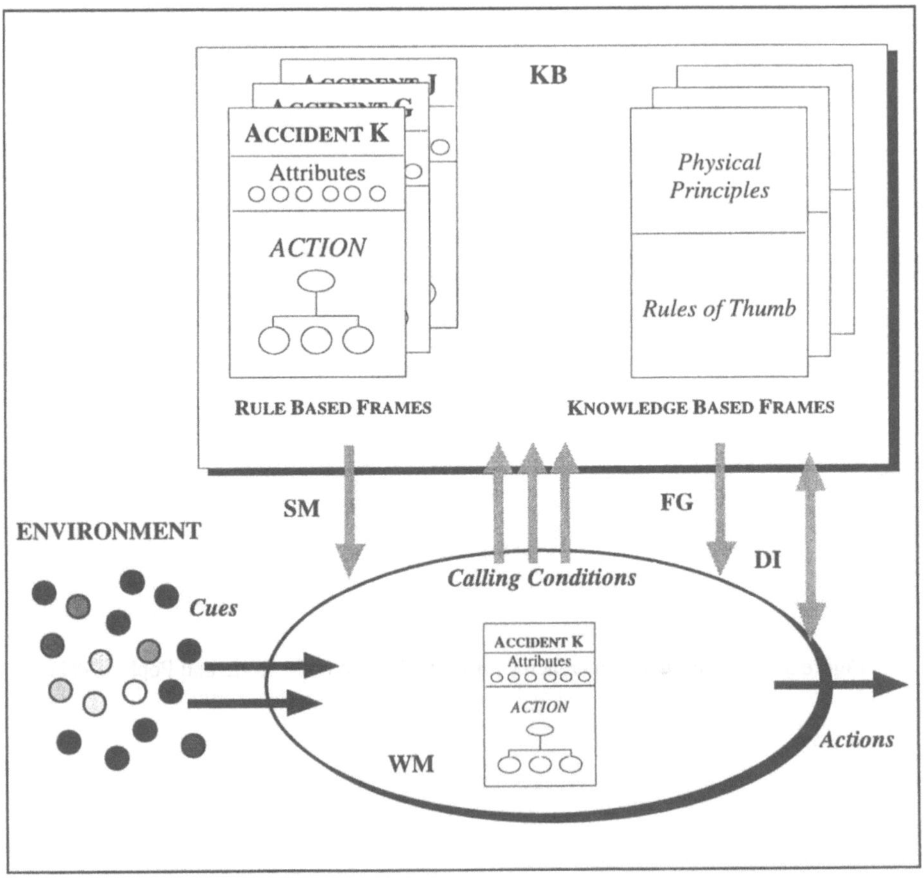

Figure 3.16. COSIMO Conceptual description (from Cacciabue and colleagues, 1992a).

The *SM* primitive matches the perceived cues coming from the work environment with corresponding diagnostic cues described in the knowledge base.

The *FG* primitive solves possible conflicts between partially matched hypotheses selected by the *SM* in favour of the more frequent, already encountered and well known incident scenario (Reason, 1987a).

By means of the "cognitive primitives" knowledge *Units* or *Frames*, stored in the *Knowledge Base* (*KB*) are selected and brought into the *Working Memory* (*WM*) for processing.

In COSIMO four inter-related cognitive activities have been modelled and implemented. The process of modelling these activities relies on theoretical accounts of operator's strategical attitudes in real situations (Bainbridge, 1989), and on what has been experimentally investigated in cognitive psychology (Reason, 1987b). These four cognitive activities are: *Cognitive Filtering*, *Diagnosis*, *Hypothesis Evaluation* and *Execution*.

The concept of *Salience* of information is used to set priorities for the perception of information during the evolution of cognitive processes.

A more detailed description of COSIMO follows, while, in a sample case discussed later in this Chapter, a specific application of COSIMO is presented for modelling a nuclear power plant operator in control of an auxiliary system during an operational transient.

Computational Implementation

The Artificial Intelligence Blackboard (BB) framework (Hayes-Roth, 1985; Engelmore and Morgan, 1988) is proposed as computational architecture to implement COSIMO.

The BB architecture supports the adaptive behaviour of operator reasoning, in particular the ability to execute parallel filtering, diagnosis or planning activities, such as, for example, a dilatory action that can be taken without being preceded by a diagnosis, when operators want to extend the time available for understanding the problem.

The BB structure also supports an explicit representation of mechanisms that enable to focus explanation and dynamic revision according to new information that continuously change the problem scenario.

Finally, the BB structure allows an incremental and opportunistic decision making development and the ability to respond to changes in the situation.

In order to implement a cognitive model in a Blackboard, three parts are considered, namely:

- The *Blackboard* itself with its internal structure.
- The *Agents*, also called knowledge sources or specialists, working on BB.
- The *Controller* governing the actions of the agents.

The similarity of this architecture with those utilised for the computerised implementation of CES, CAMEO and AIDE is quite straightforward.

Task Analysis and Work Context

COSIMO was developed in a computer program and has been applied for the analysis of power plant operators in accident management.

The theoretical foundations of the model allow the formulation of goals and intentions only with reference to well known situations, while the representation of sensory-motor activities and the application of rule based behaviour are well simulated.

The human response to unexpected conditions is carried out by the selection of well known and already "built-in" plans, while the adaptation of established control rules or the development of new plans of action, by direct inference and reasoning, have not been considered in the simulation.

In order to collect data and parameters that govern the model, COSIMO adopts a mechanisms very similar to CES. Firstly, a certain number of empirical studies are necessary in order to discover how procedures are carried out in reality and how working context and plant physical evolution affect operators' behaviour. These studies lead to the definition of *Attributes* of Frames.

Moreover, when coupled with postulated deep knowledge, these empirical studies contribute to build the formats of tasks and control procedures which make up *Rule Based* and *Knowledge Based Frames*.

3.3.5 The Operator Function Modelling - OFM

The Operator Function Model (OFM) method has been proposed as a dynamic, task-analytic architecture for designers of interfaces, procedures and support systems, that focuses on human rather than hardware function (Rubin, Jones, and Mitchell, 1988).

The OFM methodology has been utilised, for example, to develop an *operator associate* in the supervisory control of ground controlled earth-orbiting satellites (Mitchell, 1987).

Theoretical Description

The Operator Function Model combines operator functions and dynamic states of the system under control, by developing a hierarchical structure of functions and by subdividing each function into sub-functions, tasks, actions and requests for information.

The hierarchical structure is based on the representation of operator intentions in the form of a dynamic set of *goals*, associated with the state of the system. Each goal is then decomposed in *plans*, which are further organised in *tasks*, and in control (elementary) *actions*, which are the building blocks of the whole model.

Control actions, in the OFM convention, can be either *manual* or *cognitive.* While manual actions are straightforwardly explained by a single or a multiple set of commands executed in controlling or reconfiguring the plant, cognitive actions cover information processing, like assessing a state or gathering information and decision making.

The OFM structure has been developed with reference to a precise computerised artificial intelligence architecture rather than to a specific theory or modelling paradigm. This has led to the development of a very flexible structure, easy to adapt to specific cognitive and behavioural characteristics of the working context and system under control.

In the case of the application of the OFM methodology to the development of an expert system functioning as *operator associate* in the supervisory control of a satellite ground control station, the human cognitive functions have been defined by the specific application rather than the OFM itself. In particular, during these operations the main operator control functions can be organised according to a theoretical modelling, which is then imbedded in the OFM structure.

In the case of the satellite ground control system, a number of "top level" operator functions can be coupled together with the events of the system that affect transition from one function to another (Figure 3.17).

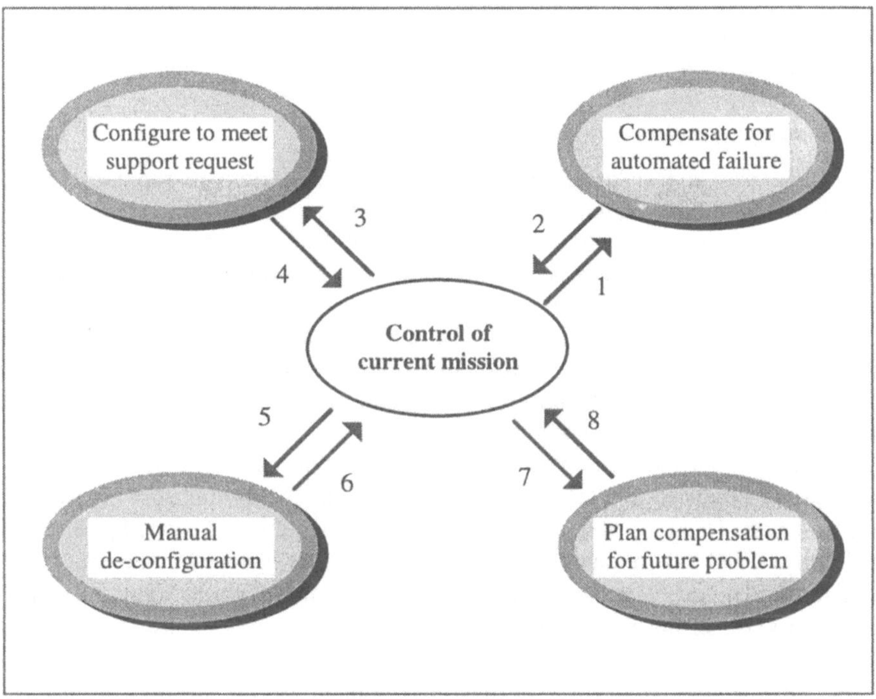

Figure 3.17. Supervisory control functions of an operator of a ground based satellite controlled station (from Mitchell, 1987).

In particular, the following supervisory functions are simulated:

- Monitoring the automatic scheduler for failures or error messages, and compensating, when possible (arcs 1-2 of Figure 3.17).
- Receiving requests of unscheduled support and "configure" system accordingly or declare inability to meet requests (arcs 3-4 of Figure 3.17).
- Receiving messages to manually "de-configure" equipment and "de-configure" as request (arcs 5-6 of Figure 3.17).
- Consulting the scheduling or mission procedures to plan compensation in case of possible expected failures and completing the compensation plan (arcs 7-8 of Figure 3.17).

In practice, the operator is involved in: a) *monitoring* data transmission during real-time spacecraft contacts, ensuring continuity and integrity of data transmission; b) *detecting faults* during transmission of data; c) *identifying* failed hardware or inappropriate configuration; and d) *compensating* according to the automated schedule or resuming other alternative system configurations or functions, if compensation is not possible.

This modelling configuration reflects almost completely the Human Problem Solver of Rouse and the Basic Supervisory Control Paradigm of Sheridan. These high level functions are then further developed into more specific sub-functions, tasks and control actions, so that the simulation can be developed in detail.

Computational Implementation

The OFM uses a Blackboard problem-solving architecture to represent operator functions and interaction with the controlled system.

The Blackboard is an ideal computerised architecture for dealing with selection, representation and management of chunks of knowledge (Hayes-Roth, 1985; Engelmore and Morgan, 1988).

As most of Blackboard models, the OFM is also organised in three major elements: *Knowledge Sources, Blackboard Control*, and *Blackboard Data Structure.* The *Knowledge Sources* contains the domain knowledge which is the representation of the system's operational and functional states.

The *Blackboard Control* is the controlling engine of the model, by which the appropriate knowledge sources are activated and "opportunistic" reasoning is implemented. This type of reasoning is representative of how knowledge is retrieved and utilised, and it is frequently applied in artificial intelligence frameworks.

The *Blackboard Data Structure* is the global database, where solutions or partial tasks are kept. In the Blackboard Data Structure the processes are carried out and solutions are developed in detail.

In the specific application of the development of the operator's "electronic associate" in the ground based satellite control station, the OFM based expert system, called OFMspert (Rubin, Jones, and Mitchell, 1988), consists of six major elements.

Each of these elements performs a specific function which is necessary for the operations of expert system. The overall system is modelled according to the Blackboard system (Figure 3.18).

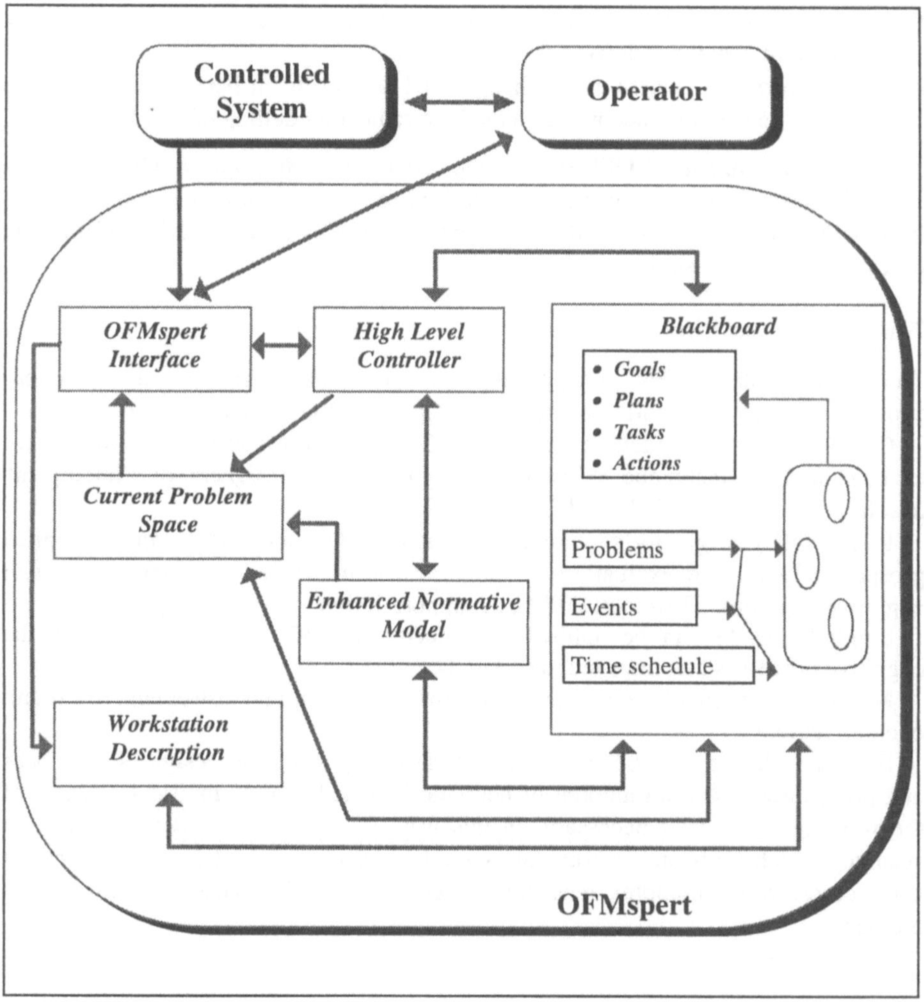

Figure 3.18. Blackboard architecture for the OFMspert model, (from Rubin, Jones, and Mitchell, 1988).

The *OFMspert Interface* decodes all messages coming from the *Controlled System*, which are then sent to the *High-Level Controller*.

The High-Level Controller manages and schedules the execution of the OFMspert by acting on three other elements of the model, namely:

- The *Current Problem Space,* where the representation of the system is updated dynamically.
- The *Enhanced Normative Model,* which contains normative information about the system behaviour derived from the OFM.
- The *Blackboard,* where event changes and new problem lists are elaborated to produce new actions according to the hierarchical modelling of OFM.

The last constituent of OFMspert is the *Workstation description* which contains semantic description of the workstation utilised for the control process and provides information on the interface and components which may be needed for information processing and goals setting or planning.

Task Analysis and Work Context

The declared role of OFM in supporting operator decision making and supervisory control, makes this simulation particularly adapted for the design of expert systems.

Consequently, the modelling structure of OFM can be proposed as an alternative to task analysis technique utilised by engineers and designers of "intelligent" interfaces (Degani, Mitchell, and Chappell, 1995).

In the OFM originator's opinion (Mitchell, 1987), the OFM can be applied prospectively for the evaluation of the information needed to carry out certain cognitive activities. Therefore, OFM can support design.

Similarly, OFM can be applied retrospectively, with the aim to explain the cognitive processes that have led to certain behaviour.

Moreover, the model can be used as an "inferential" on line instrument, to explain operator actions.

Especially in the case of prospective studies supporting the design of interfaces and procedures, a certain amount of heuristic knowledge and empirical analysis of the working contexts is necessary. In this way, it is possible to define with high accuracy which tools and instruments should to be used in real conditions and, in some cases, which complex reasoning processes could be carried out to solve a problem.

3.3.6 The Simulation of a Group - SYBORG

The Simulation of the Behaviour of a Group of operators (SYBORG) has been developed with reference to nuclear energy production, and aims at studying hypothetical severe accidents involving human factors, as well as at supporting the design of intelligent interfaces and control procedures (Takano, Sasou, and Yoshimura, 1995).

Theoretical Description

The model builds on the previously developed theories of information processing mechanisms and, in particular, on the SRK model of Rasmussen.

The simulation of a group of operators comprises a "leader" and two "followers" and builds on a model of single operator behaviour, duplicated and modified according to the specific role assigned within the group (Figure 3.19).

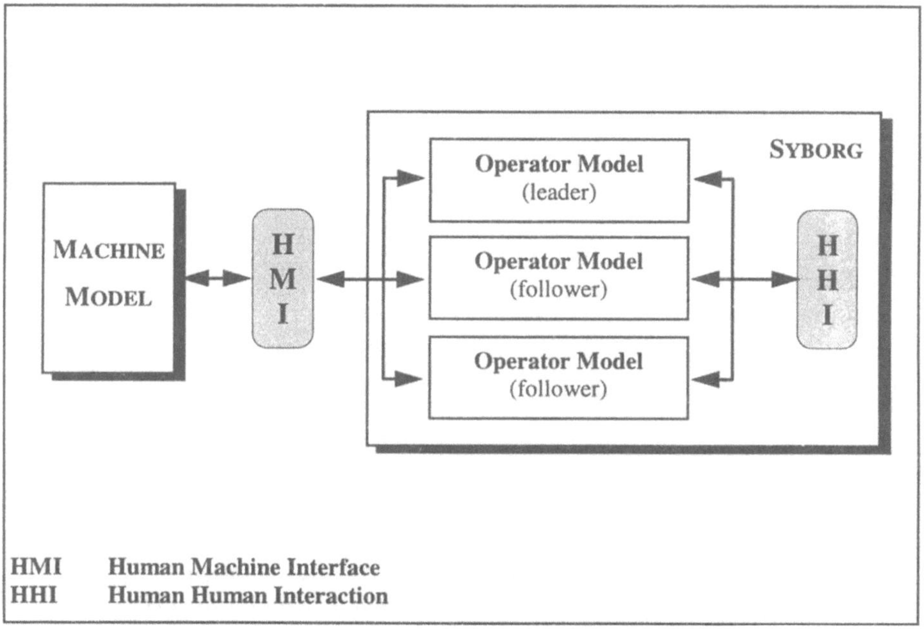

Figure 3.19. Human-Machine SYBORG architecture (from Takano, Sasou, and Yoshimura, 1995).

As SYBORG is dedicated to the analysis of human-machine interaction, the simulation architecture also covers the model of the plant and interfaces.

Moreover, SYBORG exhibits the peculiarity of simulating two interfaces, one for the interactions of humans with the machine (Human-Machine Interaction, *HMI*) and one for the group interactions (Human-Human Interaction, *HHI*).

This feature of the architecture of SYBORG, is also reflected in the model of individual operator behaviour, which contains a number of constituents, accounting for the fact that the each operator communicates with the others, as well as with the plant under control. In particular, the following elements are considered (Figure 3.20): *Attention Model, Short Time Memory, Thinking Model, Medium Term Memory, Long Term Memory, Action Model,* and *Utterance Model.*

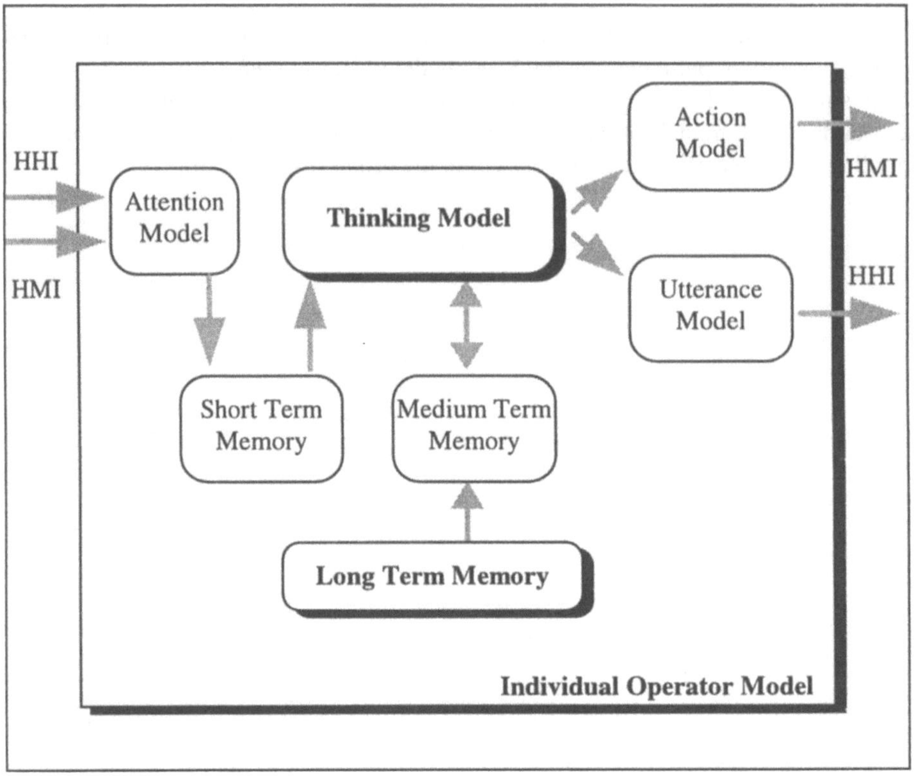

Figure 3.20. Individual Operator Behaviour Model, (from Takano, Sasou, and Yoshimura, 1995).

The *Attention Model* filters sensory information derived from machine behaviour through the *HMI* and communication between operators trough the *HHI*.

The *Short Time Memory* accumulates temporarily information from the *Attention Model*, conveying it "smoothly" to the *Thinking Model*, with a predefined time delay.

The *Thinking Model* is the core of the single operator model, as it calculates and defines the execution of procedures and actions to be carried out.

The *Medium Term Memory* serves as buffer and sustains the transfer of information between the *Thinking Model* and the *Long Term Memory*.

The *Long Term Memory* contains the knowledge necessary for the *Thinking Model*, including plant configurations, parameters, variables, dynamic behaviour, and predefined procedures.

The *Action Model* implements the control actions decided by the *Thinking Model*.

The *Utterance Model* develops the communication between operators, including typical sentences and verbalisation symptomatic of tacit knowledge.

Computational Implementation

The computerised version of SYBORG has been developed by using the C^{++} language, which is an Object Oriented programming language particularly suited for representing knowledge and cognitive behaviour.

As the model architecture develops in detail the relations between the various constituents of the model, the use of special environments, such as Blackboard or G2, are not necessary.

Task Analysis and Work Context

The analysis of the working context and the evaluation of tasks and practices of plant control are necessary for the development of the knowledge structures contained in the Knowledge Bases supporting the Long Term Memory and the Utterance Model.

The application of a formal method for task analysis and empirical evaluation of the working environment are discussed in detailed by the authors of SYBORG. These methods imply the observation of operators behaviour and communications during the management of hypothetical accidents carried out in simulator sessions (Sasou, Takano, and Yoshimura, 1993).

As these methods refer to standard Cognitive Task Analysis approaches, they bear strong similarities with the approaches already described, for example in the case of the AIDE or CES. Consequently, no further details will be given here on this subject.

3.3.7 Other Cognitive Simulations

There exist many other endeavours to develop computerised simulations of human behaviour, which are based on the Information Processing System metaphor or on some of the theoretical models discussed earlier.

However, the description of the whole spectrum of simulation approaches of human cognitive behaviour is beyond the aim of this book. What is important is that, by now, the reader has developed a basic knowledge of "which cognitive activities" are relevant for modelling purposes, "how they may be described" in theoretical conceptual forms and simulated in numerical algorithms, and "which techniques and methods" may be utilised for data collection and retrieval.

In order to complete, as much as possible, the review of simulations, we will now analyse shortly four more approaches, covering the whole spectrum from control theoretic techniques to artificial intelligence programming and fuzzy logic.

We will not elaborate a detailed discussion for these last simulations models, as the description of the approaches carried out in the previous sections is considered sufficient to show the practices and the steps of development necessary to arrive at a numerical approach, once the problem space and theoretical formalisms have been defined.

These last four simulations are:

- PROCRU (Baron and colleagues, 1980),
- MIDAS (Corker and Smith, 1993),
- Human Operator Model of the Navigator (Papenhuijzen, 1994), and
- Cognitive Model of Dynamic Ship Navigation (Hansen, and Itoh, 1995).

The aim of this last review is to show how simulation methods have evolved over time, gradually improving specially with respect to numerical techniques, while the basic theoretical paradigm has remained relatively stable.

PROCRU

PROCRU (Procedure Oriented Crew Model) is a simulation of a aircraft crew, composed of a flying pilot, a pilot-not-flying, and a second officer, for the analysis of multi-operator situations and the evaluation of the effects of communications among such operators (Baron and colleagues, 1980).

The basic reference model structure comprises the simulation of the aircraft under control and the simulation of the single operator (Figure 3.21). The simulation of the aircraft includes: *Machine Dynamics,* containing display and control variables, and *ATC/CREW* model, which comprises communications with other crew members and the external world, such as the air traffic control.

The simulation of the individual operator contains four main elements:

1. The *monitoring* process, which handles display variables and incoming communication and is affected by the situation and by psychological and external factors, such as stress.
2. The assessment of the current situation (*information processing*), which is influenced by monitored information, inherent knowledge and goals of the operator.
3. The decision of the action or other cognitive activities to carry out, which is based on a procedure oriented modelling (*procedure selector*) and is affected by the previous cognitive activities, the aims of the operator and the assessment of possible consequences.
4. The action implementation (*execution*), which implies a process of *communication* with other crew members, or the external world, and the performance of actual control activity, either by observing (*monitor requirements*) display variables or by operating the control system (*control requirements*).

The configuration of this model reflects quite accurately the basic *Reference Model of Cognition*, and the structure of the *Very Simple Model of Cognition* of the COCOM modelling architecture.

However, it has to be noticed that the work of Baron and colleagues has been developed in the early 80ies, i.e., well before COCOM, and shows very important aspects of innovation and originality, such as, for example, the structure for simulating crew behaviour.

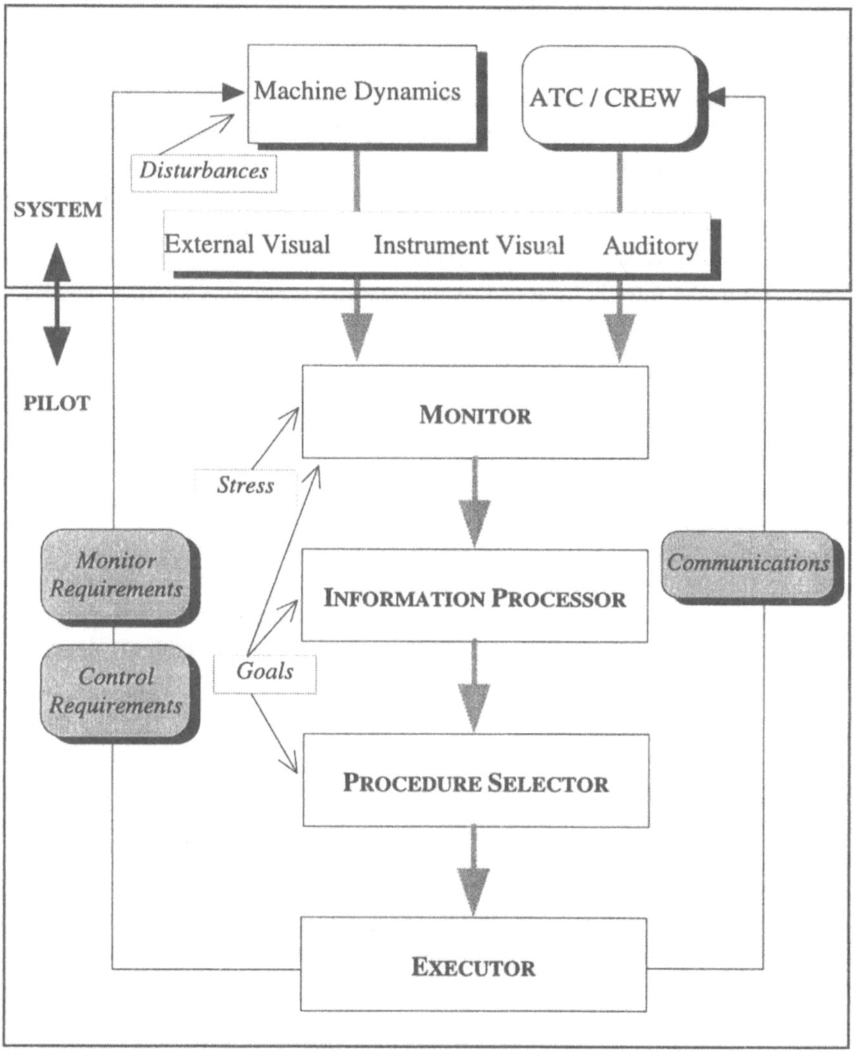

Figure 3.21. The model PROCRU for individual crew member.

The model includes, amongst the events that are considered for situation assessment, facts that are "not explicitly dependent on the vehicle state variables" (Pew and Baron, 1983). This means that one of the basic requirements for modelling cyclic cognitive processes is respected, i.e., a cognitive activity may be generated by another cognitive process and is not only the result of a machine or context stimuli. This qualifies PROCRU as a *cyclic* simulation.

The algorithms utilised by PROCRU to represent cognitive functions and processes, and control and monitoring activities are based on control-theoretic approaches, referring to the Optimal Control Model and optimal sampling techniques, already described in Chapter 1.

The simulation of communications is performed by referring to standard procedural verbal requests or responses, i.e., "callouts, requests and commands", as is required by procedures.

PROCRU presumes the use of (cognitive) task analysis for preliminary definition of procedures and actual performances carried out in the cockpit.

This feature fits very well the requirements of the *Reference Model of Cognition*, as discussed in the introduction of this chapter and represents another peculiarity of PROCRU, that shows the degree of originality brought by this approach into the domain of cognitive modelling and simulation.

It can be concluded that PROCRU, although developed in the early 80ies, remains, even today, a remarkable simulation approach worth reviewing and considering as possible means of representing pilots' (operators') behaviour, even when dealing with highly automated cockpits (control rooms) and multiple interaction processes.

In particular, while the numerical algorithms could be updated and more flexible and powerful approaches may be implemented and utilised, the architecture and structure of the model should be kept, as they match very well the requirements of the reference framework for human-machine interaction simulation, developed in Chapter 1.

MIDAS

MIDAS (Man Machine Integration Design and Analysis System) is a framework that accommodates models and knowledge structures for the simulation of human-machine interactions during safety critical conditions (Corker and Smith, 1993).

MIDAS has been conceived as a modular structure and can, in principle, be applied to study different domain environments, at different levels of complexity, as far as mission requirements, equipment and models of human performance are concerned.

In practice, MIDAS as been applied to analyse civil aviation safety situations, both from the crew (Corker and Pisanich, 1995a) and the air traffic control perspective (Pisanich and Corker, 1995b; Pisanich, Corker, and Bunzo, 1997).

The basic architecture of MIDAS (Figure 3.22) reflects the Human-Machine Interaction architecture described in Chapter 1, as it contains a model of the system under control, the *World Representation*, and the *Operator Model.* These two models interact with each other dynamically by cyclically exchanging data and information.

The *World Representation* also contains the probabilistic module, by which failure and malfunctions may be introduced on a probabilistic basis.

The *Operator Model* contains the basic modules that characterise the information processing paradigm, i.e., a module for *Perception*, an *Updateable World*

Representation (UWR), a *Scheduler,* a *Symbolic Operator Model,* and a *Motor* model.

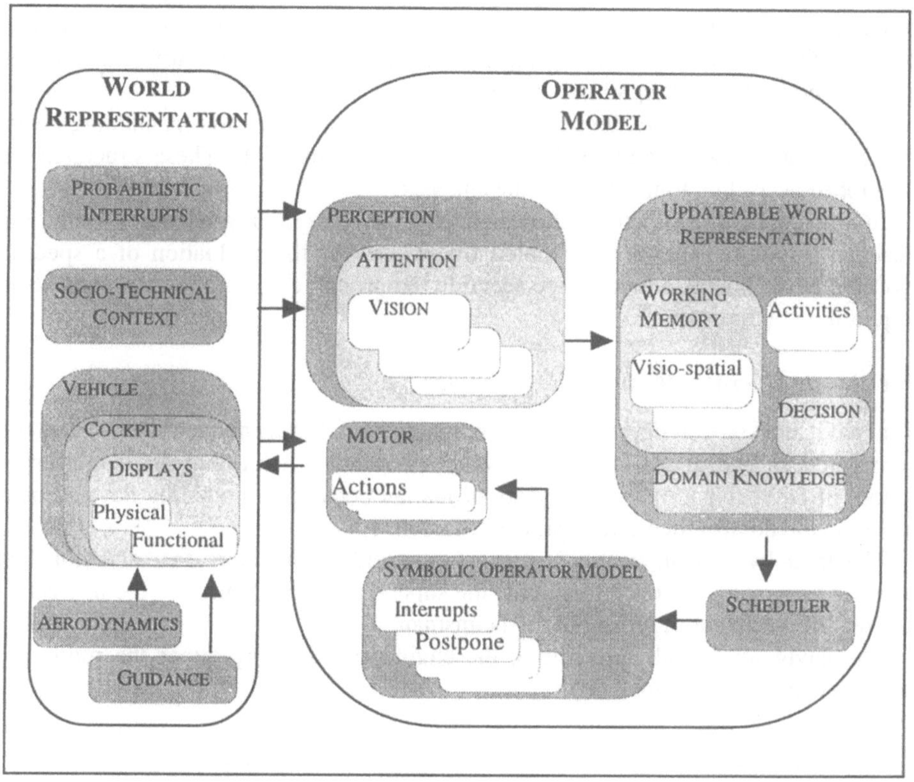

Figure 3.22. The MIDAS Model (Corker and Smith, 1993).

While the *Perception, Scheduler,* and *Motor* models perform quite simple cognitive activities, the core of the operator model resides in the *Updateable World Representation* and *the Symbolic Operator Model.* The former model contains the basic knowledge of the operators, the information concerning procedures and equipment, the activity of working memory on the information perceived from the perception module, and the known relationships between objects and system components. In particular, the decision making part of the *Updateable World Representation* is modelled in accordance to the Skill-Rule-Knowledge (*SRK*) base model of Rasmussen.

The *Symbolic Operator Model* manages the dynamic aggregation and scheduling of activities, by assigning priorities to goals, selecting current tasks, postponing or suspending tasks of lower priority, etc.

This module simulates the process of *Allocation of Resources* of the *Reference Model of Cognition* of Chapter 1.

As for numerical algorithms and programming language, MIDAS is structured in such a way as to be easily developed within typical artificial intelligence or object oriented architectures, containing "agents", "calling nodes", with associated pre and post-conditions, and "time scheduling parameters", by which the different modules can be represented and organised in sequences.

Tasks and activities are defined in a hierarchical representation, containing goals, sub-goals, tasks and several levels of attributes and constraints. These structures are developed according to the goals of the analysis.

In the case of evaluation of design of interfaces and procedures, only the designer's requirements are accounted for, while, for the evaluation of a specific working environment, a much more accurate (empirical) evaluation of the contexts is required.

The Human Operator Model of the Navigator

The navigator model described in this sub-section (Papenhuijzen, 1994) can be considered as one of the many, very valid, products generated by the engineering school of Delft (NL), over many years of human behaviour modelling development based on the control theoretic approach (Stassen, 1988).

The model of the navigator is part of a generic architecture that discriminates between two major operator functions: the supervisory one, typical of the navigator, and that of the actuator, typical of the helmsman.

The navigator's goals are: setting the engine and defining either the rudder or heading command (Figure 3.23).

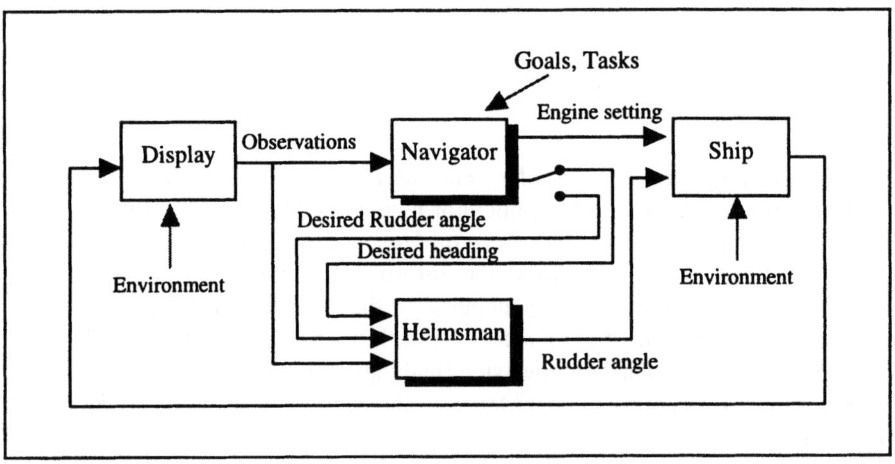

Figure 3.23. Generic navigation process (from Papenhuijzen, 1994).

Nowadays, in modern ship control systems, the helmsman role has been taken over by the auto-pilot, therefore increasing the supervisory responsibility and demands on the navigator.

The model of the navigator has been developed for differentiating between three cognitive activities: the *state estimation*, the *track planning* and the *track following* (Figure 3.24). Each of these activities is represented by a specific *sub-model* of the overall navigator model.

The sub-model for state estimation could be developed by means of a Kalman filtering approach or by a simpler explicit description of state estimation, presenting, to the other sub-models, the information exactly as shown by the display system of the boat. In the existing modelling framework, the second alternative has been selected, even if the option for a Kalman based modelling has been left for future or specific developments.

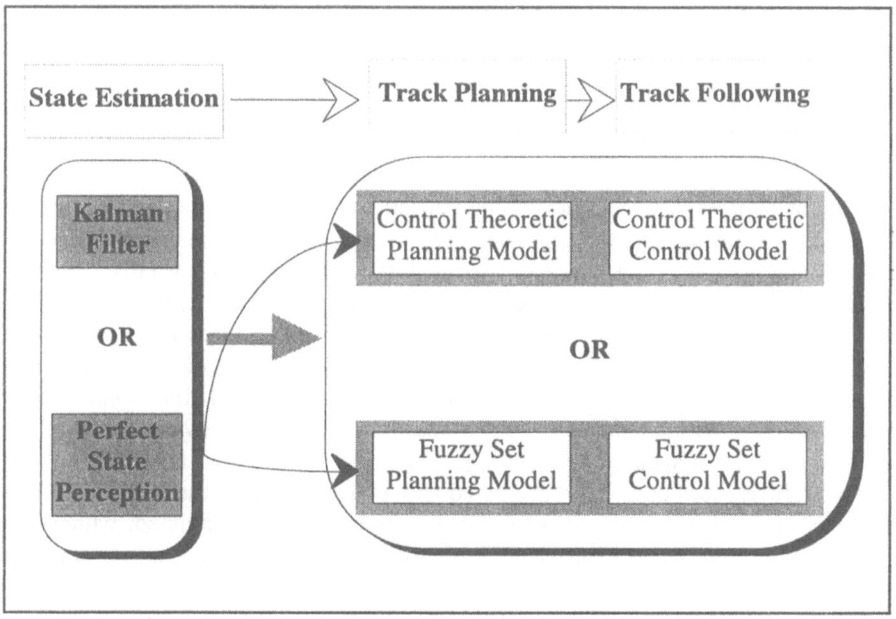

Figure 3.24. The Navigator Model (from Papenhuijzen, 1994).

The planning and control sub-models have been developed using either control theory or fuzzy set theory, with the objective to offer, to the user of the model, maximum flexibility and adaptability to simulate different situations and control strategic attitudes. A comparison of these two modelling methods has also been performed with the aim of defining advantages and drawbacks of each technique (Papenhuijzen and Stassen, 1992; Papenhuijzen and Dijkhuis, 1995).

This generic representation of basic cognitive processes, and in particular, the consideration of three levels of problem solving activities (*state estimation*, *track planning* and *track following*) relates very well to the Human Problem Solver paradigm of Rouse. However, this modelling structure does not leave very much space for the simulation of cognitive resources or for the development of memory and knowledge base content.

As this modelling approach is fundamentally linked to control theoretic approaches, and, therefore, no Artificial Intelligence methods are envisaged, the representation of knowledge and memory of the navigator is practically performed by means of an *internal model* of the ship's dynamics. This model is developed as a set of linear differential equations, which are well adapted to the sub-models of track planning and track following. In the case of fuzzy theory based modelling, some adaptation is needed.

From the theoretical viewpoint, the model of the navigator can be considered as a *sequential* model, as all human activities are carried out in reaction to a state estimation, following the dynamic behaviour of the ship. Thus, a number of limitations and drawbacks could be associated to the general modelling architecture. In particular, it could be argued that this approach is an elegant and much developed version of the Stimulus Response Organism approach of the 70ies, which has shown many limitations from the cognitive perspective and has been very much criticised for its lack of consideration for "real" cognitive flavour (Hollnagel, 1993a).

A number of experiments have been carried out with the objective of validating the ability of this simulation approach with respect to actual navigator's behaviour and to compare the performance of a control theoretic approach versus a fuzzy set based model. The main conclusions of this study have been that *"...both approaches, the control theoretic one and the fuzzy set one, may well result in a practical model that yields realistic ship tracks, given ship situations. There are some important differences between the two models, however, concerning the functionality of the models as they are existent now, or the possibility for future enhancement of their applicability....."* (Papenhuijzen, 1994, p. 127).

These differences have been discussed and analysed well in the above mentioned reference and the proposal for a *hybrid* development was formulated, which could combine the best performances of both approaches according to the task and activities that need to be simulated.

What is important to note here, is that this type of modelling, although based on an architecture of human-machine interaction simulation (the control theoretic and fuzzy logic frameworks) that may seem inappropriate or too "stiff" for representing complex cognitive processes, has been able to give relevant and satisfactory results, even for the simulation of highly supervisory activities carried out by a ship's navigator.

This is an important issue that must be considered in combination with the question of the selection of the most appropriate numerical method and simulation environment by which to implement the human behaviour modelling architecture.

In particular, the simulation of the navigator has shown that, in certain conditions and in presence of a well defined domain problem, the use of control theoretic approaches and fuzzy logic are viable means of simulation of cognitive processes.

The Cognitive Model of Dynamic Ship Navigation

The cognitive model of dynamic ship navigation has been developed over a number of years (Itoh, and Enkawa, 1991; Paulin, and Itoh 1995; Itoh, Hansen, and Nielsen, 1998), and represents a simulation approach for navigation tasks, alternative to the control theoretic approach of Papenhuijzen previously discussed.

The reference paradigm of this model is substantially the same as the one adopted by the control theoretic simulation and is based on the representation of the navigator's behaviour in three levels, or *phases*, of cognitive activities: the *monitoring phase,* the *course correction phase*, and the *execution phase* (Figure 3.25).

However, the analysis and representation of the working contexts, the identification of navigator's knowledge base and memory exploitation and the algorithms of simulation of cognitive processes are totally different.

Firstly, the model is based on very accurate task analysis of navigator's behaviour. The essential hypotheses that govern the task analysis method are based on the theories of "eye-mind assumptions" (Just and Carpenter, 1980) and "process-monitoring hypothesis" (Rayner and McConkie, 1976), which demand that an accurate evaluation of eye movements and verbal protocols is carried out by field experiments and observations.

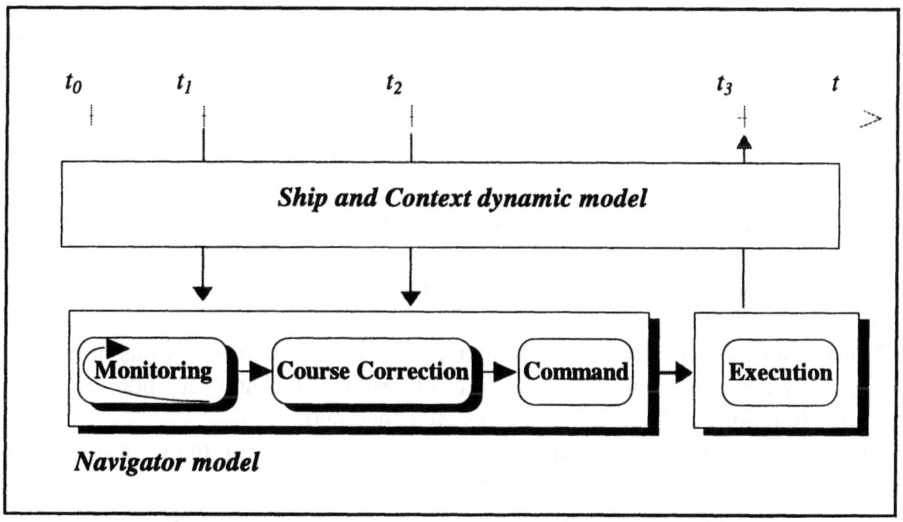

Figure 3.25. Outline of a cognitive model of ship navigation, (from Itoh, Hansen, and Nielsen, 1998).

The definition of cognitive activities and processes is carried out according to the Rasmussen (1986) *SL/SRK* modelling paradigm.

The method considers an *operation cycle* in the navigation task which is related to the eye-movement and verbal protocols. By the observation of the gaze time and the actual verbal protocols, a variety of time intervals are defined, namely:

- The time interval for perception and elaboration of information from the working context.
- The interval for the acquisition of new information and development of plan of execution
- The time delay for execution of a command, either by the helmsman or by the auto-pilot.

Another important peculiarity of this simulation approach is related to the numerical algorithms and programming environment utilised for the implementation of the model. The simulation has been developed by standard C language, which, although relatively limited in capacity to represent objects and inheritance structures typical of artificial intelligence programming, gives sufficient flexibility to represent rules and logical connections between tasks, goals and observation of ship performance.

Moreover, the consideration of a knowledge base repository of experience and its exploitation through a limited capacity short term memory (called "situation awareness memory" in this model) and long term memory, call for a simulation environment which is able to handle "production rule" types of modelling and, also object oriented programming.

The detailed studies carried out by the authors of this simulation, especially in the area of exploitation of short term memory processes, workload and risk evaluation of the navigator in performing real ship control activities, have generated a substantial knowledge base which strongly sustains the simulation.

The greatest advantage of this model lies in the richness of the knowledge base. However, this can also represent its major weakness point, if the aspects of task analysis and the development of a sound and vast amount of built-in rules are not properly considered and carried out by the user.

This observation can be considered as a generic differentiating element between "classical" engineering simulation methods, based on techniques such as control theory, and more flexible cognitive oriented approaches, exploiting artificial intelligence methods.

The former methods allow the simulation of many different situations, according to the analytical power of their numerical algorithms, but are very poor as far as cognitive functions and cyclical processes are concerned.

The latter methods are much more flexible and allow a more "human-like" representation of cognitive processes, but demand the development of an extensive work of analysis and formalisation of contextual effects and actual behaviours.

3.4 Guidelines for Development of a Simulation

In the previous sections we have reviewed and discussed the theoretical models, the algorithms and numerical techniques, and the task analysis methods by which a model and simulation of cognition can be devised.

At the end of this extensive exercise of analysis we are now in a position to summarise the major findings and develop basic guidelines which should be followed to achieve a simulation that is able to meet the requirements and needs of a user.

We have already discussed the issue of complexity of theoretical models (§ 3.2.7) and, in particular, we have shown how a complex and thorough theoretical approach may not always represent the best solution for a real problem.

We have also considered the need to develop a simulation as a recursive process between theoretical modelling, numerical algorithms, and cognitive task analysis (§ 3.3).

We will now develop further these issue, and we will describe in detail the steps of development by which a simulation can be conceived and generated.

In particular, Figure 3.26 shows how the interaction between Cognitive Task Analysis, Theory, Numerical Algorithms and Programming Environments, already discussed in general terms (§ 3.3, Figure 3.11), may be expanded into an iterative loop of 6 specific steps that should be carried out in order to achieve a sound and reliable simulation.

These 6 steps are:

1. ***Definition of Problem Boundaries and Aim of Simulation***
2. ***Cognitive Task Analysis***
3. ***Field Study of Working Context***
4. ***Selection of Theoretical Model***
5. ***Selection of Numerical Algorithms***
6. ***Implementation in Programming Language and Environment***

In general, not all 6 step are necessary in order to develop a descriptive model of cognition, as the numerical transformation by means of specific algorithms and the implementation in a programming language may not be needed.

The iteration between different steps is required when the need for refinements is identified at one step of the simulation development, and this implies a repetition of preceding steps.

After the step of *Selection of Theoretical Model*, the loop branches off into two alternative directions, so as to consider the development of either a *numerical* or a *descriptive simulation*, which represent the two forms in which a simulation may be developed for prospective and retrospective types of analysis, as already discussed in Chapter 1 (§ 1.2.1).

It is also important to point out that steps 2-3, 4-5, and 5-6 are often carried out in combination and therefore may be regarded as single steps in the loop.

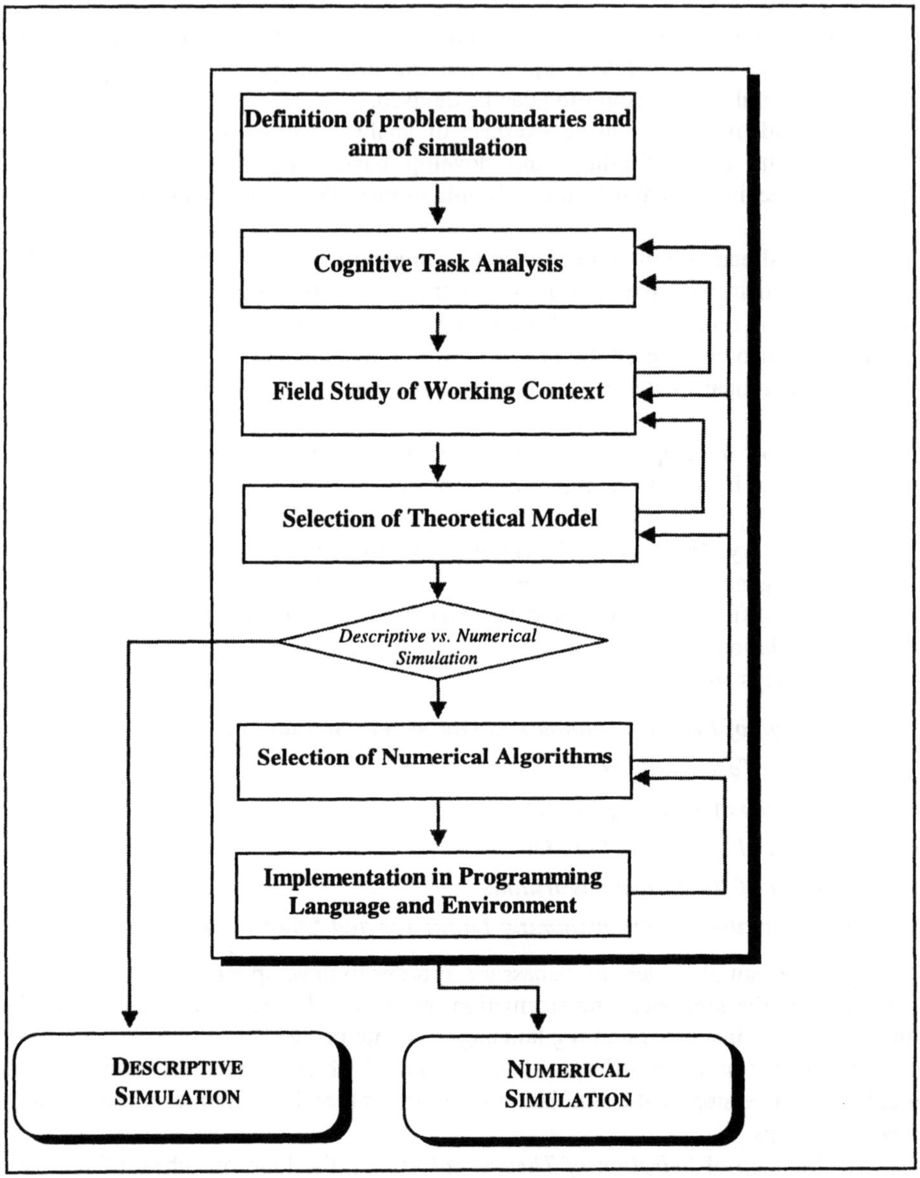

Figure 3.26. Basic Guidelines for the development of cognitive simulation.

3.4.1 Definition of Problem Boundaries and Aim of Simulation

The first step is crucial and may well be the most important one, as it can affect the whole subsequent work of development.

It consists in the definition of the problem at hand, of the goals and expected outcome desired from the simulation. In particular, it is essential to define whether the model will have to generate a *Numerical* or a *Descriptive Simulation* of cognitive functions and processes.

It is very important to define the level of accuracy required by the problem under study, as from this first steps of development, because, if this requirement is changed during the process of development, this may imply the repetition of the whole process from the beginning.

As an example, lets assume that the objective of the study is to evaluate and validate quantitatively the effectiveness of certain interfaces or control procedures, and that the model of the machine and display systems are accurate. In this case, the option of developing a detailed simulation of cognition should be applied and some specific aspects of human behaviour should be looked at and simulated as accurately as possible to match the real cognitive behaviour.

On the other hand, if only performance and external activities are included in the modelling requirement, and if the machine model also covers shallow aspects of physical plant and control system, then the option of a less sophisticated modelling development should be chosen, focusing only on most visible and generic aspect of human behaviour.

3.4.2 Cognitive Task Analysis and Field Study of Working Context

The complexity of the model to be developed affects the scope of the Cognitive Task Analysis and the amount of field studies of the working context.

As already mentioned, these two steps in the development of a simulation may be performed either independently or in strict correlation with each other.

The importance of these two steps in the overall process of generation of a simulation has already been pinpointed.

We simply reiterate on such crucial issue and we stress the fact that the complexity of the modelling approach, selected in the previous step, delimits only the scope of the task analysis and associated field and experimental studies, but not the accuracy and the formalisms that should be applied.

The accuracy of the analysis and field studies is defined by the successive steps of development. This is why a recursive procedure is foreseen (Figure 2.26), so as to emphasise the need to iterate between theoretical and practical processes in the development of the simulation. For example, as a result of the *selection of theoretical model*, further task analysis and field observations may be required in order to determine new parameters needed by the theoretical model, which were not identified in the first round of experiments.

3.4.3 Selection of Theoretical Model

The theoretical model sustaining the simulation is strictly related to the decision about the required accuracy of the simulation and to the specific working context under study, i. e., to the task to be simulated.

The selection of one of the paradigms of cognition previously discussed (§ 3.2) is only a starting point, which needs to be further integrated with the reality of the working context and with the numerical algorithms selected for implementing the simulation.

This blending process demands a further iterative loop between the task analysis and the theoretical model on the one side, and the numerical and computerised implementation on the other.

This second major loop of iterations is not needed if the objective of the development is to arrive at a *descriptive simulation*, which, by definition, contains descriptions of various cognitive functions, processes and mechanisms, but does not need numerical quantification.

3.4.4 Selection of Numerical Algorithms and Implementation in Programming Language and Environment

The steps of *Selection of Numerical Algorithms* and *Implementation in Programming Language and Environment* are essential for the development of a *quantitative numerical simulation.*

These steps are very much related to initial goals and scope of the development and are strongly linked to each other.

As we have noticed while reviewing different cognitive simulations different algorithms and programming languages exist and are able to fulfil the objectives of the model developers. These algorithms and programming environments are usually well suited to satisfy the requirements and specification for developing a reliable software simulation.

It is important to note that the iterative process between different steps of development (Figure 3.26), for the definition of best suited data, variables and parameters, can even take place at the programming environment level.

In certain cases, indeed, the numerical algorithms and programming tools offer technical solutions to a theoretical problem not previously thought of, and thus not developed in terms of data and task analysis.

At this point, a simulation developer should be in possession of a software simulation of cognitive behaviour, able to describe or calculate the mental and physical performances of a human operator during a human-machine interaction.

The practical application of such a simulation can then begin.

3.5 An Example of Application

The sample case that will be developed in this section concerns the COSIMO model, already described in the review of the existing simulations, and focuses on the process of development of the simulation and its application to a specific plant and accident scenario.

3.5.1 Definition of Problem Boundaries and Aim of Simulation - Case Study COSIMO

The simulation COSIMO (Cognitive Simulation Model) (Cacciabue and colleagues, 1992a) was conceived to explore human behaviour in simulated accident situations in order to identify important safety aspects of Human-Machine Interaction.

In particular, COSIMO can be usefully applied in a human-machine interaction architecture to:

- Analyse how operators are likely to act given a particular context.
- Identify situations that can lead to human errors and evaluate their consequences.
- Identify difficult problem solving situations, given problem solving resources and constraints (operator knowledge, man-machine interfaces, procedures).
- Identify and test conditions for error recovery.
- Investigate the effects of changes in the human-machine system.

The objective of this case study is to analyse operators' behaviour and error consequences, during the control of an emergency situation in a typical Nuclear Power Plant sub-system, i.e., the Auxiliary Feed-Water System (AFWS).

As we are only interested in the performance of the human simulation mode, we will not consider the simulation of physical phenomena, control instrumentation and procedures of the AFWS with respect to accident management. These models will be developed in a forthcoming Chapter when we discuss an example of overall Human-Machine architecture and we adopt the AFWS as reference system.

It is necessary to know, for this sample case study, that the AFWS is a crucial systems supplying cooling water to the four steam generators of a nuclear power plant in the case of an accident, and after the shut down of the main pumping and cooling systems. The AFWS is designed to provide sufficient and necessary cooling water to remove the residual heat generated in the nuclear reactor after shut down.

As a stand-by system, the AFWS is, activated automatically, at maximum capacity of water flow, in the case of an accident. There is thus a need to regulate the water flow according to the type of accident and the availability of steam generators, and to verify that the water level, the pressure and the void fraction in the steam generators are kept within established values. In some cases, the feed-water to one of the steam generators has to be stopped, as a consequence of failures or malfunctions of internal components of the AFWS.

All these control operations are performed using the control panels and the automatic control system by which the operator can assess the performance of the AFWS, regulate mass flows and isolate steam generators according to needs and control procedures.

All operations are supported by procedures and the most important activities of the operators consists in:

- The identification of the accident at hand.
- The selection and application of the appropriate procedure.
- The verification and direct observation of the evolution of the situation, performing occasional successive interventions of new regulation, if necessary.

The operator model and simulation must cover human behaviour in a number of accidental configurations based on the "station-black-out" accident, i.e., the total loss of electrical power and consequent reactor shut-down, complicated, in some cases, by small marginal local failures of individual components or parts of the AFWS. These latter failures, irrelevant in most cases, can become crucial for the human-machine interaction, if they are combined with certain scenarios and contextual conditions.

More detailed objectives of the study by means of the human model are:

1. To enable the simulation of control procedures and regulation actions, so as to cover at least one hour of accident.
2. To evaluate the consequences of generic error types, and not single erroneous actions.

Examples of these types of errors are: cognitive collapse, resistance to change views, cognitive lock-up, channel visioning and other cognitive erroneous attitudes, that manifest themselves in different ways according to the time and contextual situation.

In order to study this type of situation it has been deemed necessary to develop a sufficiently complete model of cognitive behaviour, that is at least able to consider cognitive functions and cognitive processes exploiting knowledge base and memory as well as an adequately articulated mechanisms of allocation of resources.

3.5.2 Cognitive Task Analysis and Field Study of Working Context - Case Study COSIMO

The task analysis of the control process of the AFWS in several accidental configurations, related to the *procedures* in case of "station-black-out" accident, has been carried out.

In particular, the "station-black-out" accident has also been studied in combination with other types of failures, such as limited availability of steam generators, unexpected leakage from a steam generator, malfunctioning of pumps and valves of the AFWS during operations, etc.

Each *procedure* has been decomposed in goals, sub-goals, task, sub-tasks and individual actions. For each action, preconditions and post-conditions have been identified. The overall procedure has been graphically represented according to the criteria of the Cognitive Task Analysis, discussed earlier.

Several procedures have been constructed in this way and have been organised in the form of "frames" of structured sequences of actions, verifications and decisions.

Following this theoretical and formal activity, the evaluation of the "working context" has been carried out, with the objective to evaluate practical aspects of tasks performance and to identify important variables, parameters and data, which are significant to operators from the empirical viewpoint.

In our case, as we are performing a sample case study, based on a simulation of a real feed-water system, we have performed only laboratory experiments, by observing the behaviour of a number of subjects, knowledgeable in nuclear energy and sufficiently familiar with the type of feed-water system, controlling the simulator of the AFWS in a variety of conditions.

The control strategies and behaviour of these subjects have been confronted in 5 different accidents:

1. "Station-black-out" accident with no other component failure.
2. "Station-black-out" accident, with break in the inlet line to Steam Generator number 1.
3. As case 2 with a further malfunction on the pumping system of Steam Generator number 1.
4. As case 2 with a further malfunction on the regulation system of Steam Generator number 1.
5. As case 2, with a further failure on the steam line feeding the control system of Steam Generator number 1.

While the statistical analysis of the responses of the subjects is outside the scope of this work, the most relevant outcome of this study consists in the identification of the relevant physical cues and cognitive parameters used by operators in controlling the AFWS in different circumstances.

These data and parameters have been utilised for the development of the simulation.

3.5.3 Selection of Theoretical Model - Case Study COSIMO

The theoretical paradigm adopted to develop the simulation COSIMO is based on the "fallible machine" model of Reason (1987a, 1987b, 1990), already described in a previous section of this Chapter (§ 3.2.4).

We recall that this theoretical model is based on the mechanisms of Similarity Matching (*SM*), Frequency Gambling (*FG*) and Direct Inference (*DI*), and on the concepts of Working Memory (*WM*) and Knowledge Base (*KB*). The *KB* is made-up of "frames" containing known procedures and accident configurations (Figure 3.6).

We will firstly concentrate on the detailed description of the content of knowledge base and on the cognitive functions that govern the model.

The basic constituents of the Knowledge Base are *Rule Based Frames* (*RBF*) and *Knowledge Based Frames* (*KBF*). The *RBF*s were originally conceived as static frames for diagnosing and recovering from an accident, i.e., snapshots of the configuration of the process controlled by the operator (Cacciabue, Mancini, and Bersini, 1990; Bersini, 1993).

Rule Based Frames

Each *RBF* contains the following elements (Figure 3.27):

- An index label identifying the *type of accident* (e.g., loss of electrical power, steam generator tube rupture).
- A *subjective frequency tag* related to the number of times the operator has encountered this accident in the past as well as his familiarity with the accident.
- A set of properties or *attributes*, and their associated *attribute-values* describing expected symptoms which characterise such an accident (e.g., pressure of water in the steam generator: high and increasing).
- Each *attribute* and associated *attribute-value* is also coupled with a *Diagnosticity* value, which is a measure of the importance of the attribute-value with respect to the accident scenario identification. Further details on the concept of *Diagnosticity* are given in the forthcoming sub-section on the Entity Attribute Matrix.
- A set of appropriate *actions* for the management and the performance of the selected tasks to deal with the current situation.

Two types of symptoms (*attributes*) can be described in a *RBF*: physical symptoms and logical symptoms. Physical symptoms are related to values of plant parameters (e.g., high mass flow to steam-generators, decreased pressure in steam-generator) and logical symptoms related to the Boolean indicators (e.g., pump indicator, regulation valve state indicator).

A symptom consists of the following elements:

- time(sec),
- plant component (e.g., SG1),
- material (e.g., water),
- variable (e.g., pressure),
- numeric value (e.g., 55),
- numeric derivative (e.g., + 4 u/t),
- numeric value history, trend (e.g., 200 300 350),
- semantic value (e.g., high),
- semantic derivative (e.g., increased),
- physical salience,
- cognitive salience.

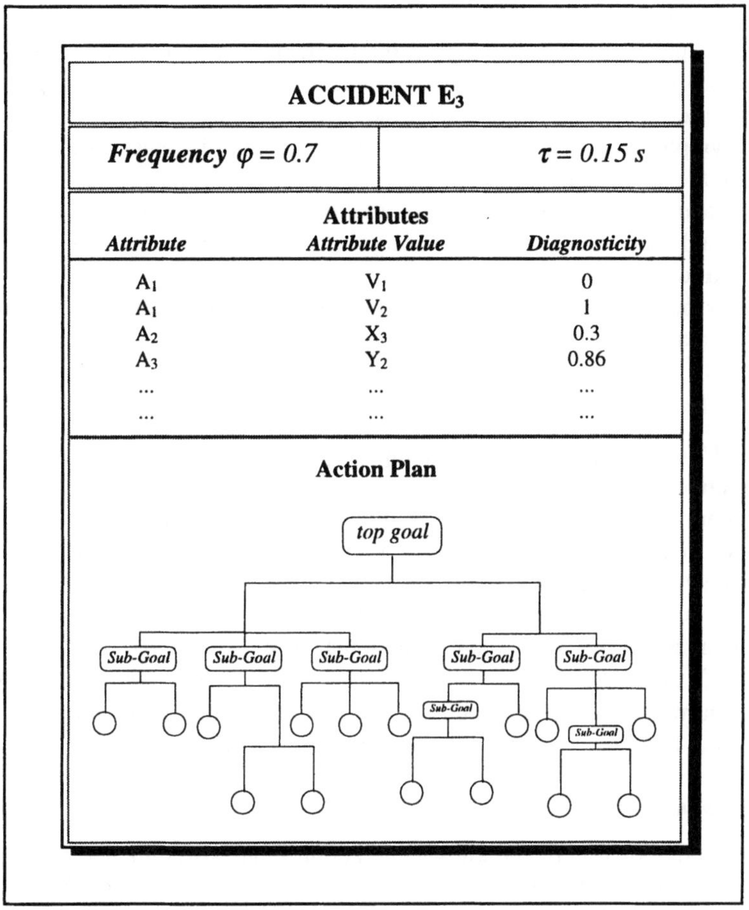

Figure 3.27. Prototype Rule Based Frame of COSIMO.

Physical salience reflects the physical properties of a symptom (e.g., type of indicator, position in the control panel, light or sound intensity of alarms).

Cognitive salience represents the subjective familiarity and significance of a symptom. For instance, a particular indicator may be checked by the operator because that indicator is of particular relevance for the current task.

Semantic values and *semantic derivatives* are expressed by means of fuzzy expressions and fuzzy functions according to the empirical knowledge, elicited and analysed during the phases of Task Analysis and experimental data collection.

In order to represent the dynamic evolution of the process a further knowledge structure is defined - a dynamic frame - as a time-dependent extension of a static *RBF*. A dynamic frame describes the evolution of a particular process configuration, i.e., the various steps of the cognitive process carried out by the operator while controlling the plant.

Each dynamic frame is associated with a particular instance of the overall plan of action, when, for example a sub-goal is reached, and a number of expected new conditions are met. These are identified with the newly appearing symptoms, changed symptoms or *attributes* and a time stamp characterising the specific snapshot of the accidental situation.

All data and values included in a *RBF* are obtained from the process of knowledge elicitation and analysis carried out during the previous phase of task analysis and experimental data collection.

Iterations between these two step of simulation development may be necessary when new data are needed for a better definition of frames.

Knowledge Based Frames

*KBF*s are *Units* of knowledge containing only heuristic rules of thumb as well as general engineering and physical principles on the operation of the plant, usually developed during training, experience and theoretical background.

Contrary to the *RBF*s, *KBF*s do not contain organised sets of recovery and control actions and symptoms associated with well defined situations.

*KBF*s are called into play in the *WM* when a new planning process has to be developed, as no *RBF* is available to handle the current situation.

By a *KBF* a totally new strategy of actions is developed by adapting, to the novel situation, known parts of plans and heuristic and theoretical knowledge. This part of the modelling architecture is not considered for this sample case study.

Entity Attribute Matrix

A synthetic composition of the knowledge about diagnosis contained in the *KB* can be developed by combining several *RBF*s. This structure is the *Entity-Attribute Matrix* (EAM), which takes the form of a two dimensional matrix, with entities as row heads and attributes as column heads (Reason, 1986, 1987a).

Each entity is associated with a *RBF* and is characterised by the set of attributes/attribute-values, i.e., symptoms, and the frequency tag of that *RBF*. An element of the matrix represents therefore the value of a particular attribute in a particular accident situation. The example of EAM given in Figure 3.28 represents the general EAM, which contains the *RBF* describing the (Event) Entity E_3 previously shown in Figure 3.27.

Since the set of attributes describing each entity differs from one entity to another, empty spaces are left in the EAM concerning value of attributes not present as attribute of the corresponding entity.

The number of attributes, and thus the number of empty spaces, of an entity is related to the knowledge that the operator has about the accident associated with that entity.

This knowledge depends on the number of times the accident has been encounter by the operator: the higher the accident frequency, the more numerous the descriptive attributes. This refers to Reason's experimental finding "*the more frequent in the world, the more will be known about it*" (Reason, 1987b).

ENTITY / ATTRIBUTE	E_1	E_2	E_3	E_4	E_5	E_6	E_7	$E_{N=8}$
A_1	V_1	V_1	V_1	V_1	V_1	V_1	V_1	V_1
A_1	V_1	V_1	V_2	V_1	V_1	V_1	V_1	V_1
A_2	X_1	X_3	X_3	X_3	X_3	X_1	X_3	X_3
A_3	Y_3	Y_1	Y_2	Y_5	Y_2	Y_1	Y_1	Y_1
...	...	...	...	...	...	...	...	...
A_i	Z_1	Z_5	Z_2	Z_1	Z_1	Z_4	Z_1	Z_3
...	...	...	...	...	...	...	...	...
A_M	U_1	U_5	U_2	U_1	U_1	U_4	U_1	U_3
FREQUENCY	0.8	0.5	0.7	0.3	0.8	0.5	0.2	0.6

Figure 3.28. Entity Attribute Matrix - The values of E_3 are found in Figure 3.27.

Each attribute of the EAM is characterised by a domain of variation, i.e., the set of different values taken by the attributes over the entities. The number of these domains varies from one attribute to another. A particular value v of an attribute A_j is characterised by its *Diagnosticity* $D(A_j, v)$, which is an easy-to-compute information measure given by:

$$D\left(A_j, v\right) = \frac{(N-n)}{N-1}$$

where N is the total number of entities and n $(0 < n \leq N)$ is the number of entities where the value of A_j is v.

The *diagnosticity* of Entity E_3 for some of its *attributes* can be evaluated applying the above expression. The following results are obtained (Figure 3.27, 3.28):

A_1, *value* V_1, i.e., $D(A_1, V_1,) = 0$; A_1, *value* V_2, i.e., $D(A_1, V_2,) = 1$;
A_2, *value* X_3, i.e., $D(A_2, X_3,) = 0.3$; A_3, *value* Y_2, i.e., $D(A_3, Y_2,) = 0.86$

The diagnosticity of an attribute value reflects the level of discrimination provided by this value over the entities. The more a value is present, the less "informative" it becomes and the lower is its Diagnosticity value. In particular, the limit conditions are: $n=N \Rightarrow D=0$ and $n=1 \Rightarrow D=1$).

This simple equation has the objective to facilitate the understanding of the concept of pertinence of information.

A more formally complete and mathematically elegant framework for developing the notion of *diagnosticity* may be given, for example, applying the Information Theory, which provides an appropriate and more complete set of algorithms (Shannon and Weaver, 1949). However, for the sample case under development, the above analytical expression of the diagnosticity is considered sufficient.

Primitives of Cognition

As discussed during the description of COSIMO (§3.3.4), the model is governed by the primitives of cognition of *Similarity Matching*, *Frequency Gambling* and, less frequently, Direct Inference.

The *SM* primitive matches information (*cues*) coming from the work environment with diagnostic *attributes* described in the *KB*. More precisely, the *SM* compares external cues, i.e., data usually perceived because of their high physical salience, and internal cues, i.e., data usually investigated because of cognitive process, with previously experienced symptoms contained in *KB*.

The result of the matching is either a unique hypothesis or a set of partially matched hypotheses, depending on the discriminating ability of the considered cues. These hypotheses are characterised by a *matching value* which represents the result of the *SM* primitive.

The *FG* primitive resolves the conflict, which may occur between partially matched hypotheses selected by the *SM*, in favour of the most frequently encountered and well known accidental situation.

These two primitives operate so as to minimise cognitive strain and maximise the chances of immediate information processing, which is often recognised as the most common problem solving strategy of operators in emergency situations.

The problem solving based on the *SM-FG* primitives is therefore defined as a process of immediate information processing in contrast with deep reasoning. Switching from immediate information processing to deep reasoning is assumed to be possible only when sufficient time and attention resources are available. In this sample case, only immediate information processing have been considered.

Functions of Cognition

Even if the theory of the "fallible machine" does not explicitly discuss the simulation of *cognitive functions*, these must be considered in the simulation, as they cover the basic cognitive activities like diagnosing, monitoring, scheduling, etc.

In COSIMO, four interrelated *cognitive functions* have been described, namely *Filtering*, *Diagnosing*, *Hypothesis Evaluation* and *Execution*. These functions match well the four cognitive functions of the *Reference Model of Cognition*, i.e., Perception, Interpretation, Planning, and Execution. They will now be briefly discussed.

Filtering Function

The communication between the plant and COSIMO is done through a "cognitive filter". The filter selects among the large number of variables, parameters and data produced by the environment those which are actually perceived.

This perception is guided by a *salience* criterion based on the physical and cognitive salience associated with the incoming environmental data.

In order to be perceived, variables and data need to reach a salience value greater that a pre-defined *filter threshold* value, which is one of the input data of the model (see table 3.13 for further details concerning COSIMO Parameters and Default values).

After the filtering process, data are interpreted: numerical data are translated in semantic information (e.g., nominal, low, on, off, etc.). Only filtered data are interpreted according to the psychological argument that only the information "perceived" is actually interpreted.

Diagnosis Function

Once the cues have been filtered and interpreted, they are matched with the content of the Knowledge Base (*KB*) using the *SM* mechanism in order to identify a *RBF* to be executed.

The number, form and quality of these cues determine the matching value: the more pertinent the calling conditions, the less numerous and the better supported are the hypotheses selected from the *KB*.

Once a frame is activated in the *WM*, it will govern a hypothesis driven information processing, by which a confirmation process is initiated according to the attributes which better qualify the selected frame.

In the case of ambiguity, either in the interpreted data or in the symptoms represented in the *KB*, more than one explicating hypothesis may be brought to the operator's attention by the *SM*. The *FG* resolves the conflict between partially matched hypotheses in favour of the one that has been more frequently encountered in the past.

Matching and frequency values are combined to deliver a support score for each hypothesis. This combination may take various forms according to the level of expertise and current *cognitive state* of the operator, i.e., the mental and physical conditions characterising the operator as a consequence of his/her interaction with the working environment and the socio-technical context. For instance, while an expert operator may always favour the frequency value, a novice may give higher priority to the similarities between situations. The hypothesis which is eventually selected is the one obtaining the largest support score.

The diagnosis is meant to be a dynamic process which can be carried out over an extended period of time, interrupted by the implementation of some alternative plan or new diagnosis and possibly resumed at a later stage. A diagnosis can also be triggered off at any time during the execution of another function just to confirm the initial choice of a *RBF*.

Hypothesis Evaluation Function

This function aims at deciding whether a hypothesis can be trusted or has to be rejected.

If the hypothesis selected after the diagnosis function is not supported with sufficient evidence, the hypothesis is rejected and a new diagnosis is initiated utilising newly perceived cues from the environment and the known information already elaborated previously.

The level of confidence required for a hypothesis to be considered - *evaluation threshold* - can be dynamically modified by the model according to the current operator's cognitive state.

Execution Function

Once a hypothesis has been selected, the *WM* is cleaned out and receives an instantiation of the *RBF* associated with the selected explanation. Therefore only one *RBF* is present in the *WM* at one time.

This *RBF* is called the *Currently Instantiated Frame* (*CIF*). The control and recovery actions contained in the *CIF* are executed over the time evolution of the accident. Note that a control action does not necessarily imply a physical action on the plant (e.g., a monitoring process).

We have shown that the *cognitive functions* are not necessarily sequentially concatenated. On the contrary, depending on their level of expertise, operators are able to take short-cuts. In particular, operators may cycle forward and backward by focusing first on a diagnosis, and performing some actions and, then, if unsatisfied with the results, by modifying their initial diagnosis and starting an entirely new plan of actions.

Sometimes a "dilatory action" is performed to delay the evolution of a sequence when the operator wants to extend the available time for understanding the current situation. As a consequence of this, the *cognitive functions* cannot be sequentially ordered along the temporal axis but should intervene in an opportunistic way.

Such an interaction between *cognitive functions* is guided by higher level cognitive functions which are contextual to the specific work environment of application of the model. In particular, such functions have to take into consideration the modes of control and the specific competencies which are developed by the operators in different working environments. These competencies are consequently a consistent part of the "culture" of those individuals. These special functions have to be accounted for in the implementation of the model as the controlling element of the simulation (Amalberti and Deblon, 1992; Hollnagel, 1993a).

However, if the evolution of the accident follows the expectations of the operator and thus the information contained in the *CIF*, the interaction is substantially sequential and follows the order Filtering, Diagnosing, Hypothesis Evaluation and Execution at predefined time instants, in agreement with the plan of actions contained in the *CIF*.

3.5.4 Selection of Numerical Algorithms and Implementation in Programming Language and Environment - Case Study COSIMO

The COSIMO model has been implemented in a Blackboard framework (Hayes-Roth, 1985; Engelmore and Morgan, 1988).

This architecture was been chosen for three main reasons:

1. It supports the adaptive behaviour of reasoning of an operator.
2. It allows an explicit representation of mechanisms which enable the dynamic revision of human behaviour.
3. It sustains an incremental and opportunistic decision making development.

These three reasons will now be discussed in detail in combination with the algorithms that they entail, as to illustrate the quantification aspects and computer implementation that have been developed for COSIMO.

The Blackboard Paradigm

The Blackboard (BB) architecture is often described as being composed of three parts (Nii, 1986), namely:

1. *Blackboard* itself, with its internal structure.
2. *Agents*, also called knowledge sources or specialists, working on BB.
3. *Controller*, governing the actions of the agents.

Blackboard

The Blackboard structure consists of different *levels* of information, which correspond to the various kind of structured-objects that may be dealt with by the model.

These levels create a hierarchy along which the reasoning process takes place while the properties of the data describe the level at which objects are intended to be placed. Thus, the BB constitutes a unique global database accessible and viewed by all the agents of the system.

The set of all the properties is the restricted to well-defined vocabulary by means of which the model expresses its various actions and results

Agents

The know-how of the domain is contained in small independent units, or *agents*, each of which is specialised in a particular "inference" scheme. Each agent only focuses on the BB levels of interest to its particular specialisation and communicates indirectly with other agents through the BB. Agents are independent of each other.

An agent is mainly composed of three parts, or bodies: the *triggering* body, the *precondition* body and the *action* body. In the *triggering* body, the initial conditions are evaluated by which the agent can be animated. An agent, even if triggered by some event, may not be able to perform directly its action body.

Supplementary data have to be acquired in order to fulfil the conditions of execution. These required conditions form the *precondition* body of the agent.

The *action* body contains the knowledge on how to perform the inference mechanism on the data that match its triggering and precondition parts.

The only way to modify the BB is during the execution of the action phase of an agent, either by creating new objects on the BB or by modifying previously existing objects.

Each time new incoming data of the BB matches the triggering part of an agent, this agent generates a task. Such a task contains a set of data which is used during the schedule process as well as during the possible execution of the action body of the agent. This task informs the Controller of the potential activation of the agent. It is then up to the Controller to decide which agents are called into play, according to certain rules and conditions, as we will see in the next subsection.

Thus, the execution of an agent takes place in three steps:

Step 1. The agent is triggered and generates a new-task, called "triggered task".

Step 2. The triggered task remains dormant until its preconditions are satisfied (these tasks are called "invocable tasks").

Step 3. If the invocable task is selected, according to the criteria generated by the Controller, the action part of the agent is executed.

Controller

Since more than one agent can be triggered by the same object of the BB and more than one object can be generated by a single task, a control mechanism is required in order to manage the interaction between agents and the sequence of tasks.

The control mechanism is able to select amongst various invocable tasks the one to be activated next.

It is obvious that this mechanism is of major importance in a BB system. Indeed, the tasks are due to *agents* dealing at different *levels* of information and, therefore, have completely different meanings. Since the results, as well as the execution of the tasks are highly different, the product of the problem-solving process is directly related to the selection mechanism of the Controller: the more pertinent the selected task the faster the plant will converge towards accurate behaviour.

We will now discuss in more detail the implementation of the functions of cognition and the process of Allocation of Resources within the Blackboard Architecture.

In the BB structure that sustains COSIMO, as in most Blackboard systems, two different BBs coexist and interact with each other: a *Domain Blackboard* (DBB) and a *Control Blackboard* (CBB).

These two BBs will be associated to a number of agents and levels according to their role and objectives. In particular, these are described in a condensed form in Tables 3.9 - 3.12, which contain:

- Table 3.9 Control Blackboard *Levels.*
- Table 3.10 Control Blackboard *Agents.*
- Table 3.11 Domain Blackboard *Levels.*
- Table 3.12 Domain Blackboard *Agents.*

Table 3.9. Control Blackboard *Levels* for the COSIMO sample case.

LEVELS	**Definition**	**Example**
PROBLEM	problem to be solved	maintain the safety levels
STRATEGY	strategies currently used	data acquisition, diagnosis, confirmation, ...
FOCUS	temporary agent selection criteria	prefer tasks at signal level
POLICY	permanent agent selection criteria	prefer recent tasks vs. old tasks
AGENDA	set of invocable and triggered tasks	get-environment, decode, update-focus, ...
TASK	current task to be executed	update-focus

Table 3.10. Control Blackboard *Agents* for the COSIMO sample case.

AGENTS	**Triggering**	**Preconditions**	**Body**
PROBLEM-ON-BB	none	none	initialise BB activities
STOP-PROBLEM	new problem	solution of trigg. problem	stop BB act. on current problem
START-STRATEGY	new problem	none	activation of initial strategy
UPDATE-STRATEGY	new strategy	activation of new strategy	selection of a strategy
INITIALISE-FOCUS	new strategy	none	first focus of triggered strategy
UPDATE-FOCUS	new focus object	satisfy triggered focus	selection of next focus

Table 3.11. Domain Blackboard *Levels* for the COSIMO sample case.

LEVELS	**Definition**	**Example**
ENVIRONMENT	Untreated representation of physical world	300, 9.7, 11, ... 10.8, 0.1 ...
SIGNAL	Data decoded, set of indicators and alarms	• indicator 5 = 10.8 • alarm 23
SIGN	Data cognitively and physically filtered	• indicator 5 • alarm 23
PERCEIVED CUES	Significant data interpreted (perceived cues)	Perceived problem relative to steam generator 1
SET OF HYPOTHESES	Potential hypotheses related to perceived cues	• SG1 inlet break • Valve V21 blocked open
CURRENT HYPOTHESIS	Hypothesis currently being used	SG1 inlet break
ACTION	Actions to be executed on the plant	• close valve v031 • open valve v035 • regulate SG1 at 30

Table 3.12. Domain Blackboard *Agents* for the COSIMO sample case.

AGENTS	**Triggering**	**Preconditions**	**Body**
Get-Environment	none	none	Acquisition of new set of data
Decoder	New raw data	none	Decoding new raw-data
Filter	New object on signal level	none	Filtering on basis of physical and cognitive salience
Interpreter	New object at level Sign	none	Semantic interpretation of filtered data
Similarity Matcher	New object at level perceived or new object at level *CIF*	none	Matching perceived cues and knowledge in Dynamic Frames
Confirmation Matcher	New object at level perceived or new object at level *CIF*	none	Matching perceived cues and *CIF*
Frequency Gambling	New object at level Set of Hypotheses	none	Conflict resolution on set of hypotheses
Executor	New object at level *CIF*	Evaluation signal activated	Execution of actions

Implementation of Allocation of Resources

The *Control Blackboard* implements the Controller part of the BB and, therefore, manages the sequence of tasks and agents related to the DBB as well as its own tasks and agents.

In particular, the CBB generates tasks that are added to the tasks of the DBB, and then, given that the CBB governs the scheduling mechanism by selecting the task to be executed next, it may occur that it will choose one of its own tasks in alternative to others. In this way the CBB may influence its own scheduling mechanism. The control *agents* and control *levels* of the CBB are shown in Tables 3.9 and 3.10.

It is important to note how, by means of the CBB, we are able to consider the mechanisms of *Allocation of Resources*, which have not been explicitly modelled in the theoretical development of COSIMO, but are necessary for a complete description of the processes of human-machine interaction. In this case, these important characteristics of human behaviour are brought back in the simulation by the computational architecture, and lead to a more complete simulation framework.

Moreover, the implementation of the CBB requires certain data, variables and parameters that may have to be searched for and identified by further iterations, i.e., by repeating some of the steps preceding computer implementation, including model development, task analysis and experimental studies (Figure 3.26).

Implementation of Cognitive Functions

The implementation of *Cognitive Functions* is carried out by various agents and levels of the *Domain Blackboard* (DBB), described in Tables 3.11 and 3.12.

Five correlated activities are simulated by the Domain Blackboard, namely: *data acquisition, diagnosis, confirmation, planning* and *execution.* These five activities will now be discussed in more detail.

<u>Data Acquisition Activity</u>

This scenario implements mainly the *Perception* cognitive function and consists of the following stages (Figure 3.29):

1. Uninterpreted input data from the environment (i.e., external cues), are represented by objects of the *Environment Level* (Table 3.11). This task is performed by the *Get-Environment Agent* (Table 3.12).
2. The *Decoder Agent* associates these raw data (a list of numbers) to different physical and logical variables, creating new objects on the *Signal Level.*
3. The *Filter Agent*, is, then, applied to the *Signal Objects* in order to evaluate their expected significance and create new objects at *Sign Level.* In this way, only objects that are important because of their overall salience in the given context are kept for further evaluation.
4. The *Interpreter Agent* translates objects at sign level to semantic values and creates objects at *Perceived Level.*

In Figure 3.29 the *agents* and the *levels* that make up the *data acquisition activity* are shown and their correlation is graphically described.

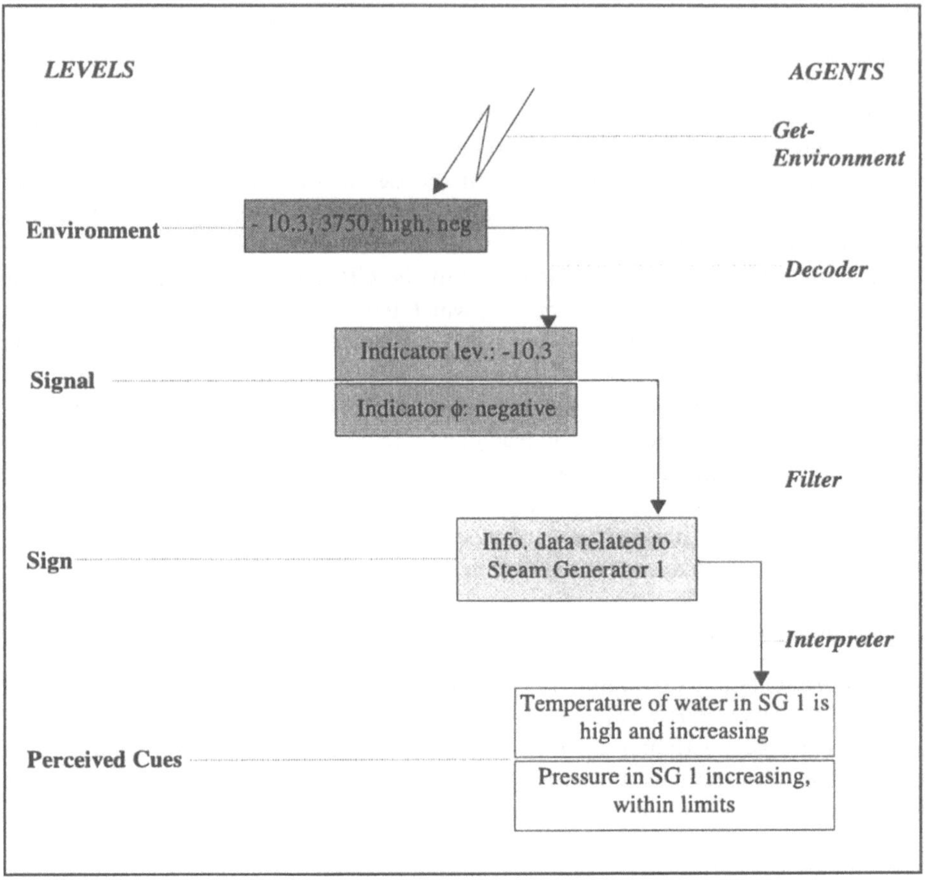

Figure 3.29. Data Acquisition Activity of COSIMO.

Diagnosis Activity

This scenario implements the *SM* and *FG* primitives and thus the *Interpretation* cognitive function.

A diagnosis process may be initiated following a set of perceived objects and takes place in two stages (Figure 3.30):

1. The *Similarity-Matcher Agent*, triggered by new cues of the *Perceived Level*, generates a set of potential diagnoses which constitute an object of the *Set of the Hypotheses Level*.
2. From that object, one hypothesis is selected by the *Frequency-Gambler Agent*. This hypothesis becomes an object of the *Current Hypothesis Level*.

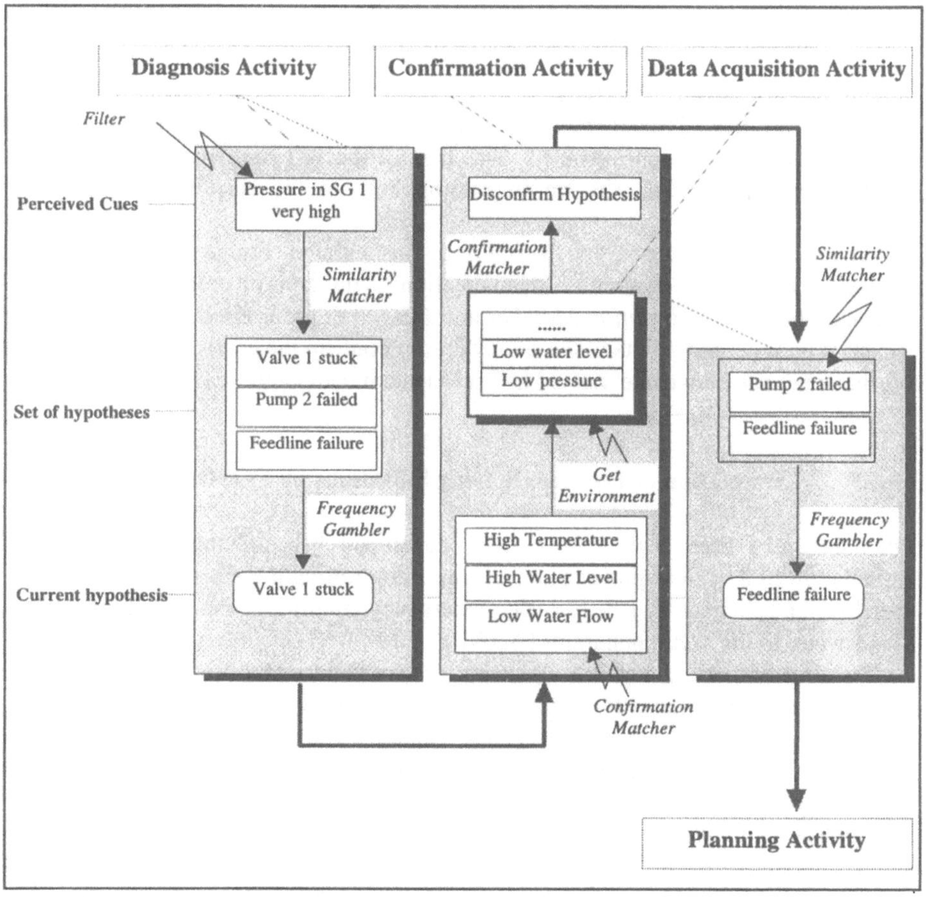

Figure 3.30. Diagnosis and Confirmation Activities in COSIMO.

Confirmation Activity

A confirmation stage may be started by searching into the environment for pertinent information (Figure 3.30).

The confirmation strategy can be exploited to assess either the results of a diagnosis or of an execution activity. It consists of three steps:

1. Selection of data to be considered for confirmation (*Confirmation Matcher*), at the level of *Current hypothesis*.
2. Acquisition of these data strategy, executed on the basis of the requirement of the first step of the confirmation. This activity is performed by tuning the parameters of the filter agent, so that the appropriate selection is performed.
3. Actual stage of confirmation, executed by comparing expected and perceived data, exploiting again the *Confirmation Matcher Agent.*

Note that this scenario of *Confirmation Activity* comprises the execution of another activity, i.e., the *Data Acquisition Activity*. This type of coupling between activities is quite common in Blackboard structures.

Planning and Execution Activity
When a hypothesis is confirmed, the Action Plan associated with the hypothesis is firstly selected and then executed. The first action is taken from the Rule Based Frames of the operator *KB*, corresponding both to the current time and to the diagnosed situation.

The actions contained in the Current Instantiated Frame are performed in sequence and, as the dynamic implementation of the control procedure takes place, the activities of *Confirmation Matcher* and *Diagnosis* are utilised, at predefined time instants or at certain points in the procedure, to confirm the correct execution and the correspondence between expected and real system behaviour.

3.5.5 Simulation of Problem Solving Situations by COSIMO

A number of case studies have been developed by applying COSIMO to the simulation of a nuclear power plant operator controlling the transient behaviour of a feed-water system, only active during emergency conditions: the system of auxiliary feed-water to the steam generator (AFWS).

As already discussed (§ 3.5.1), such a system is normally in a stand-by condition and becomes automatically operative when the power station is shut-down for an operational transient or in an emergency situation. The AFWS needs to be regulated at the beginning as well as during the evolution of the transient in order to adapt the heat removal capacity of the AFWS to the progressive decrease in the amount of decay heat being produced in the plant.

The initiating event considered in all sample cases has been a loss of electrical power, or "station black-out", i.e., an emergency situation. Such an accident induces the nuclear reactor emergency shut-down and the consequent automatic starting of the auxiliary systems, including the AFWS.

A model of the physical behaviour of the AFWS has been developed accounting for a variety of faults which may occur on the plant while performing its functions as well as on the control and instrumentation system associated with the AFWS and the steam generators.

The AFWS and COSIMO models have been run in an interactive mode in order to create the control simulation of the operator model acting on the auxiliary feed-water system. An accident period of one hour has been simulated. This allows sufficient time for the decrease of the decay heat generated in the plant and thus for a number of control actions to be simulated by COSIMO on the AFWS.

As previously discussed, the COSIMO input data concerning the Knowledge Base have been developed from the expert elicitation on the management of the AFWS and by running the simulation of the AFWS for many different initial conditions and transient configuration in order to build a sufficiently rich number of *RBFs* for inclusion in the *KB*.

The remaining input data of COSIMO are the parameters, by which it is possible to tune the behaviour of the main parts of the model, like, for example, the interpretation and cognitive filtering activities. The adjustment of these parameters, either as input data or during the dynamic simulation, allows to reproduce subtle variations in the overall model performance. A detailed description of the COSIMO parameters and default values utilised in this case study can be found in the final part of this-section (Table 3.13).

The case studies performed with COSIMO have been obtained by modifying one part of input data concerning the model parameters while maintaining untouched the other part related to the content of the *KB*.

In particular, four cases have been studied. The first one describes the expected behaviour of the operator in the management of the AFWS, with the physical system and the control instrumentation performing optimally without failures or malfunctions during the accident evolution.

The other three cases simulate failures in the AFWS mechanical components occurring during the evolution of the accident, after the initiating event of station-black-out. These three cases aim at the study of deviations from the appropriate control strategy and their consequences on the overall accident management. By deviations of problem solving we intend a progressive de-coupling between the plant evolving in time and the operator ability to perceive and interpret an accident, to integrate new events which modify dynamically the scenario and then to act appropriately on the system.

De Keyser and Woods (1989) have shown that deviations frequently observed in dynamic and complex environments are due to fixations, which are failures to perform appropriate situation assessment, as new evidence is produced, and are difficult to identify.

Sometimes, certain actions performed by an operator, or some verbal exchanges with other operators, may indicate that there is a fixation on one type of explanation for a given situation. Often, such fixations become manifest later on in the evolution of the accident, when evidence indicates that operators have not succeed in updating their views on the world (Masson, 1991, 1994).

Another type of deviations that operators are likely to show are "premature closures", which result in the termination of the problem solving without generating all the alternatives and without seeking all the available information. Moreover, the effect of workload and stress, induced by the working environment, may also contribute to reducing the operator's problem solving capacities This effect is particularly noticeable when the problems to be dealt with comprise of numerous competing features, such as control actions, verification, monitoring and supervisory activities.

Three main aspects of deviations of problem solving have been examined in the framework of COSIMO. These are the core of the case study. They have been defined as: *cognitive collapse*, *unadapted change* and *cognitive lock-up*.

In summary, four applications of COSIMO have been performed:

Case A. This case describes the *nominal* expected behaviour of the model controlling the AFWS during the evolution of the "station black-out accident", when no new particular events occur in the AFWS and no problem solving deviations affect human cognitive behaviour.

Case B. This case describes a situation of a *cognitive collapse,* in which the operator loses the capacity of reasoning clearly. This situation is typical of sharp stress conditions, which may be generated by the complexity of the working environment or by the sudden demand for more control and supervisory activity due to novel incidental events occurring during the evolution of the station black-out sequence.

Case C. This case describes an *unadapted change*, whereby, in presence of a new event occurring during the accident evolution, the model changes its view and diagnosis, but it fails to account for the effects of past events.

Case D. This case describes the case of a *cognitive lock-up*, which causes the persistence of the first explanation, even after a new plant failure has occurred during the evolution of the accident. This generates what we may call a fixation error.

Figures 3.31 and 3.32 indicate respectively the distribution in time of the amount of filtered variables and the similarity matching values during the four simulations. These cases and the related figures will now be described in some details.

Case A: Expected behaviour

In this case, no human error nor plant failure occur following the initiating event of the station black-out. The actions and decision making of COSIMO are always appropriate and lead to a successful control sequence and compensation for the system demands in case of a station black-out accident.

The parameters and input data of COSIMO for this case are shown in Table 3.13. These parameters and data allow the simulation of a generic operator with no effects on problem solving deviations.

In particular, the results of the simulation show the following characteristics:

1. The model deals appropriately and according to procedures (*Currently Instantiated Frame*: *Station Black-Out accident*) with an expected set of indicators either logical (position of valves, pumps, alarms, etc.) or physical (temperatures, pressures, etc.). A limited number of control panel indicators (Filtered variables) are utilised to carry out the planned task (Figure 3.31).
2. The value of *Similarity Matching* throughout the simulation is constant and equal to 1 (Figure 3.32). This means perfect correspondence, at all times, between the COSIMO expectations regarding the plant behaviour and the incoming cues, which describe and represent the effective process evolution.
3. The corresponding actions are executed according to an optimal solution of the control procedure.

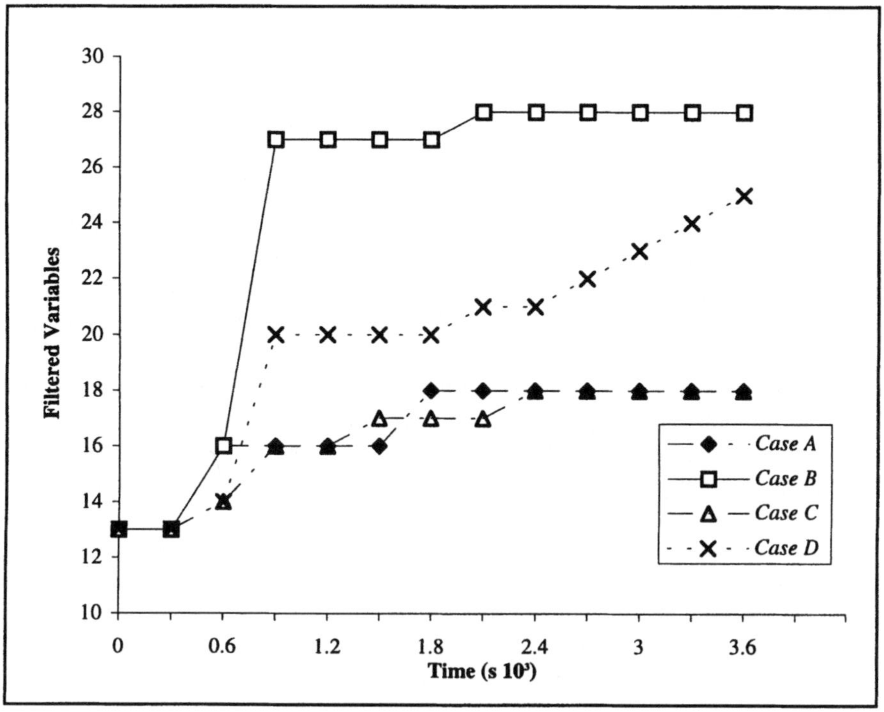

Figure 3.31. COSIMO filtered variables for Cases A - D.

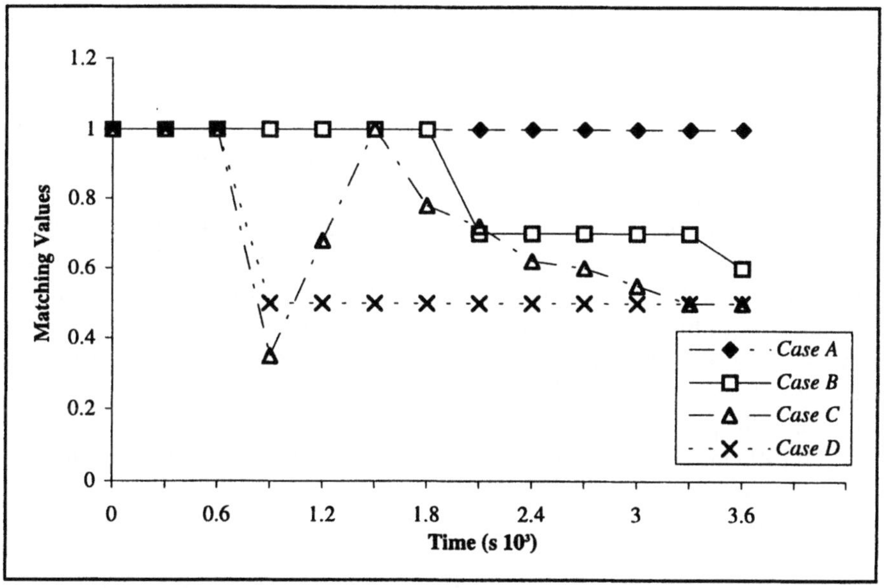

Figure 3.32. COSIMO distribution of matching values for Cases A - D.

Case B: Cognitive Collapse

In this case, the accident progression in the AFWS consists in the same station black-out transient as for the case A, but a further event of inlet-line-break occurs in one of the steam generators connected to the AFWS at approximately 2100 s after the initiating event.

This new event requires an immediate control reaction by the operator in identifying the failed inlet-line and isolating it together with the combined steam generator. This would limit the propagation of the effects of the new failure to the other parts of the AFWS. Following this further diagnosis and intervention of isolation, an action of regulation of mass flow should be carried out, so as to ensure appropriate feeding to the remaining steam generators.

However, in the case of cognitive collapse, the operator is not able to recognise the new accidental scenario and situation. The model enters in a stage of confusion with more and more cues being filtered (Figure 3.31), many of which may not appear related to the current problem, and with no execution plan being selected and carried out.

The lack of understanding of the current process evolution leads to a search for always larger sets of indicators. But, these extended amounts of collected (filtered) data are only poorly and badly exploited.

Indeed, when the model is in this cognitive state, it is unable to exploit novel incoming information or changed cues, which would have normally allowed a change of views.

COSIMO does not succeed in performing an accurate diagnosis. In particular, COSIMO is well "aware" that its current view of the world is insufficiently supported and that its problem representation is unrealistic. However, it does not possess adequate resources to succeed in finding the appropriate situation assessment.

It tries to take into consideration all possible known explanations, jumping from one to another, without being focused on the new current situation. It is therefore stuck in a kind of endless diagnostic stage and does not switch to the recovery activity by taking appropriate action.

After 2100 s, the *matching value* with a certain Rule Based Frame is always lower than the minimum necessary to bring the Frame into the Working Memory (Figure 3.32). This means that the *matching value* always results lower than the *evaluation threshold* adopted by the evaluation agent, i.e., 0.7 from Table 3.13.

The number of activities called into action becomes drastically smaller in comparison to case A and they are repeated without obtaining a satisfactory resolution of the problem.

In COSIMO the simulation of this problem solving deviation was obtained by:

- Expanding the model field of attention, i.e., decreasing the *filter threshold*.
- Decreasing the *cognitive salience weight* and increasing the *physical salience weight*.
- Reducing the interpretation of small variations by increasing the *interpretation threshold*.

- increasing the *weight of the frequency* relative to the *weight of the similarity matching*.
- increasing the *evaluation threshold*.

It is interesting to observe that, in Figure 3.31, there is a noticeable difference between the set of filtered variables of case A and B, even before 2100 s when the new event occurs in case B. As a matter of fact, given that the two accident evolutions are the same for the first 2100 s, one would not have expected differences in the COSIMO output before that time.

However, the modification of the parameters, in particular the decreased value of the filter threshold, implies that a much greater amount of cues are actually filtered by the model at all times. This results in a greater work of matching, which is performed by the model before reaching the maximum value 1 for the time period between 0 and 2000 s (Figure 3.32).

Moreover, this considerable amount of matching work implies that, if the AFWS simulation had included "noisy" and "imprecise" data coming from the information system, instead of precise and correct data, the tendency to a cognitive collapse may have induced a problem solving deviation even before the event at 2100 s.

Case C: Unadapted change

The accident sequence described in this case is similar to the one of case B, but the failure event on the AFWS, i.e., the feed-line break on one of the steam generator, occurs at a very early stage of the accident (300 s) and consists in a very small break of the line with delayed effects on the physical behaviour of the system.

The parameters setting which has been selected in order to simulate the unadapted change of the operator during the sequence consisted in the reduction of the *evaluation threshold* from the default value and in the increase of the *weight of the frequency versus similarity* for the process of frame retrieval from the knowledge base. The remaining parameters have been kept at their default value.

As consequence of this input data configuration, at the beginning of the accident COSIMO succeeds in selecting the correct scenario through the *SM* and *FG* primitives, i.e., the plant has experienced a station black-out and the AFWS has automatically started.

The second major failure event takes some time to generate relevant effects on the plant. These are actually reported by the instrumentation system and are consequently only noticed by the operator with a certain delay.

This explains why the matching value remains high, i.e., 1, up to 600 s (Figure 3.32), even though the failure has occurred at time 300 s.

At a later time instant, i.e., 900 s, the new accident pattern can be objectively distinguished from the initial sequence and, at that time, the hypothesis of nominal station black-out, which represents the Currently Instantiated Frame (*CIF*) at 900 s, badly matches the data filtered by the model (Figure 3.32): the *Matching Value* is approximately 0.3.

Given the mismatch, COSIMO tries to update its current view of the world by searching for a better candidate in the knowledge base, but, following the high bias of the *FG* primitive, the wrong scenario is selected. The situation is misdiagnosed and the inappropriate sequence of actions is initiated. The model does not succeed in performing the expected task of appropriately cooling the system in the one hour of transient.

A number of further observations can be made. Firstly, there is a substantial amount of symptomatic or surface similarity between the incorrect hypotheses selected by the model (for time 900 < seconds < 3600) and the effective state of the plant. Indeed, as an example, the similarity matching value is 1 at 1400, and it remains high, above 0.7, between 1400 and 2100 s. This stems from the slowly evolving effects of the failure event on the AFWS and from the fact that the initial diagnosis was good and consequently the plant was properly regulated during the initial part of the accident.

The two lowest values of the matching are found at 900 s and during the final phase of the analysis. This is due to the fact that up to 900 seconds the *CIF* was the one initially developed, and no different configuration was attempted given the high value of the matching. At 900 s, the model "realises" (matching value very low) that a new event has occurred and that a different configuration has to be evaluated and this causes an improvement in the matching value. Even if the good solution is never obtained, an initial amelioration of the matching is found.

At the end of the analysis the persistent failure in diagnosing correctly the situation leads to a progressive degradation of the plant and therefore to a continuous decrease of the matching value.

The continual misdiagnoses result from the slight difference in frequency between a number of possible cases, which favour the selected erroneous frame. The selection of the most effective schema of the basis of a frequency criterion, which compensates for the difference in matching, is a commonly encountered and proven deviation of cognition in accidental circumstances such as the one considered here (Reason, 1990). Moreover, the low value of evaluation threshold allows a weakly supported hypothesis to be trusted even in presence of a decreasing matching with the cues of the environment.

Case D: Cognitive lock-up

In a cognitive lock-up condition, the model is fixed on a selected hypothesis and continues acting on the system in the light of an obsolete explanation, even in presence of strong disconfirming evidence. A progressive de-coupling is then observed between the model understanding and the evolution of the plant.

The accident scenario simulated in this case is totally similar to the previous case and, thus, as explained above, the failure event on the AFWS takes some time to manifest its consequences on the information system, because of the slow dynamics of the process. Consequently, as in the previous case, the similarity matching of the Rule Based Frame "Nominal Station Black-Out", which is the *CIF*, with the actual cues of the environment remains at level 1 up to approximately 600 s. At this time it is not yet possible to acknowledge that a new failure has occurred.

At 900 s the accident patter of the *CIF* does not satisfy the actual plant evolution, but COSIMO does not succeed in noticing the change from the indications of the AFWS. The nominal station black-out accident scenario, on which COSIMO is fixed, has a greater subjective frequency than the inlet break scenario, which is the dominating event of this sequence.

Even if the abnormal behaviour of AFWS demands a greater cognitive effort of variables collection and interpretation (Figure 3.31), the high satisfaction of the model in regard to its current hypothesis does not allow any attempt to update the reasoning to be made in the light of the disconfirming cues.

This leads to a poor matching of the selected cues with that of the expectation (Figure 3.32). Moreover, even with a very low similarity matching value, COSIMO remains “confident” in its explanation and does not succeed in revising the situation assessment. It carries out the initially selected strategy, step by step, according to the scheduling of actions contained in the "Nominal Station Black-Out" frame with no attempt to perform a revision.

The sequence of control actions does not obviously reach the expected goals by the end of the one hour transient simulation.

Technically, this fixation pattern has been obtained mainly by modifying two parameters of the model: the *weight of the frequency gambling* and the *evaluation threshold*. The *weight of the frequency gambling* has been increased with reference to the similarity matching for the computation of hypotheses. This gives more importance to past experience than to the observation of actual cues of the world.

The *evaluation threshold* has been increased. This has made the model less aware of possible mismatches between the hypothesis driven expectations and the interpreted cues of the world.

Therefore, even if the matching value becomes very low from 900 s onward (Figure 3.32), the support to the initial *CIF* "AFWS nominal station black-out" remains substantially high and, in any case, sufficient for the initial hypothesis to be trusted and maintained throughout the sequence.

COSIMO Parameters and Default Values

Table 3.13 shows the parameters that are used for defining and selecting the sequence of cognitive activities.

These parameters are the input data of COSIMO, by which intra-individual differences and special deviation mechanisms of problem solving are simulated. A set of default values for these data is built-into the program by which the behaviour of an average subject is matched.

Data necessary for the construction of the Knowledge Base, which defines the level of expertise of different subjects, must be compiled for each case of human-machine interaction.

Filter threshold

The filter threshold governs the attentional level of the model.

Decreasing the filter threshold extends the field of attention and thus more cues of the environment are collected. This does not mean that only pertinent cues are taken into account: indeed, the subject can be, for instance, focused on unimportant details.

Increasing the filter threshold means narrowing the field of attention by which only very salient cues are taken into account. The selection of information for interpretation also depends on the salience content of each cue.

Table 3.13 COSIMO Parameters and default values

PARAMETERS	DEFAULT VALUE	DEFINITIONS
Filter threshold	**0.75**	Threshold value for the salience of cues in order to gain acceptance through the filter
Physical salience weight vs. Cognitive salience weight	**2**	Relative influence of physical vs. cognitive factors in the computation of the salience of a cue.
New cue weight	**1**	Weights used to qualify either new cues firstly appeared, or changed cues or persistent cues
Changed cue weight	**0.9**	
Symptom weight	**1**	
Cognitive inhibition	**nil**	Set of quantities to be inhibited.
Interpretation threshold	**10/60**	Sensitivity (relative vs. absolute) in accounting for the variation of quantities to be interpreted.
Similarity weight	**1**	Weight of *SM* and *FG* primitives in selection of hypotheses.
Frequency weight	**1**	
Extended search	**All KB**	Amount of KB and time frames considered in selecting and asserting a hypothesis
Time search	**0**	
Evaluation threshold	**0.7**	Degree of confidence for a hypothesis

Cognitive vs. physical salience

The salience is the parameter which measures the attention grabbing power of a cue.

Two types of salience are defined in the model: the cognitive salience and the physical salience, which refer respectively to the subjective significance and the physical properties of cues.

Greater importance may be given to the physical or to the cognitive salience, by adjusting the relative weights assigned by the two parameters "physical salience weight" and "cognitive salience weight".

New and changed cues

New and changed cues describe respectively the new events occurring during a dynamic human-machine interaction and the cues which modify their values during the evolution in time.

In COSIMO, different weights can be given to newly appeared cues or to changes that have occurred. At certain moments it could be crucial for the operator to interpret the new information that is arriving, while in other circumstances it may be more important to be able to interpret changes in the system.

Cognitive inhibition

The cognitive inhibition is a parameter by which part of the system may be inhibited from reaching attentional level. For example when a part of a system does not work because of a previous incident or because it has been excluded from the loop for maintenance, and therefore the operator knows that this system can be neglected. It can be said that this part of the system is, from the operator's point of view, cognitively inhibited and other events related to it are not taken into account.

Cognitive inhibition can be a very economic process because it allows to neglect some unimportant or not pertinent events and to limit the time allowed for the information search. But it can be a source of errors too.

Interpretation threshold

The interpretation threshold regulates the sensitivity of the model to changing cues. The greater the interpretation threshold, the greater is the variation of the reading on the control panel which is interpreted as a stable state.

The interpretation threshold is assigned in percentages or fractions, with reference to the value of the observed cue at the previous time step.

Similarity vs. frequency

Similarity matching and frequency gambling are the two main primitives of cognition which govern the decision making of the model.

Differential weights can be given to the similarity matching or to the frequency gambling in order to privilege either of the two primitives.

During the simulation of the human-machine interaction, each frame selected from the knowledge base shows a matching value and a frequency tag: the final selection is based on a weighted combination of these two factors. Consequently, the relative weights are assigned as input for simulating different tendencies of human behaviour.

Extended and temporal search

These two parameters are used to improve the dynamic search process in the knowledge base. The first parameter, i.e., extended search, delimits the amount of the knowledge base, in terms of the number of frames, over which the search is performed for the selection of the *CIF*.

The second parameter, i.e., temporal search, defines the amount of time history or time periods which can be revised for information retrieval and/or for evaluating trend indications.

Evaluation threshold

The evaluation threshold specifies the lowest support value which can be associated to a hypothesis in order to be trusted and accepted by the model. The higher the value of the evaluation threshold, the more cautious or even suspicious the simulated behaviour becomes.

An evaluation stage aims at deciding if a potential solution can be trusted or if it must be rejected. The hypothesis selected has the greatest support, but this can be insufficient to give the model any confidence in this explanation.

3.6 Summary

Review of Models of Cognition

In this Chapter we have firstly reviewed the *theoretical models* and the psychological standpoints which sustain the analytical representation of cognitive behaviour (§ 3.1 and § 3.2).

The theories and models examined, have been developed over the time period that goes from the end of the 40ies until the end of the 90ies.

Six theoretical models have been reviewed. They cover a large spectrum of theories and modelling architectures of varying complexity and completeness. Some of them focus on specific aspects of cognition, which are modelled in great detail, while others look at cognitive behaviour from a more general perspective, with the aim to offer a comprehensive and balanced representation of different processes and functions of cognition.

These models have been compared with the requirements of the *Reference Model of Cognition* defined in Chapter 2, as well as with each other.

From such comparison, a model developer can identify and choose the theory that suits best the specific problem and the Human-Machine Interaction architecture that he/she is dealing with.

Review of Simulations of Cognition

In the second main section of this Chapter (§ 3.3), we have tackled the problem of how to implement theories of cognition into working simulations.

In so doing, we have considered a variety of numerical algorithms, programming languages and computer architectures that are applied to theories.

Ten different simulations have been analysed. These simulations were selected as they represent important stereotyped applications which use different algorithms, such as fuzzy sets, control theory, analytical expressions, production rules, and different computer languages/architectures, such as AI-Expert-Systems formalisms, Object Oriented Programming, Blackboard, etc.

These models have been developed in different domains of applications, namely, Nuclear Power Plants, Maritime Transport, Civil and Military Aviation, Space and Air-Traffic Control. This variety of domains has contributed to show similarities of operators' behaviour working in different technical environments. In particular, two important aspects of human behaviour have been enhanced, that are common to all these technologically developed domains and need to be simulated.

These are: the supervisory role of operators and the extended cognitive activity carried out during system management.

Guidelines for Development of a Simulation

The list of steps that need to be carried out for the development of a simulation have then been summarised (§ 3.4).

These *Guidelines* cover the whole process of simulation development, which comprises the following steps: 1- Definition of the problem boundaries and aims of the simulation; 2- Cognitive Task Analysis; 3- Field study of working context; 4- Selection of theoretical model; 5- Selection of numerical algorithms; 6- Implementation in programming language and environment.

This section combines in a single structure, the methods and techniques discussed in Chapter 2 and Chapter 3.

An Example of Application

Finally an example of application has been presented, by choosing one of the simulations described previously, and discussing in detail its application to the simulation of an operator's behaviour in controlling an auxiliary system during a major accident in a nuclear power plant (§ 3.5).

This application has shown the complexity of the process of simulation development and the spectrum of possible results that can be obtained, once the whole simulation is complete and running.

One interesting feature shown by the specific application has been the application of the *Guidelines* in the development of the simulation, specifically dedicated to the case study. In particular, the iteration between computer architecture and data retrieval and modelling, has shown the importance associated with the development of each element of the simulation.

The results of the application have been focused on the ability of the simulation to reproduce the cognitive processes that generate the manifestations of behaviour.

The consideration for the dynamic interaction between humans and machines and the effects of human behaviour on plant control will be considered in the next Chapter, so as to complete the Human-Machine architecture discussed in the first Chapter.

CHAPTER 4
MODELLING MACHINE AND INTERACTION

4.1 Introduction

In the previous Chapters we developed a set of guidelines and we have revised a number of models and techniques for the elaboration of a simulation of cognition in a human-machine interaction architecture.

As these methods are mainly focused on the human aspect, we have privileged the variables and results of the simulations concerning cognitive and behavioural performances. Very little attention has been paid to the overall dynamic evolution of the plant and working context.

Therefore, in this Chapter, we will start by reviewing the basic requirements for modelling and simulation of "non-human" constituents of the human-machine interaction architecture. These include the *Machine Model* and *Interaction Model.*

We will focus on the *Interaction Model,* which plays a very important role in regulating the whole flux and exchange of data and parameters between models and data architectures.

As we have already discussed in Chapter 1 the requirements for these models, we will only briefly reconsider them in the view of practical applications, mainly for prospective types of analysis and quantitative numerical simulations.

The main part of this Chapter, will be devoted to the description of two detailed applications aiming at showing how the proposed architecture of human-machine interaction can be exploited for safety assessment studies and for the evaluation of the design of interfaces and procedures.

Even though the two sample cases that will be discussed refer to the control and emergency management of a nuclear power plant, the architectures and models that will be developed may easily be extrapolated to other technologically similar domains, for example, chemical and process control plants, aviation, air traffic and space management.

The reader can find in § 1.3.2 the discussion on the *Elements and Mechanisms of Machine Model* and in § 1.4.1 and 1.4.2 the description of the *Interaction Model* and its correlations with the Human-Machine Interaction Simulation and Data Management.

4.2 Models and Simulations of Machines

As we discussed in the first part of this book, machine and human models need to show the same level of complexity and flexibility, so as to reach a balanced simulation of the human-machine system.

A Machine Model, as already discussed in general terms in Chapter 1, must include at least three main constituents characterising the behaviour of a plant and associated environment (Figure 4.1):

1. A model of the *physical behaviour* of the most important plant state variables and parameters.
2. A model of the response of *interfaces* and *actuators*, which include, indicators, control panels, Video Display Units, decision support tools and automatic systems.
3. A model of the *context and working environment*, that may affect the overall performance of operators and plant. These include external parameters, such as temperature, noise, humidity etc., but also factors related to the performance of a task, such as communications, rules, operating standards, etc.

Physical models are applied for the evaluation of dynamic behaviour of state variables and parameters and can be obtained by applying the basic conservation principles and physical laws that govern any technical system. This leads to the development of sets of analytical and differential equations.

The use of numerical techniques and iterative methods, coupled with appropriate simplifying assumptions, can be utilised for solving these equations, and for generating computerised simulations.

The models of *interfaces* and *actuators* are essential for the whole human-machine architecture, as they offer the real connection with the human model.

The interfaces provide the information about the plant state and evolution, while the actuators are the instruments that the operator uses to implement tasks and procedures.

Interfaces and *actuators* are, in principle, very simple to model, as they relate directly to physical variables and human actions.

However, they are substantially hardware components and, like all other components of the plant, they may be affected by reliability and stochastic behaviour.

Moreover, they are affected by intrinsic characteristics, mainly derived from their design, which pay an important contribution to the generation and development of human errors and to human factors related events in general.

While the human errors are mainly dealt with by a specific reliability model, the human factors aspects of the *interfaces* and *actuators* are more delicate and require special attention.

The best viable solution to the problem is by performing very accurate Task Analysis, by which it is possible to identify data and correlations to be utilised for the development of the simulation.

The model and simulation of the *context* and *working environment* present substantially the same sort of features as the *interfaces* and *actuators*.

On the one hand, the variables and parameters of the *context* and *working environment* that affect plant and human behaviour can be represented quite straightforwardly by analytical expression as well as by numerical algorithms.

However, the most important issues are related to the definition of "which" are these variables and parameters and "how" they affect plant performance and human behaviour. The answer to these questions comes, once again, primarily from the Task Analysis and from the field examination of the working context.

The importance of the Task Analysis, which appeared very clearly in the discussion about the simulation of cognition, returns here. This is not surprising, as the indicators/actuators and the context/working environment are the two elements that affect primarily human-machine interaction and through which the interaction actually takes place.

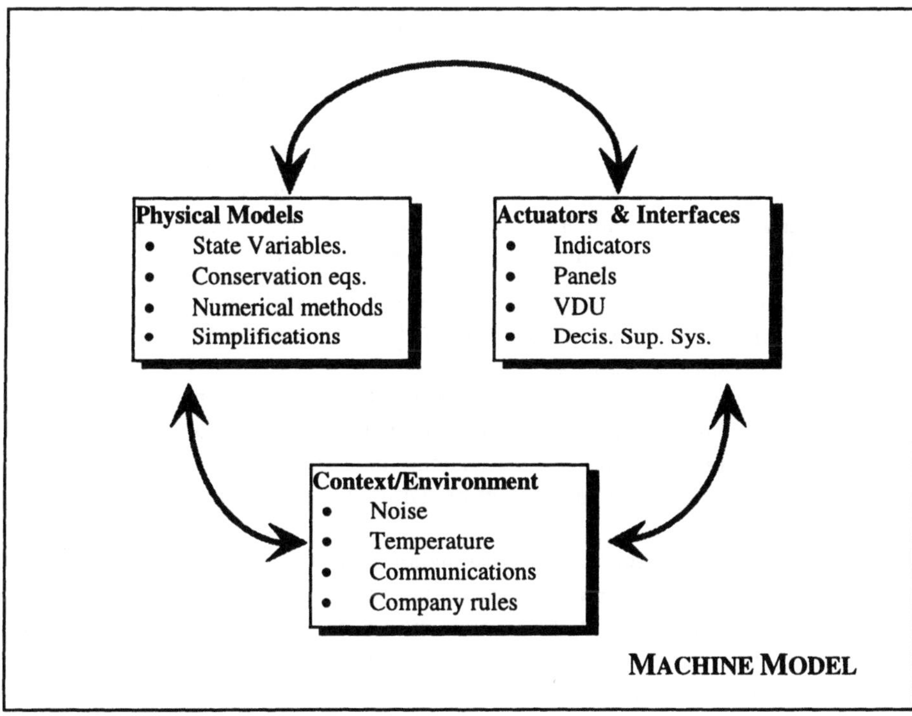

Figure 4.1. The constituents of the machine model.

The models utilised, in practice, to represent the elements of the Machine Model are based on well established principles and techniques that will not be discussed here. The reader con refer to other text books and monographs on modelling and simulation physical systems, and interfaces (Ordys and colleagues, 1994) and on reliability performance of components and systems (Andrews and Moss, 1993), to develop further knowledge on these subjects.

We will apply these physical principles and numerical and analytical techniques in the forthcoming sample cases, under the assumption that they are known and familiar to the reader, as we have already done with other principles and techniques for the development of previously discussed sample examples.

4.3 Interaction Model

The *Interaction Model* plays a very important role in the Human-Machine Interaction architecture taken as reference throughout this book, as it manages the connection between machine and human models, and controls the retrieval of data from the databases.

As was discussed in general terms in Chapter 1, the connection between machine and human models is governed by the Interaction Model by means of three modelling algorithms (Figure 4.2):

- The time step evolution and synchronisation between the machine and human models (*Time management*).
- The preservation of logical correlation and exchange of data between machine and human simulations (*Logical human-machine interconnection*).
- The reliability of the components of the machine and of the human behaviour during the time dependent interaction (*Dynamic reliability*).

In performing its role in the overall simulation framework, the *Interaction Model* manages the flux of *dynamic data* between the HMI Architecture and the Data Architecture (see also Figure 1.14).

The data that are utilised by the *Interaction Model* to perform its search in the database are the Human-Machine variables and parameters calculated by the human and machine simulations during the dynamic interaction, such as *temperatures, pressures, alarms,* etc. These are all factors that may foster or initiate malfunctions, failures or human errors.

The retrieval of data from the database mainly concerns the reliability and logical behaviour of machine and human constituents, which are usually in the form of *failure modes, error modes, failure rates,* etc. These data are necessary for the calculation of the operating or malfunctioning dynamic state of components of the machine and human model.

Even though the general aspects of an Interaction Model have been discussed in Chapter 1 (§ 1.4.1 and § 1.4.2), we will now consider in greater detail the theoretical background and the numerical algorithms necessary for an Interaction Model.

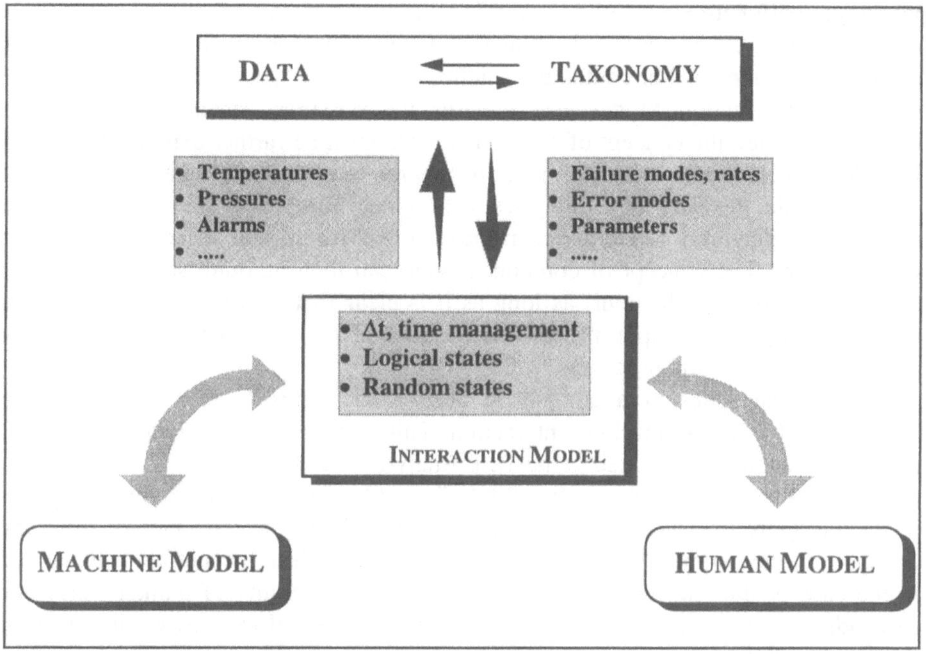

Figure 4.2. The Interaction Model in the Human-Machine Architecture

4.3.1 Algorithms for Interaction Models

Time Management

The management of the time evolution of the whole simulation and the data retrieval from database are performed by algorithms specifically developed for each case at hand. These algorithms depend on the complexity of the models and on the richness of the databases.

The most important aspect of the time management criteria and methods concerns the synchronisation between the simulations of human and machine, however, it is not possible to develop a set of general rules for these algorithms. In many, cases they are the result of some numerical process of adaptation of the machine and human simulation to the outcome of experimental analysis and evaluation of working processes and performance of procedures.

Logical Human-Machine Interaction

The Interaction Model controls the "operating state of a component", or simply the "component state", as it develops during the dynamic interaction of the human-machine system.

The "component state" represents one of all the possible operating modes, such as working, malfunctioning, or non-working, which a component can reach as a consequence of either a particular human-machine behaviour, or a random event, or a certain environmental or contextual condition.

As the Interaction Model also governs the possible "states" of the human behaviour model, the concept of component state must be further expanded, so as to include possible human errors, or, more in general, inappropriate behaviours.

In principle, the same type of causes that regulate component states, i.e., machine or human behaviour, random occurrences or environmental effects, also affect human errors. The concept of component state can then be quite straightforwardly extended to human behaviour, as long as it is clear that, when considering human error, this includes inappropriate actions as well as incorrect performance of cognitive functions and processes.

All possible states of a component need to be defined *before* performing a simulation of a human-machine interaction. This is part of the analytical, empirical and experimental work related to the Task Analysis activity.

The number of possible operating "states" must be finite, for the obvious purposes of making more manageable simulation and calculation. The relevance of the Task Analysis on the output of the simulation becomes very clear. This effect is enhanced in the case of the Human Model, as the enormous variability of human behaviour demands an important and accurate analysis of the tasks and working environment in order to highlight the most significant parameters and variables.

The Interaction Model manages the search in the database for components' states in relation to the dynamic behaviour of the machine and human and to the evolution of the context. The state of a component can also be modified by pure random events. This other change in operating state must be managed by the Interaction Model, with appropriate analytical and numerical algorithms.

Dynamic Reliability

The real complex element of the Interaction Model is the calculation of the dynamic behaviour of components that are governed by probabilistic and stochastic response, in addition to logical and deterministic behaviour.

This is a typical problem of reliability analysis which goes beyond the classical approaches such as Fault Trees (FT) and Event Trees (ET) methods, that are well developed and consolidated techniques (Feller, 1968; Henley and Kumamoto, 1981; Høyland and Rausand, 1994).

A number of approaches have been developed to permit the evaluation of the probabilistic behaviour of system unavailability versus time or dynamic reliability, when the plant is decomposed in a reasonable number of super-components. These dynamic approaches are based upon Petri Nets theory and Markov chains (Vesely and Goldberg, 1977; Jeong, Chang and Kim, 1987; Aldemir and colleagues, 1994).

Although these methods deal very well with the stochastic behaviour of components, they seem somehow limited for handling the actual interaction between the process of dynamic behaviour of physical variables and the dynamic change of components' operating state (Siu, 1994).

This particular domain of logical/reliability problems has been tackled with a variety of special approaches. Hassan and Aldemir (1990) have developed a method that separates the physical and probabilistic analysis and assesses event sensitivity to uncertainty on the component failure data. Another approach is based on the Continuous Event Tree method, which links the system model to probabilistic treatment using Markovian or semi-Markovian chains in a complete theoretical analysis (Aldemir, 1987). The Dynamic Event Tree method (Siu and Acosta, 1991), and the DYLAM approach (Cojazzi and Cacciabue, 1994) combine a quantitative dynamic process model and probabilistic analysis in dynamic conditions to evaluate both reliability and physical behaviour in an integrated architecture. The GO-FLOW method enhances the time-order analysis of the system at hand (Matsuoka and Kobayashi, 1988).

We will now focus on the DYLAM (Dynamic Logical Analytical Methodology) methodology, which was originally developed some years ago (Amendola and Reina, 1981) for analysing the dynamic stochastic and logical behaviour of plants. DYLAM dedicates particular attention to the need of combining physical simulation and probabilistic evaluation in a manageable computing architecture, without requiring a too complex analysis and/or endless calculation of possible scenarios and sequences.

These are generated either by random events or by physical and environmental changes experienced by the machine and working context as a consequence of the accident dynamics.

For systematic studies integrating Human-Machine simulation, the DYLAM method seems to be ideal, as it combines the stochastic nature of system failures and human (erroneous) behaviour (Cacciabue and Cojazzi, 1994).

The DYLAM method will now be briefly described and it will then be utilised as the basic structure for the Interaction Model, to be applied in a number of sample applications of the reference architecture of human-machine interaction.

4.3.2 The DYLAM Methodology

The DYLAM methodology, used as a tool for dynamic logical/reliability analysis, permits the evaluation of time dependent behaviour of systems when the boundary conditions and failure modes of its components are defined.

DYLAM generates and follows a series of possible incidental sequences, which can arise during the running time, due to failures or inappropriate behaviour of some components of the plant. Failures occur at time instants which are unpredictable at the beginning of the analysis, as they are calculated either by probabilistic algorithms or by logical correlations between events and occurrences. Indeed, failures are affected by the physical and stochastic behaviour of the system, but also contribute to changing its performance (Cojazzi, Cacciabue, and Parisi, 1993).

For each sequence, DYLAM controls the physical simulation and the evaluation of dynamic probability of occurrences, recording the attainment of failed configurations and/or of possible recovery conditions, i.e., the conditions which lead to mission failure and/or mission recovery. These conditions are chosen or defined by the user at the beginning of the analysis.

The application of the DYLAM methodology is only possible when a simulation of the machine is provided for the deterministic analysis and it is linked to a probabilistic/logical model. The use of DYLAM as Interaction Model of human-machine interaction architecture is ideal in this case.

There are no limitations to the type of plant, as, in principle, DYLAM can be applied both to simple hardware structures, i.e., small systems constituted by a limited number of components, and to very complex plants, including control instrumentation, and human behaviour models.

In practice, the application of DYLAM can be carried out by a number of iterative formal steps, namely (Figure 4.3):

1. Identification of Main System Components.
2. Failure Mode and Effect Analysis.
3. Selection of Data and Parameters.
4. Selection of Type of Analysis.

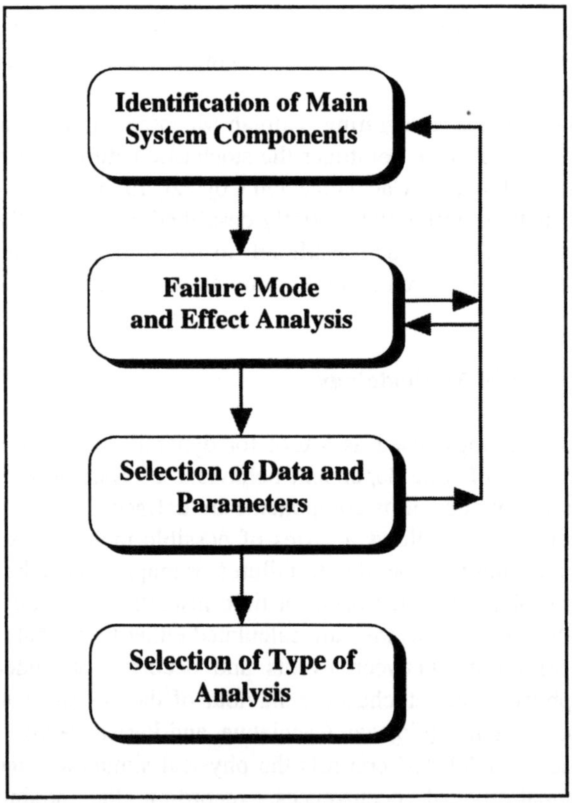

Figure 4.3. DYLAM application procedure.

Step 1: Identification of Main System Components

In the case of simple plants, all system components can be considered for the failure and reliability analysis, by studying all possible combinations of their working states.

In the case of complex plants, a preliminary analysis is usually required in order to identify the components, or the parts of plants (macro-components), which can become critical during accidents.

The identification of these components may be completed by the analyst through a qualitative analysis which involves implementing the experience and knowledge of the system or, alternatively, through a more formal method, i.e., Task Analysis for human performance, and standard analysis based on Fault Trees and Event Trees methods, for machine and hardware components.

Step 2: Failure Mode and Effect Analysis

Having identified the components or macro-components which may fail and affect the plant safety, the application of the DYLAM approach requires the performance of a Failure Mode and Effect Analysis (FMEA) for each component and macro-component.

This analysis aims at identifying three main elements (Figure 4.4):

1. The complete spectrum of "working" *states* of the components of the system.
2. The *transitions causes* between states.
3. The *physical and logical behaviour* of the components in each state.

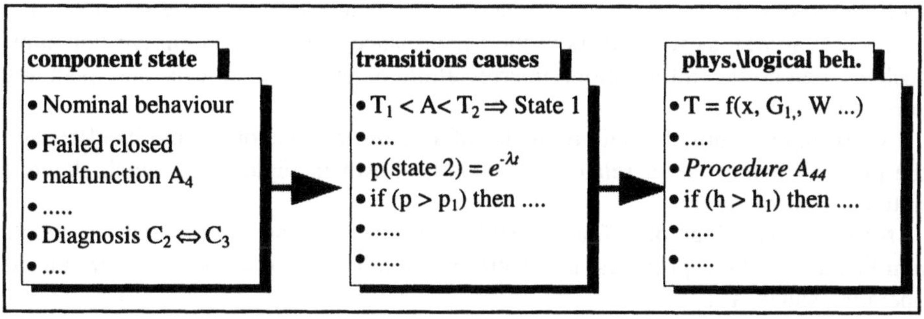

Figure 4.4. Outcome and elements of Failure Mode and Effect Analysis.

In case of hardware components, for example, the number of *states* can be very high, with a spectrum of possible performances that go from the perfect response according to design, called *nominal state*, to several failure modes.

The *transitions causes* between states can be related to physical system behaviour, or environmental conditions or simply stochastic events. In all these cases, as we will see in the next sub-section on DYLAM data, a number of failure conditions can be defined, such as: 1) failure to demand probabilities, 2) transition rates between states, 3) conditional transition matrices between states, in the case of dependence from physical variables and 4) conditional failure matrices, due to common cause failure with other components.

The *physical and logical behaviour* of a hardware component in a certain state is usually represented by equations. However, it is also possible to describe physical and logical states by expression and production rules.

In case of human "components" the definition of the three elements resulting from the Failure Mode and Effect Analysis is even more difficult and certainly requires very accurate studies in order not to impoverish excessively the human model.

The possible *states* related to human behaviour can be identified as performances adhering perfectly to procedures, or inappropriate performances of manual and cognitive activities. Examples of inappropriate performance could be: "the systematic omission of checks following regulation actions" or "the exchange between two similar control procedures", etc.

The *causes of transitions* between states of human behaviour can be related to environment and contextual conditions as well as to random events. Personal attitudes and internal factors affecting behaviour are obviously causes of individual erroneous performance.

However, these are seldom included in the type of study discussed in this book, as they are too specific to single human behaviour, while we are interested in large groups and populations, for which only average performances and characteristic data should be accounted for.

The *physical and logical behaviour* of a human "component" can be associated with specific manifestations of erroneous behaviour, during the execution of a procedure, as well as with errors in the performance of specific cognitive functions, such as perceiving information, interpretation of sign and signals and planning control procedures.

An example of possible failure modes of a hardware component and its effects on the physical behaviour of related variables is shown in Figure 4.5, which shows a notional control valve.

Nominal behaviour plus four possible failure modes are considered. A simple non-linear equation provides the representation of the control valve in all its operating states, as :

$$Q_v = K_v \, y \sqrt{\Delta p_v} \tag{4.1}$$

where:

Q_v = Mass flow through the valve v

K_v = Constant parameter for the specific valve v

Δp_v = Pressure drop across the valve v

y = parameter for definition of component state

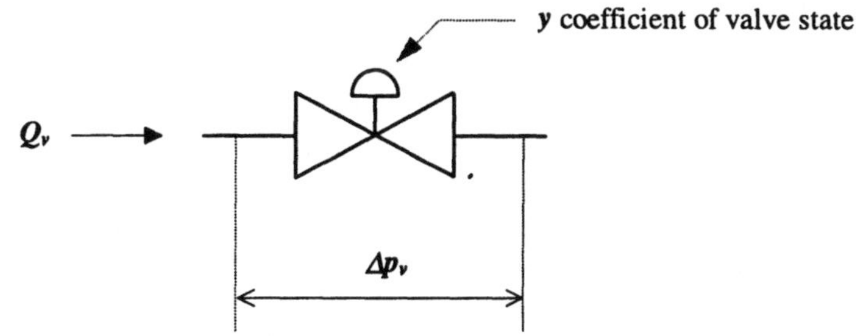

State	Description	*y* value
0	nominal behaviour	any value (0 -1)
1	failed closed	0
2	failed open 100 %	1
3	failed open 50 %	.5
4	failed in previous condition	previous value

Control Equation:

$$Q_v = K_v y \sqrt{\Delta p_v}$$

Figure 4.5. Example of DYLAM modelling of a control valve.

Finally, two important considerations have to be made with respect to the FMEA in DYLAM:

1. After the Failure Mode and Effect Analysis, some important components that have not been included in the first selection of main components could be identified. Therefore, an iteration with the step "Identification of Main System Components" may be needed, so as to ensure that no critical elements have been omitted from the study (Figure 4.3).
2. The Failure Mode and Effect Analysis leads to the development of equations like Eq.(4.1) for modelling component failures. These equations have to be strictly correlated to, or integrated within, the machine and human models, so as to give a balance between the dynamic physical behaviour of the system with the dynamic generation of failures and human errors.

Step 3: Selection of Data and Parameters

The step of identification and selection of data is very important and it relates closely to the Task Analysis and to the availability of databases of failure modes and causes in different environmental conditions.

As we have already discussed the Data Architecture (§ 1.5) and the Task Analysis method (§ 2.4), we will now focus on the identification of data and parameters with respect to DYLAM.

The data used by DYLAM in a human-machine simulation study can be subdivided into two classes:

1. Probabilistic and logical data for machine behaviour; and
2. Probabilistic and logical data for human behaviour.

The first class of data are utilised in DYLAM for performing the probabilistic/logical analysis of sequences related to hardware components, and have to be supplied in different ways, according to the probabilistic behaviour and dependencies chosen for the components.

As previously stated, components may be described by *initial state probabilities*, *constant failure rates*, *conditional transition matrices* between states, due to dependence from physical variables, or *conditional failure matrices* due to common cause failures. These data have to be introduced as INPUT to the model.

In general, these types of data may be found in traditional Reliability Data Banks already used for methodologies such as Fault Trees and Markov Chains.

However, in the case of failure probability dependent on physical conditions, problems can arise in assigning the values to (conditional) transition matrices between states, which give the failures or state transitions of components with respect to physical variable behaviour. This is because, generally, literature supplies only failure rates (related to different ranges of a physical variable) and the state transition probabilities have to be defined according to specific studies and accurate evaluation of the behaviour of the system in hand.

As an example, with reference to the Control Valve described in Figure 4.5, the user must be able to assign probability of valve blockage, either in a fully open or in a fully or partially closed state in accordance with temperature thresholds of the fluid that flows through the valve. However, in the transition matrix the probabilities of reversing to normal behaviour or to further changes of state also have to be evaluated in a very careful and logical way.

Table 4.1 shows a typical transition matrix, which could apply to the Control Valve of Figure 4.5 for the threshold temperatures 80° and 120°. The "arrows" indicate that the threshold value is passed increasing (↑) from lower to higher temperatures or decreasing (↓) from higher to lower temperatures.

In order to assign the probabilistic and logical data for human behaviour the methodology requires the availability of some matrices containing conditional error rates, which correlate the mode of error that the human operator can make with contextual and environmental conditions that affect performances.

For example, one or more functions of environmental variables could be defined and then, these functions could be correlated, on the basis of threshold values, to the probability of error making.

During the simulation, these error rates could be used to evaluate, in a dynamic fashion, different accident evolutions, corresponding to different error modes.

It is particularly important that transition matrices, like the one of Table 4.1, are devised in accordance to the specific human functions and cognitive processes that are simulated and to all possible modes of errors that can be considered in the study of human-machine interaction.

In the sample cases that follow, some special instances of such types of transition matrices are discussed.

Table 4.1 Transition Matrix: Probabilities of component state transition vs. temperature. Component: Control Valve (Figure 4.5); * Probabilities • 10^{-3}.

T Threshold (°C)		80 ↓	80 ↑	120 ↓	120 ↑
Comp. State Change		T dec. < 80 °C	T inc. > 80 °C	T dec. < 120 °C	T inc. > 120 °C
0	1	0.1^*	0.9^*	2.9^*	5^*
0	2	0.1^*	0.9^*	2.9^*	5^*
0	3	0.9^*	1.9^*	4^*	9^*
0	4	1.5^*	3^*	9^*	12^*
0	0	$1-\Sigma p_{0i}$= .9974	$1-\Sigma p_{0i}$= .9933	$1-\Sigma p_{0i}$= .9812	$1-\Sigma p_{0i}$= 969
1	0, 2, 3, 4	0	0	0	0
1	1	$1-\Sigma p_{1i} = 1$	$1-\Sigma p_{1i} = 1$	$1-\Sigma p_{1i} = 1$	$1-\Sigma p_{1i} = 1$
2	0, 1, 3, 4	0	0	0	0
2	2	$1-\Sigma p_{2i} = 1$	$1-\Sigma p_{2i} = 1$	$1-\Sigma p_{2i} = 1$	$1-\Sigma p_{2i} = 1$
3	0	5^*	1^*	0.3^*	0
3	1	2^*	8^*	12^*	25^*
3	2	5^*	1^*	0.3^*	0
3	4	0	0	0	0
3	3	$1-\Sigma p_{3i}$= .988	$1-\Sigma p_{3i}$= .99	$1-\Sigma p_{3i}$= .9874	$1-\Sigma p_{3i}$= .975
4	0	5^*	1^*	0.3^*	0
4	1	2^*	8^*	12^*	25^*
4	2	5^*	1^*	0.3^*	0
4	3	0.9^*	1.9^*	4^*	12^*
4	4	$1-\Sigma p_{4i}$=.9871	$1-\Sigma p_{4i}$=.9881	$1-\Sigma p_{4i}$=.9834	$1-\Sigma p_{4i}$=.963

As a consequence of the Selection of Data and Parameters, a repetition, or revision, may be necessary of the previously performed steps of Failure Mode and Effect Analysis, and, possibly, Identification of Main System Components (Figure 4.3).

This is usually the case when data related to some critical components can not be found in the literature or are too difficult to collect in the field. Consequently, the analyst has to review the initial hypotheses and has to carry out a new FMEA.

Step 4: Selection of Type of Analysis

The last step in the application of the DYLAM methodology, is the selection of the type of analysis that is carried out, and the assignment of data that are specific to DYLAM for the management of the machine or human models.

Two types of analysis are possible:

1. *Systematic* analysis, in which DYLAM generates all the possible sequences resulting from the combination of faults and logical conditions occurring during a simulation.
2. *Interactive* analysis, in which the analyst assigns an initial configuration to the plant and a limited number of descendants are studied.

In both cases DYLAM triggers dynamic failures in the plant components according to probabilistic/logical criteria.

In Figure 4.6 an example of 18 sequences, a *nominal sequence* (S_0) and seventeen *descendants* (S_1 - S_{17}), are shown.

These sequences are generated at various times and are distinguished from each other because of different failure modes of components. The "nominal sequence" is the sequence generated by the initiating event with all components behaving as expected, i.e., in *nominal state*.

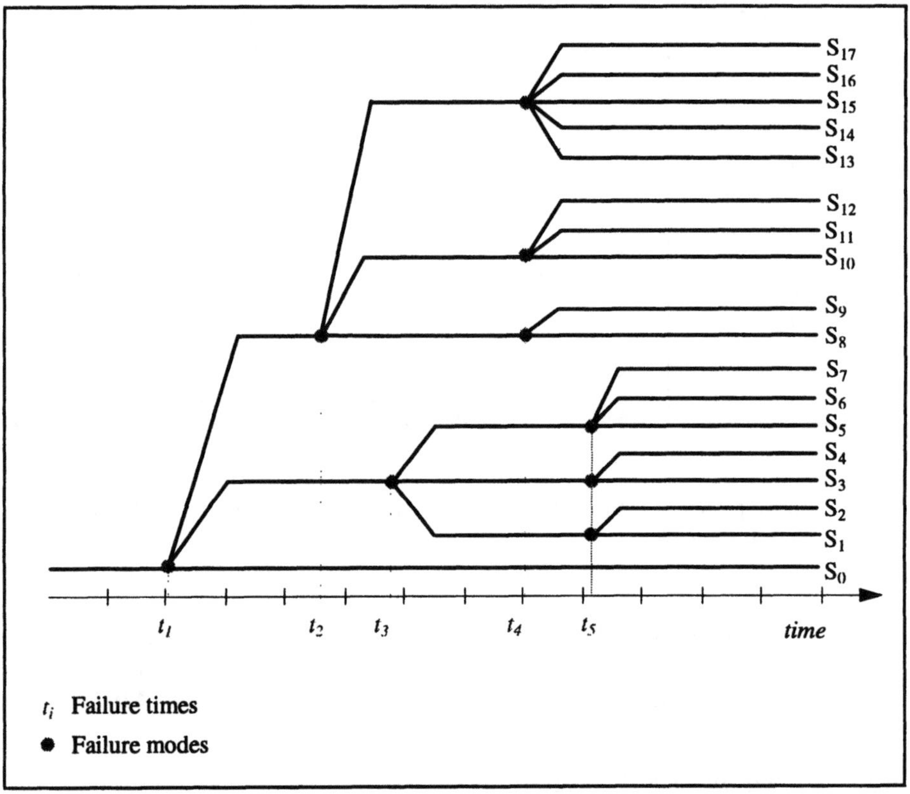

Figure 4.6. Generation of Sequences and descendants by DYLAM.

In the remaining part of this Chapter, a number of case studies will be performed, where the models and simulations described in the previous Sections and Chapters will be applied in order to show some complete applications and to demonstrate the potentiality of the proposed architecture for studying Human-Machine Interactions.

4.4 The Chemical and Volume Control System Case Study

The objective of this study is the analysis of human-machine interactions in the case of management of a complex control system of a large plant, which performs constant on-line, partially automatically controlled functions, and stand-by emergency activities.

4.4.1 Plant Description and Accident Scenario - Case Study CVCS

Plant Description

The system under study is a typical Chemical and Volume Control System (CVCS) of a large Chemical, Process or Nuclear Power Plant (NPP). In this sample case the NPP type is a Pressurised Water Reactor (PWR).

The CVCS is an auxiliary system which controls the boron concentration and coolant volume in the primary circuit and ensures the seals of the primary pumps (Figure 4.7).

As we intend to maintain the level of complexity within manageable boundaries for our case study, only the main components and hydraulic lines of the CVCS are shown in Figure 4.7.

The system is normally connected to one of the Primary (Coolant) Circuit loops and it is essentially composed of four parts: the main *let-down line*, the *make-up*, the *charge line* and the line to the pump seals (*seal line*).

The main *let-down line* is connected to the cold leg of the primary circuit, between the steam generator and the main pump. The water drained from the circuit is firstly cooled and de-pressurised before entering the regeneration, decontamination and deionisation system, where it is purified by means of ionic-exchange resins and then finally enters into the Volume Control Tank (VCT).

In the VCT an appropriate quantity of boron and other chemical additives are added to the purified water by the *make-up line* which also provides an additional supply of water to the VCT.

The treated primary coolant water is then fed-back partly to the reactor, through the *charge line*, and partly to the seals of the primary pumps, through the *seal line*.

The CVCS is provided with an automatic control and protection system. In alternative to automatic control, manual operation is also possible.

Accident Scenario

Given the characteristics of the methodology and the complexity of the plant under study, we have concentrated our attention on a well defined accident scenario and control procedure:

- The loss of electrical power to the primary pumping system, which implies reactor shut-down and very heavy intervention of many auxiliary systems, including the CVCS.

In such a case, the CVCS is available for controlling the system, manually as well as by protection and automatic mechanisms. On the Primary Circuit there is a rapid and important decrease of the most important physical quantities, namely pressure, temperature and level of the Pressurizer. This is the macro-component that is designed to control the water level and pressure variation within the Primary Circuit.

During the evolution of the accident, a number of specific interventions of the operator on the CVCS are demanded to maintain the main functional goals of preserving sealing of pumps and water purification in the Primary Circuit.

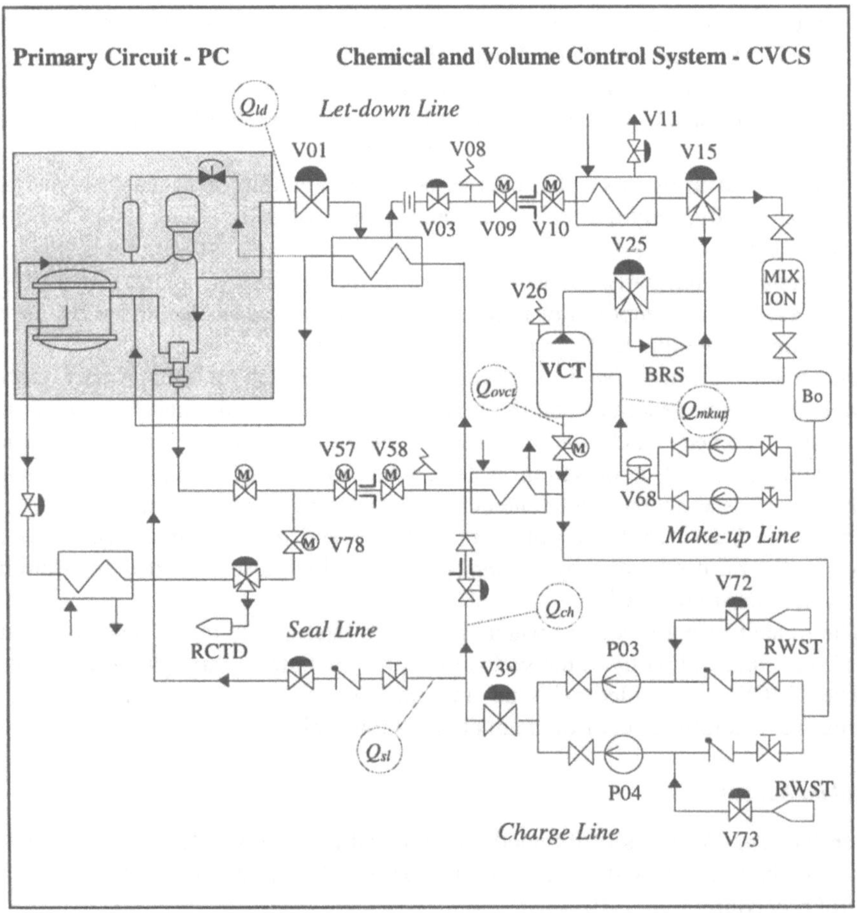

Figure 4.7. Schema of a typical Chemical and Volume Control System

4.4.2 Machine Model and Simulation - Case Study CVCS

The model of the CVCS has been developed mainly considering the physical behaviour of plant variables and hardware components. Little attention has been paid to the interfaces for controlling the system.

This choice has been made deliberately, as the objective of the study is to consider the human-machine interaction problem in the case of management of a complex system from the functional and physical behaviour viewpoint. In other words, in this case we are less interested in the evaluation of interfaces, but rather in the way in which the operator can control the CVCS during an incidental condition, with different parallel physical processes ongoing at the same time.

A simulation code was developed to describe the thermo-hydraulic behaviour of the CVCS plant both in nominal and abnormal conditions, resulting from possible faults of a selected number of components. In this application the chemical aspect related to the boron regulation is not treated, while the operator activity with respect to system control and protection are carefully considered for plant management studies (Carpignano, 1989; Vivalda, 1989).

The simulation developed for CVCS allows the dynamic analysis of the plant and accounts for automatic as well as for manual control.

The thermo-hydraulic behaviour of the CVCS plant is evaluated by successive equilibrium states for what concerns the hydraulic network and by time dependent equations for the Volume Control Tank (VCT) response. The physical quantities of the primary circuit represent the boundary conditions for the CVCS and they are described by a predefined dynamic behaviour in terms of pressures and temperatures for the whole period of the transient.

The simulation of the CVCS has been developed into a computer program using standard FORTRAN 77 programming language.

Hydraulic Behaviour of the CVCS

The model and simulation of the hydraulic behaviour of the CVCS is based on a number of simplifying assumptions.

In particular, the following hypotheses are adopted in the model:

- The fluid, i.e., water, is assumed to always remain single phase liquid with constant physical properties.
- The totality of the pressure drops in the network are concentrated in the components.
- The potential energy, derived from differences in height of the different sections of the network, and the convection forces, generated by temperature differences, are not taken into consideration.

The direct consequence of these hypotheses is, firstly, the complete independence of the thermal and hydraulic calculation. This implies a considerable simplification of the solution mechanism, as the thermal evaluation of the CVCS plant can be performed consecutively to the solution of the hydraulic equilibrium. In other words, temperature differences have no effect on the evaluation of the mass flow and pressure distribution within the network.

Moreover, in order to reduce further the complexity of the solution method, two other hypotheses have been adopted:

- The pressure in the VCT is assumed to remain at constant, or to vary according to a predefined law.
- The pumping system can be described by symmetrical behaviour of the pumps when they are both operative.

The first hypothesis is rather logical, in consideration of the fact that a number of inlet and outlet lines of gas, such as nitrogen, are specifically designed to allow the operator to manage the pressure in the VCT.

In this way, the hydraulic analysis is divided in two: the *let-down line*, running from the primary circuit to the VCT, and the *charge line* and *seal line*, downstream to the VCT towards the primary circuit and the pumps seals.

The second hypothesis is mainly oriented to simplifying the simulation.

From the structural viewpoint, the CVCS network has been subdivided in a number of sections and nodes. A set of differential equations are obtained by applying the equilibrium condition of mass flows at nodes and the pressure balances in each section of the network:

$$\frac{dm}{dt} = \sum_{in} w - \sum_{out} w = 0 \tag{4.2}$$

$$\frac{1}{S}\frac{d(mv)}{dt} = \sum \Delta p \tag{4.3}$$

where:
m = mass, v = speed, w = mass flow, p = pressure, and S = surface area.

The solution algorithms adopted to represent the dynamic behaviour of the CVCS from the hydraulic viewpoint are rather flexible and capable of handling networks of quite high complexity. The algorithm chosen is based on the Cross method (Douglas, Gasiorek, and Swaffield, 1983), which has been generalised by considering passive components, i.e., valves, as well as active components, i.e., pumps, and elements of discontinuity, such as relief and check valves.

The set of equations (4.2) and (4.3), for all nodes and sections of the network, are solved by successive iterations. After having chosen the initial tentative values of pressure distribution within sections, the method evaluates the corresponding mass flows and pressure distributions by solving the equilibrium equations in all nodes of the network. The calculated values of pressure distributions are then confronted with those assumed or calculated in the previous iteration: the difference between calculated and prior values is used to make the appropriate corrections for the successive iteration.

The hydraulic simulation of the VCT has been obtained using the mass balance equation:

$$\rho S \frac{dL}{dt} = \sum w_{in} - \sum w_{out} \tag{4.4}$$

where:
ρ = density; L = water level in the VCT; S = VCT surface.
The solution of equation (4.4) has been obtained by the Euler (explicit) method.

Thermal Behaviour of the CVCS

The thermal analysis of the CVCS is also based on a number of simplifying assumptions.

In particular, no heat dispersion to the atmosphere is considered along the lines of the network nor within the components. The temperature distribution is evaluated by:

- Assuming the conservation of the internal energy of the fluid in the nodes:

$$\frac{dU}{dt} = \sum c w_{in} T_{in} - \sum c w_{out} T_{out} = 0 \tag{4.6}$$

- Applying the energy balance equation to the VCT:

$$c \rho S \frac{dLT}{dt} = \sum c w_{in} T_{in} - \sum c w_{out} T_{out} \tag{4.7}$$

- Modelling heat exchangers by the following equations:

$$\begin{cases} \varphi = c w_c (T_{oc} - T_{ic}) \\ \varphi = c w_h (T_{ih} - T_{oh}) \\ \varphi = AH \, \Delta T_{lm} \, F \end{cases} \tag{4.8}$$

where:
φ = exchanged power; F = geometric factor; T = temperature; c = specific heat; AH = thermal exchange coefficient; U = internal energy;
subscripts:
lm = logarithmic ratio; c/h = cold/hot leg; i/o = input/output

As in the case of the hydraulic system, the solution of the thermal analysis requires an iterative algorithm, because there is a thermal coupling between the *let-down line* and the *charge line*, due to the presence of the VCT and a regenerative heat exchanger between them.

4.4.3 Human Model and Simulation - Case Study CVCS

Definition of Problem Boundaries and Aim of the Simulation

The problem boundaries of the human behaviour simulation are summarised in the identification of those elements which may affect the behaviour of operators and may generate biases in the decision making process, during the interaction with a complex plant like the Chemical and Volume Control System, for managing and controlling a major accidental condition.

The human model is particularly focused on the decision making, with associated actions, and on the biases and mechanisms of errors occurring during the human-machine interaction.

The analysis of human errors related to safety and reliability studies needs to be developed at three dependent levels, namely: the *error causes*, the *error modes* and the *error consequences*.

In an architecture where interactions of human and machine are explicitly modelled, the *consequences* of erroneous actions are the result of the simulation and, therefore, they do not need to be discussed on the basis of special methodological hypotheses.

On the contrary, *error modes* and specially *error causes* are key features to be considered during the definition of the aims of the human model, and, once the boundaries and limitations of the model are defined, *error modes* and *error causes* need to be analysed and further defined during the Task Analysis and field observation.

In this case study no specific modelling paradigm, like those described in the previous Chapter, has been followed, as the objective of the study is the evaluation of a control procedure in complex environmental conditions.

The following hypotheses have been made:

- Only one operator acts on the system during the accident.
- The accident under study is familiar to the operator.
- The operator performs a correct diagnosis of the event.
- There is no delay in the execution of the actions of the procedure.

Cognitive Task Analysis and Field Study of Working Context

The optimal procedure for managing the accident scenario has been derived from the task analysis which aims at the definition of the main goals for the operator and the effects of the working context on error making and inappropriate performances, as discussed in Chapter 2 (§ 2.4.3).

Performance of Procedures

In principle, in order to manage the CVCS appropriately, three main goals that need to be fulfilled at different times of the accident scenario have been identified. These goals are:

1. ***Safeguard the CVCS.*** To fulfil this goal, the operator has to:
   a. *Increase* the VCT level during the early phases of the accident; and
   b. *Isolate* the regeneration and deionisation system to prevent damage.

2. ***Support the Primary Circuit.*** To fulfil this goal, the operator has to:
   a. *Guarantee* the isolation of the let-down line; and
   b. *Increase* the mass flow in the charge line.

3. ***Management of the CVCS.*** To fulfil this goal, the operator has to:
   a. *Control* the VCT level during the evolution of the accident;
   b. *Divert* the charge line suction to the Refuelling Water Storage Tank (RWST); and
   c. *Isolate* the return line from the primary pumps seals.

The actions related to each goal have been distributed in tasks. The tasks are performed at different time instants during the evolution of the accident. The actions included in each task have been arranged in sequence of execution but they may have different weight and importance within the task. This weight needs to be identified and distinguishes one action from the other.

Forms of Erroneous Behaviour

In the present model development we have selected a simple error taxonomy and error/recovery reference model, which assumes that human errors and recovery stem from two different sources: random occurrences and/or direct consequence of the human-machine interaction and environmental conditions (Figure 4.8).

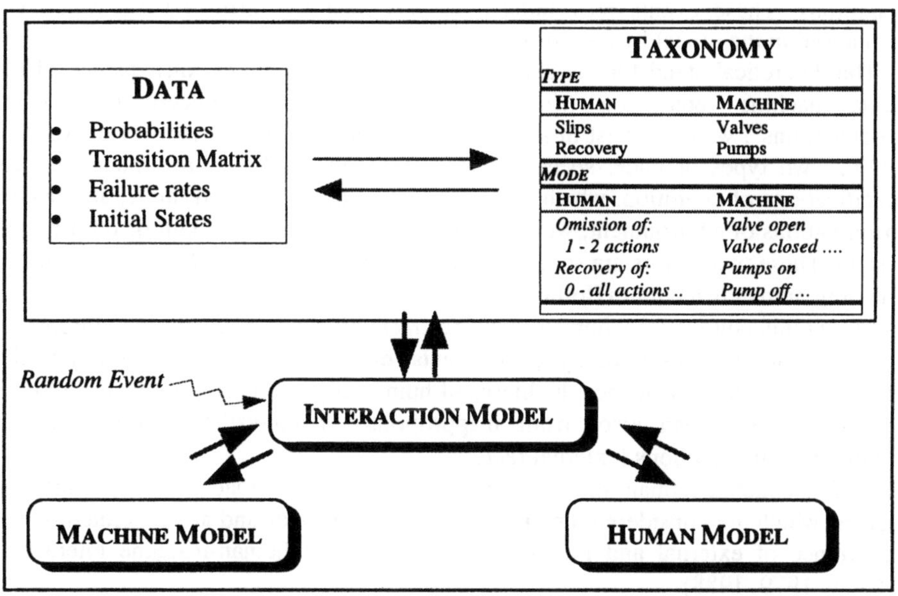

Figure 4.8. The error generation mechanisms in the HMI Architecture.

For this sample case, it has been assumed that the only types of error made by the operator during the execution of a procedure, are *slips,* or omission of a number of actions or verifications. The possibility of recovering from slips at a later stage of the accident is also considered in the model.

More precisely, at each task of the procedure, several possibilities of error exist: one or more actions belonging to the same task can be forgotten by the operator. Forgotten actions can be recovered by the operator at a later time, and, when recovery occurs, then all omitted actions, during a task, are performed.

The event of error is strictly related to the development of an accident and it is deterministic or probabilistic, and this may not be defined "a priori". More precisely, the model does not choose at the beginning of the analysis *which* are the actions omitted by the operator, but only their *mode* or *manifestation.* This limitation aims at reducing the wide spectrum of possible inappropriate behaviours which could, in principle, be considered but would make rather cumbersome and extended the amount of calculation required to perform the safety study.

Factors Affecting Behaviour

In principles, errors may be generated by random events or may be caused by the human-interaction process.

The generation of errors on the basis of a purely *random event* is justified by the consideration that, in many circumstances, the reasons for human erroneous behaviour are either too deeply rooted in the cognition of the individuals to be deterministically modelled, or are actually the results of a purely stochastic occurrence. In the model, random error events are omissions of procedural steps during the execution of a task, independently of the human-machine system conditions. These events may indeed be described by "error rates", which are formalised in the model by a set of input data.

The theoretical stand for the generation of errors as a direct consequence of the human-machine interaction (*direct event*) follows the concept of "under-specifications", which consist in an incomplete representation of the world (Reason, 1990). Two types of under-specifications are accounted for in the model: under-specifications of cognition, which are the origin of human related faults; and under-specifications of environment, which are related to faults of the plant or information system. These two factors are sometimes also called "internal" and "external" factors affecting human behaviour (Hollnagel, 1991).

The architecture of the human-machine interaction generates the conditions for error making as a result of dynamic events, either by propagating automatically through the accident sequence the biases of human cognition (internal factors), or by producing the human error from inappropriate or unavailable information or environmental conditions (external factors).

In this context, human stress and mental-physical workload are very important factors, which are considered a major cause of human error and are generated by the occurrence of external and internal events during the human-machine interaction (Moray, 1979, 1988).

Indeed, stress and workload can derive from the deterioration of a plant situation in terms of critical physical parameters as well as from the establishment of unfavourable cognitive and social working conditions.

Therefore, according to the level of stress and workload reached during the accident, the probability of errors increases in some directly proportional measure.

In this case study, a function has been defined, called *stress-in-action*, which characterises the working environment of the operator (Cacciabue, Carpignano and Vivalda, 1993). The *stress-in-action* function, $F_{stress}(t)$, is dynamically evaluated during the human-machine simulation by an empirical correlation, and also affects the recovery, which is defined by a different correlation, $F_{recovery}(t)$.

The stress-in-action function, considered here, has been formulated on rather general factors, related to the level of workload, the time pressure and the quality of the information system. It is a dynamic function because it varies according to the operator physical and cognitive state, to the complexity of the control system and information processing flow between operator and plant control.

This function should, in principle, also be able to account for the condition of mental relaxation of operators due to boredom and lack of activity, derived from the apparently undemanding role of supervision assigned to humans in the control loop of highly automated plants.

The probability of erroneous actions and of recovery is thus a function of a generic "stress-in-action" correlation, which represents the effect of the internal-external worlds on the operator behaviour.

For the case study of the Chemical and Volume Control System, only external factors have been considered. In particular, it has been assumed that the operator is affected by the following five main factors:

1. The *time* of the sequence;
2. The *number of alarms* occurring during the accident;
3. The *number of actions to be performed* according to procedures;
4. The *behaviour of the pressure head* of the main feed loop of the CVCS; and
5. The *water level* in the VCT.

The first three are quite commonly encountered factors affecting system management.

The fourth and fifth factors are associated with the most critical physical quantities to be controlled during the management of the CVCS.

These quantities have been highlighted by task analysis and the quantification of their effect on stress and workload have been evaluated by experimental work and field examination.

Theoretical Model of Operator Behaviour and Numerical Algorithms

A simple model of operator behaviour, related to stress and workload, has been considered and it has been based upon the optimal procedures for managing the accident scenario.

Model of the Control Procedure

The procedure implemented by the operator in controlling the accident under study, is based on the Task Analysis previously performed.

The detailed sequence of control actions is shown in Figure 4.9. The actions related to each goal have been distributed in tasks. The tasks are performed at different time intervals in accordance to the physical evolution of the accident.

The actions included in each task have been arranged in sequence of execution and they carry a numerical tag which defines the order of importance within the task.

During the execution of the procedure the operator can make errors, as discussed during the analysis of the operator's control task.

- ***Task 1***

The *Safeguard of the CVCS* is performed during the first 2000 seconds of the accident, and is completed by the actions and verifications of "TASK 1".

During this first phase of the accident the operator may not be totally sure about the type accident, and, in order to protect the Volume Control Tank (VCT) and the regeneration, decontamination and deionisation system, the operator will start the make-up system and will isolate the ionic-exchange process. However, in order to avoid filling the VCT with non-purified water, a certain amount of let-down flow is re-directed to the Boron Recycle System (BRS).

This regulation of the led-down flow and the flow of make-up water have to be closely monitored with respect to the water in the VCT, as to maintain the level within normal values, i.e., 2 - 3 meters, and avoid either emptying the tank or reaching its maximum level of 4 meters.

Task 1 is then fulfilled by the following actions, to be carried out in sequence and tagged according to their relative importance:

a. *Regulation of the VCT level*[1]
   This first regulation is obtained by starting the make-up pumping system and by opening valve *V68*.

b. *Increase of flow to BRS*[2]
   This increase of flow to the boron recycle system is obtained by acting on the three ways valve *V25*.

c. *Bypass the Regeneration and Deionisation systems*[4]
   This third action is carried out by acting on the three ways valve *V15* and implies that all let-down flow is sent directly towards the VCT without passing through the ionic-exchange resins for regeneration.
   This action is the less important in the Task, as it does not really affect the safety of the system. It simply decreases the amount of work of regeneration, decontamination and deionisation and, thus, reduces the overall cost of performance.

d. *Conservation of VCT level*[3]
   This activity does not imply a direct action of the operator, but rather the perception and identification of inappropriate system performance, that may be reflected in a deviation of the VCT level from its normal level (~ 2.5 meters).

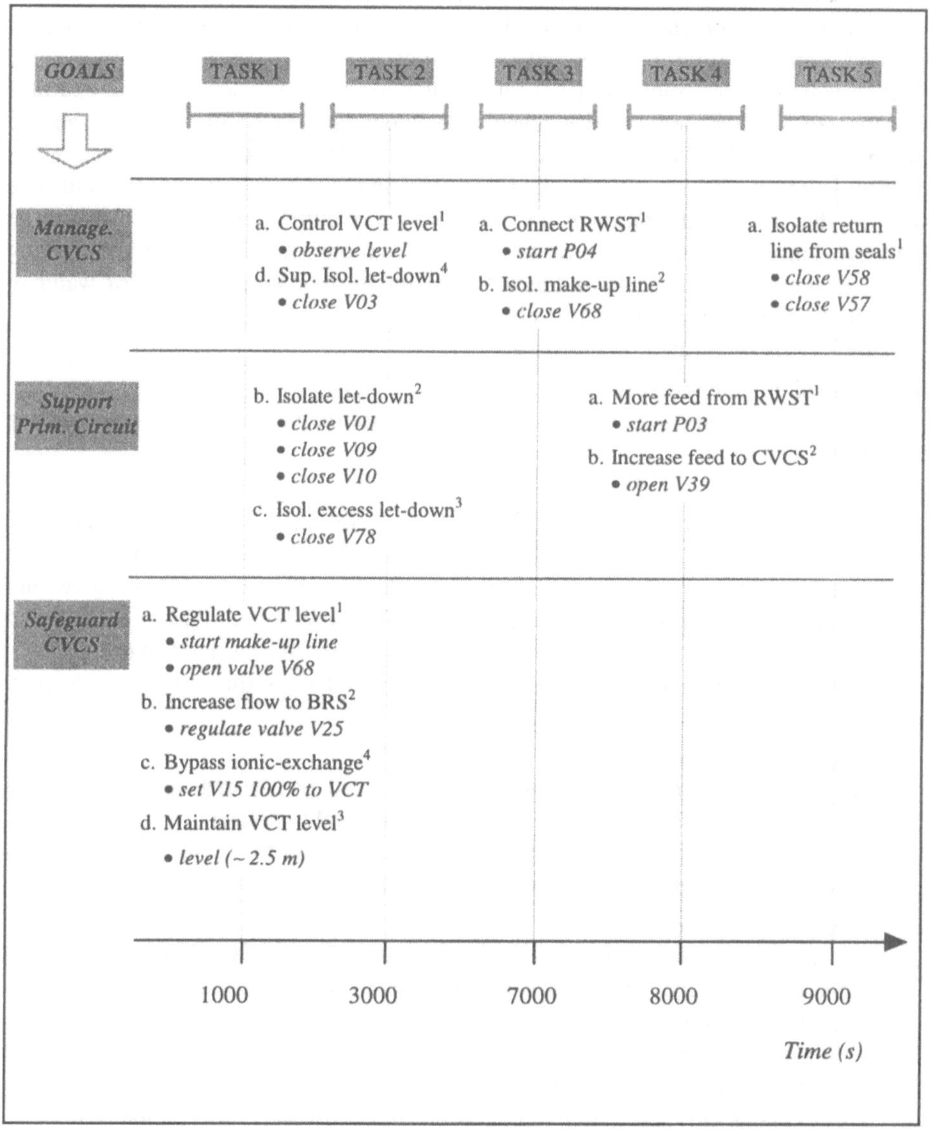

Figure 4.9. Operator Tasks and Goals in the Management of the CVCS

- ***Task 2***

Task 2 contains actions and verifications carried out at approximately 3000 seconds of accident development, aiming at fulfilling two goals, namely the *Management of the CVCS* and the *Support of the Primary Circuit.*

At this time of the transient, i.e., almost one hour inside the accident, the operator is expected to have developed a reasonable understanding of the situation and to begin to manage the CVCS system and to support the Primary Circuit.

The Primary Circuit is indeed, de-pressurising and cooling rapidly and this implies that the fluid contracts and reduces its volume. This effect has to be counterbalanced by spilling as less as possible from the Primary Circuit and feeding back as much as possible, as to attempt to maintain the overall volume of water in the Primary Circuit at a constant level.

This Task is performed by 4 main actions:

a. *Control of VCT level*[1]
   This activity pairs and continues the same operation carried out for safeguarding the CVCS.

b. *Isolation of Let-down*[2]
   The isolation of the Let-down line is obtained by three combined actions. The closure of the main valve V01 and the successive closure of V09 and V10.
   It is important to observe here that the order in which these three valves are closed is very important, as an inversion between V01 and V09 or V10 would cause the rapid and strong increase of pressure in the let-down line with consequent opening of the relief valve V08 and consequent loss of Primary Circuit water.

c. *Isolation of excess let-down*[3]
   The isolation of the excess let-down is obtained by closing valve V78. This action is not strictly necessary, as this line is normally isolated.
   However, this action is aimed at preventing its opening by the protection system, with the consequent loss of water from the Primary Circuit.

d. *Supplementary Isolation of the Let-down line*[4]
   The action considered here is a supplementary and, usually, redundant isolation of the let-down line, and is performed by closing the valve V03. This contributes to the reduction of the high pressure of the Primary Circuit water.
   This action is made redundant by the isolation carried out in support of the Primary Circuit, which is performed next in time, in this same Task (Figure 4.9). This explains why this action has been given the lowest priority in the task.
   However, in the case of inappropriate performance of the components for the isolation process, this action further ensures the successful execution of the isolation procedure.

- ***Task 3***

Task 3 is dedicated to the *Management of the CVCS* after approximately 2 hours of transient (7000 seconds), when the source of make-up water begins to be severely reduced and compensation is necessary.

The system that is utilised is the Refuelling Water Storage Tank (RWST), which contains a vast repository of water, specifically stored for supporting this kind of activities. The following actions are foreseen:

a. *Connection of the RWST*[1]
   In order to activate the feed flow form the Refuelling Water Storage Tank, the operator has to start pump PO4, as the system is in a stand-by condition with all remaining components in open and operating state which ensure maximum flow from the RWST.

b. *Isolation of the Make-up line*[2]
   The isolation of the make-up line is obtained by closing valve V68.

- ***Task 4***

Task 4 is carried out at 8000 seconds in reaction to the need to support as much as possible the loss of volume in the Primary Circuit.

This Task is then characterised by the goal of the *Supporting the Primary Circuit.*

To fulfil the task, the operator uses all the remaining means possible, and in particular:

a. *More feed from the RWST*[1]
   This extra feed is obtained by starting the stand-by pump P03. As in the case of pump P04, the other components on the line connected to P03 are in stand-by fully operative condition, and this ensures maximum flow as soon as the pump starts functioning.

b. *Increase feed-flow to CVCS*[2]
   In order to increase the water flow to the CVCS to its maximum, the operator opens fully valve V39. This valve controls the charge-flow to both the Primary Circuit and the seals of the primary pumps.

- ***Task 5***

Task 5 is the last task considered in this procedure and consists of the final control action aimed at the optimal *management of the CVCS.*

This task is carried out following the start up of the feed flow from the Refuelling Water Storage Tank in order to avoid the return of water in the VCT from the seals. This could well be the case, as the source of water is the RWST, which feeds directly the charge-line and all the system feeding water to the VCT have been isolated. In this case the water returning from the seals of the primary pumps could be fed back to the VCT through its outlet. The actions carried out to fulfil this task are:

a. *Isolate return line from seals*[1]
Close valves V57 and V58, which are located inside and outside the reactor building. The closure of any of these two valves fulfils the needs of this task.

Model of human errors

The model of error making and recovery has been developed by means of two distinct correlations, which take into consideration the five main contextual and environmental factors affecting workload and operator stress, identified by the Task Analysis.

The *stress-in-action* function, $F_{stress}(t)$ is mainly applied to evaluate the probability of error making. While, in the case of recovery, a slightly more complex formulation has been considered, called *recovery* function, $F_{recovery}(t)$.

Indeed, the *recovery* function depends on the *stress-in-action* function itself, as well as on the most important consequences of errors, e.g., the number of forgotten actions and the state of the plant following the intervention of protections and alarms.

These two functions will now be described in some details, also showing how their values and thresholds may be correlated to the probabilities of error making and recovery in dynamic conditions.

The Stress-in-action Function

The *stress-in-action* function is written as:

$$F_{stress}(t) = \frac{1}{5}\left[f(\tau) + f(actions) + f(alarms) + f(\Delta p_{seals}) + f(L_{VCT})\right] \quad (4.8)$$

Each of the elementary sub-function represents the contribution to the *stress-in-action* (function) of a specific external factor.

These five sub-functions have been developed as follows.

- $f(\tau)$ characterises the dependence of stress on time. It is assumed that stress increases as time passes:

$$f(\tau) = (1 - e^{\alpha\tau}) \quad (4.9)$$

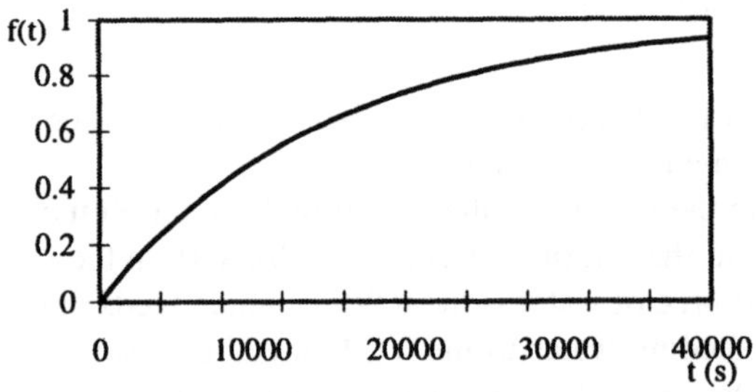

Where the constant $1/\alpha$ has been assigned the value of 15000 seconds, which implies that the stress function $f(\tau)$ reaches the value 1 after approximately 12 hours of accident management.

- ***f(actions)*** implies that the operator performs n_0 actions without any difficulty, with respect to a maximum number of n_{max} actions. If the number of actions being carried out (n_{ac}) is greater than n_0, there is a positive contribution to the stress function. While if the number of actions being carried out is smaller than n_0, the contribution is then negative, and represents a measure of boredom:

$$f(actions) = \left(\frac{n_{ac} - n_0}{n_{\max ac} - n_0} \right) \tag{4.10}$$

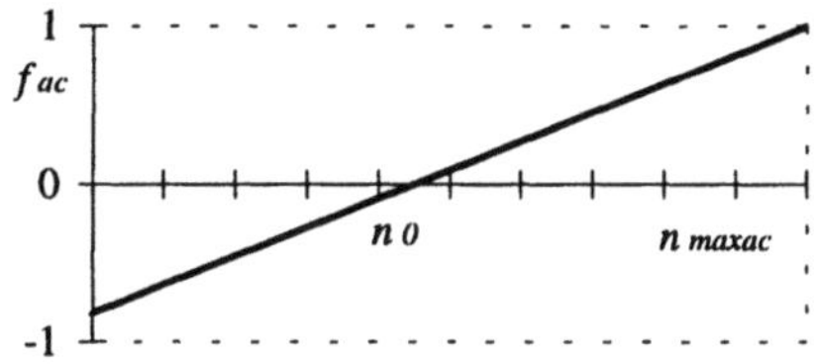

- ***f(alarms)*** aims at considering an increasing stress in the operator, as the number of active alarms increases. Thus a simple linear correlation has been considered:

$$f(alarms) = \left(\frac{n_{al}}{n_{\max al}} \right) \tag{4.11}$$

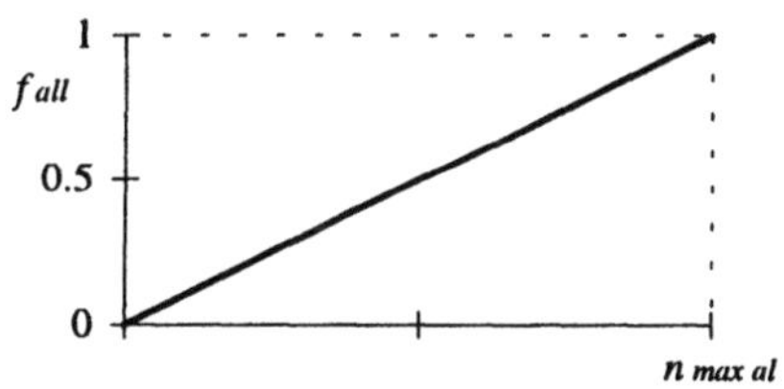

- ***f(Δp_{seals})*** and ***f(L_{VCT})*** depend on the pressure and water level in the steam generator. These are very important physical quantities that give a measure of the correct functioning of the system. They have thus been selected for inclusion in the stress function. In particular, their trend, or rate of variation over time, is one of the most important significant characteristics that allow the operator to develop a "feeling" of how well the system is operating. High trends, either in the positive or in the negative direction, are symptoms of possible problems and thus should be avoided as much as possible.

The function *f(Δp_{seals})* has been assigned the following equation:

$$f(\Delta p_{seals}) = \frac{\left|\frac{d\Delta p_{seals}}{dt}\right|}{\frac{d\Delta p_{seals}}{dt}} \left[1 - \exp\left(- \frac{\left|\frac{d\Delta p_{seals}}{dt}\right|}{\left.\frac{d\Delta p_{seals}}{dt}\right|_{t=0}} \right) B_1 \right] \tag{4.12}$$

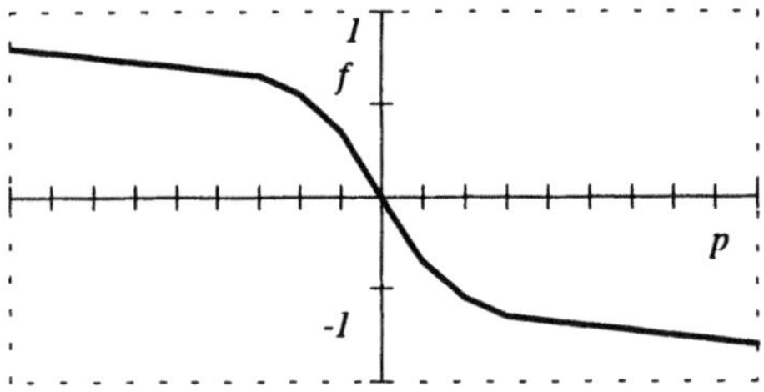

The function $f(L_{VCT})$ has been assigned the following equation:

$$f(L_{VCT}) = \left[1 - 2\exp\left(- \frac{\left|\frac{dL_{VCT}}{dt}\right|}{\left.\frac{dL_{VCT}}{dt}\right|_{t=0}} \right) B_2 \right] \tag{4.13}$$

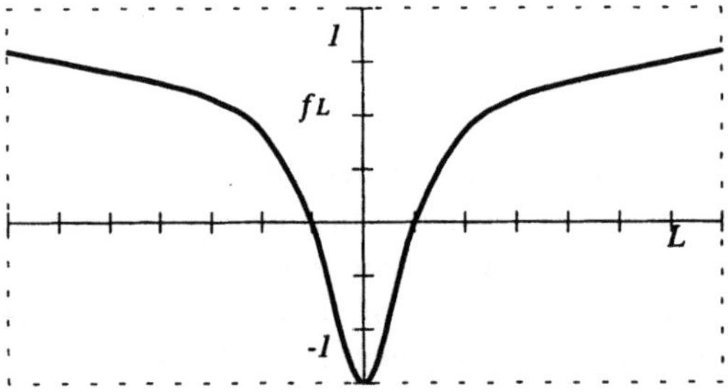

In Eqs. (4.9) - (4.13) the following meanings are associated with the parameters:
τ = time since beginning of the accident;

α, B_1 and B_2 = empirically evaluated parameters;
L_{VCT} = level in the Volume Control Tank;
n_{al}, n_{maxal} = number of triggered alarms, maximum number of alarms in the VCT;
n_{ac}, n_{maxac} = number of actions to be performed, maximum number of actions;
Δp_{seals} = pressure difference between pump *seals* and *reactor.*

The Recovery Function

The function for modelling recovery has been considered somehow dependent of the stress level and the number of *protection sub-systems* that intervene during the accident evolution.

The activation of a protection mechanism is a clear symptom of some inappropriate management of the system and thus it alerts the operator. It represents a sort of notification to revise previously performed actions, and offers a further possibility of recovering from errors.

On the other hand, a recovery process can take place only if the stress level is not too high and allows sufficient attention to be allocated to the mental process of reviewing past actions and decisions. Starting from these hypotheses, the following recovery function may be derived:

$$F_{recovery}(t) = f(stress,\ protections) = \frac{c_1 R_s + c_2 R_p}{c_1 + c_2} \tag{4.14}$$

where the following definition of parameters and correlations apply:

$$\begin{cases} R_s = 1 - c_3 \left| F_{stress}(t) \right| & \text{if} \quad \left| F_{stress}(t) \right| < 1/c_3 \\ R_s = 0 & \text{if} \quad \left| F_{stress}(t) \right| \geq 1/c_3 \end{cases} \tag{4.15}$$

$$\begin{cases} R_p = c_4 n_p & \text{if} \quad n_p < n_{p0} \\ R_p = 1 & \text{if} \quad n_p \geq n_{p0} \end{cases} \tag{4.16}$$

and

c_1, c_2, c_3, and c_4 = empirically evaluated parameters;
R_p, R_s = recovery level due to *protections,* and to *stress*;
n_p, n_{p0} = number of *triggered protections*, and *maximum* number of perceived protections.

The model presented is very simple, but is able to show how a dynamic methodology may work.

In general, for a real case study, the structure of the stress-in-action function will not only be strictly related to the operator reaction but also to the plant design: the control room design, the hazard related to the different processes, the critical physical variables. This function, as it includes cognitive and hardware information, and strongly affects the human behaviour model, needs to be very accurately developed in combination with the Task Analysis and empirical observations of the working environment.

Programming Language and Environment

The model of the operator described in the previous sections has been developed in such a way that a simple programming language and environment could be applied.

The algorithms and the simplifying assumptions have created the conditions for the development of the model by standard programming languages like C or even FORTRAN.

As the latter is the language that has been adopted for the development of the simulation of the CVCS, the use of FORTRAN for the implementation of the operator model ensures maximum compatibility between the two simulations.

An immediate consequence of the use of these simple languages for the simulations, is that the resulting System Simulation Architecture is quite portable on the most commonly utilised software operating environment, and computer types.

The only requirement is to have the appropriate FORTRAN or C compiler.

4.4.4 Interaction Model - Case Study CVCS

As already mentioned, the Interaction Model technique adopted in this sample case study, is the DYLAM methodology.

In order to develop the specific data and models for DYLAM, the results of the Task Analysis and of the Failure Mode and Effect Analysis have to be exploited in order to define the most important components and human activities affecting the CVCS performance.

Since the correct functioning of the CVCS is mainly related to the water level in the Volume Control Tank (VCT) and to the primary pumps seal pressure, a further qualitative analysis was performed to identify a limited number of components and operator activities whose faults may affect the behaviour of the VCT and/or the management of the CVCS system.

The objectives of this analysis were to define, according to the DYLAM requirements, the complete spectrum of *possible states,* the *causes of transitions* between states, and the *physical and logical behaviour* of "components".

The discussions on the application of Interaction Model to components and to human behaviour are performed separately, as they play a very important role on the accident evolution and on the type of interaction that are studied for safety purposes.

Interaction Model and Machine Components

The choice of relevant hardware components has been performed by checking the role assumed by every component on the management of the VCT level and selecting the most remarkable and significative ones for their functions.

The set of components obtained in this way has been further reduced enhancing those connected in series to the same pipeline and then analysing the most critical configurations, usually those operated by the control system. This has led to the selection of some components to be analysed in their possible failure modes, while all the others are always kept in nominal conditions in all sequences analysed.

A first set of five components are selected because of their importance in the hydraulic balance of the system. These are: the control and isolation valve V01, the three way valve V25, the regulation valves V68, and V39, and the motor-operated control valve V73 (Cacciabue, Carpignano, and Vivalda, 1992).

The control and isolation valve V01 can be operated for isolating the let-down line in case of need and for controlling the water flow from the Primary Circuit.

- This valve has been assumed to be in two possible states: *nominal state*, i.e., working according to design, or *failed state*, in a stuck open position.

The three way valve V25 is driven by the automatic control to maintain a correct level in the VCT, diverting a fraction of the letdown flow to the Boron Regeneration System (BRS); in the case of very high level of the VCT, the valve V25 is operated by the protection system to isolate the tank.

- This valve may be in *nominal state* or in *failed state*, stuck at a certain fraction of opening.

The valve V68 is driven by the automatic control and regulates the make-up flow. It ensures, together with V25, the preservation of the correct VCT water level.

- This valve is modelled in three possible states: *nominal state*, *failed state closed* and *failed state stuck* in a certain position.

The regulation valve V39 is driven by the automatic control to set the flow rate in the charge line according to the Pressurizer level.

- This valve is modelled in three possible states: *nominal state*, *failed state open* and *failed state stuck* in a certain position.

The motor operated control valve V73, together with its twin valve V72, has the role of connecting the CVCS to the Refuelling Water Storage Tank. We recall that the RWST is a source of water that may be needed in certain cases to make up the feed for the Primary Circuit.

- This valve is modelled two states: *nominal state* and *failed state*, stuck in closed position.

For the reliability analysis of the CVCS, the probabilities associated with each state of these five components, and the transition rates relative to random event and logical/physical conditions, have been identified. These data, not reported here, have been obtained from existing databases or by means of expert judgement.

Interaction Model and Human Behaviour

As in the case of the hardware components, the human behaviour has also been modelled, with reference to performance "states", in accordance with DYLAM requirements.

Firstly, certain erroneous and recovery behaviours have been identified and associated with a number of possible "states" of the operator. This satisfies the first DYLAM requirement of identification of *possible states* of components.

Then, the *causes of transitions* between states, already recognised in the *stress-in-action* function, $F_{stress}(t)$, and the *recovery function*, $F_{recovery}(t)$, have been associated with the transitions between different states of behaviour.

Finally, the probabilities of the transitions between states have been assigned.

Error States and Transition Probabilities

The erroneous behaviour of the operator has been schematised by the DYLAM variable *ERR*.

It has been assumed that the variable *ERR* is associated with the number of actions that the operator has to carry out during the various phases of the accident, in accordance with the control and management procedure, described by the Task Analysis. The following possible operating "states" for *ERR* are envisaged:

- $ERR=0 \Rightarrow$ Nominal state. The operator performs all activities according to the Tasks of the procedure.
- $ERR=1 \Rightarrow$ Failed State 1. The operator does not perform *one* action for each Task carried out. It is assumed that the less important action is omitted.
- $ERR=2 \Rightarrow$ Failed State 2. The operator does not perform *two* actions for each Task being carried out. In particular, the two less important actions are omitted (see § 4.4.3, Figure 4.9).

We have limited our analysis to the state *ERR=2*, but this simple error mode could be extended to the omission of more actions in each Task of the procedure.

The association of the states of the variable *ERR* with the *stress-in-action* function has been made according to the following rule:

- threshold values of the *stress-in-action* function, $F_{stress}(t)$, have been identified, for which transitions between states of *ERR* may occur:

  Threshold 1 $F_{stress}(t) = |0.08|$

  Threshold 2 $F_{stress}(t) = |0.16|$

This means that every time that the value of $F_{stress}(t)$ crosses one of the thresholds during the dynamic evolution of the accident, the Interaction Model (DYLAM) considers the possibility of a transition between error modes of the operator, i.e., a change in the value of the variable *ERR*.

The *initial state probabilities* and *conditional transition matrices* between states of the variable *ERR* due to dependence on *stress-in-action* function are introduced as INPUT to the model, in agreement with the DYLAM requirements, and are shown in Tables 4.2 and 4.3.

It is important to note here that the transition matrix, as already mentioned, is quite complex to develop, as it must contain all probabilities of change from one state to the other, every time a threshold is exceeded.

Moreover, it must be remembered that the probabilities of state change have to be assigned according to whether the threshold of the reference variable is reached increasing from lower values, or decreasing from higher values.

In Table 4.3 this difference is enhanced by having separate columns for thresholds of the reference variables reached either "increasing from lower values" (↑) or "decreasing" from upper values (↓).

Table 4.2 Initial state probabilities for variable *ERR* vs. F_{stress}.

F_{stress} Thresholds / *ERR* Initial State	$F_{stress} < \lvert .08 \rvert$	$\lvert .08 \rvert \le F_{stress} \le \lvert .16 \rvert$	$\lvert .16 \rvert < F_{stress}$
0	0.9	0.2	0.03
1	0.07	0.6	0.07
2	0.03	0.2	0.9

Table 4.3 Transition Matrix for variable *ERR*: Probabilities of state transition vs. F_{stress}.

F_{stress} Thresholds / *ERR* State Change		$\lvert 0.08 \rvert$↑	$\lvert 0.08 \rvert$↓	$\lvert 0.16 \rvert$↑	$\lvert 0.16 \rvert$↓
0	1	0.6	0.05	0.35	0.75
0	2	0.35	0.05	0.6	0.2
0	0	$1-\Sigma p_{1i} = .05$	$1-\Sigma p_{0i} = .9$	$1-\Sigma p_{0i} = .05$	$1-\Sigma p_{0i} = .05$
1	0	0.2	0.6	0.05	0.05
1	2	0.1	0.05	0.75	0.65
1	1	$1-\Sigma p_{0i} = .7$	$1-\Sigma p_{1i} = .35$	$1-\Sigma p_{1i} = .2$	$1-\Sigma p_{1i} = .3$
2	0	0.2	0.65	0.02	0.1
2	1	0.75	0.3	0.08	0.8
2	2	$1-\Sigma p_{2i} = .05$	$1-\Sigma p_{2i} = .05$	$1-\Sigma p_{2i} = .9$	$1-\Sigma p_{2i} = .1$

Recovery States and Transition Probabilities

Following the same process applied for the development of the operator error variable *ERR*, the definition of a variable for recovery, called *REC*, allows the assignment of "recovery states".

In particular, it requires the identification of functional dependencies of the variable *REC* from the recovery function, $F_{recovery}(t)$, and the assignment of the initial state probabilities and transition matrices.

A very simple hypothesis has been made: it has been assumed that the variable *REC* is associated with either the possibility to recover or not from previous errors.

The following operating "states" for *REC* are envisaged:

- *REC=0* ⇒ The operator does not recover from errors.
- *REC=1* ⇒ The operator recovers all omitted actions within the task being carried out (§ 4.4.3, Figure 4.9).

The association of the states of the variable *REC* with the *recovery* function $F_{recovery}(t)$, has been made, as in the case of the variable *ERR*, according to the following rule:

- a number of threshold values of the *recovery* function have been identified, for which transitions between states of *recovery* may occur:

Threshold 1 $F_{recovery}(t) = 0.3$

Threshold 2 $F_{recovery}(t) = 0.5$

The *initial state probabilities* and *conditional transition matrices,* introduced as INPUT to the model, are shown in Tables 4.4 and 4.5. For the transition probabilities, similar considerations apply to those developed in the case of the error function *ERR*.

Table 4.4 Initial state probabilities for variable *REC* vs. $F_{recovery}$.

$F_{recovery}$ Thresholds	$F_{recovery} < 0.3$	$0.3 \le F_{recovery} < .5$	$0.5 \le F_{recovery}$
ERR Initial State			
0	0.9	0.5	0.1
1	0.1	0.5	0.9

Table 4.5 Transition Matrix for variable *REC*: Probabilities of state transition vs. $F_{recovery}$.

$F_{recovery}$ Thresholds		0.3 ↑	0.3 ↓	0.5 ↑	0.5 ↓
REC State Change					
0	0	0.7	0.9	0.1	0.7
0	1	0.3	0.1	0.9	0.3
1	0	0.9	0.9	0.1	0.3
1	1	0.1	0.1	0.9	0.7

4.4.5 Results of Case Study CVCS

In order to introduce the analysis of the Chemical and Volume Control System within the entire architecture of Human-Machine Interaction, based on the models described in the previous sections, we will start by recalling the boundary conditions and the main objective of the study.

The CVCS is one of the most relevant systems expected to support reactor performance during an accident condition, following shut-down. The system is particularly complex as it controls more than one process, namely coolant volume, boron concentration and pump seals on the Primary Circuit, and demands a number of diversified interventions by the operator, both at manual and planning level.

During the accident, CVCS operations have to be performed for several hours, and this adds to the overall workload on the operator.

The aim of the study is mainly to evaluate, by means of the simulation of Human-Machine Interaction, the effects of component failures, human errors and environment conditions on the safety of the nuclear system.

In this way we expect to show the potential scope of analysis of a simulation environment, such as the one developed.

The basic assumptions are that:

- As a major accident has occurred, the reactor has been shut-down by the protection system.
- The electrical power, necessary for the operation of the CVCS components, is ensured, either by the grid or by the Diesel backup systems.
- The nuclear core of the reactor is generating a decay power and exhibits a physical behaviour which is known and described by a simple time dependent correlation.
- All components, except those for which specific failure types and modes have been envisaged by the Interaction Model, behave according to design specifications.
- The CVCS is expected to operate according to procedures for approximately 4 hours.

In the following we will discuss two specific situations:

1. The *Nominal Case*, in which all components and the operator behave according to expectations, i.e., in nominal condition. This first case has been studied with a dual objective:
    a. To compare the response of the system with the operator management activity vs. the case with no operator interventions.
    b. To show the overall human-machine behaviour during the accident evolution, including physical variables, i.e., pressures, temperatures, mass flows, etc., and operator characteristics, i.e., mainly stress and workload.
2. The *Integral Case*, in which hardware components are allowed to fail, either at time 0 or during the accident, and the operator can make errors and recovery according to the modelling characteristics described earlier. In this way it will be possible to analyse all combinations of errors, recoveries and failures.

Nominal Case of CVCS Management

One of the "design basis" accident conditions of the CVCS demand that the safety and protection measures ensure system operability even in absence of operator intervention.

However, the presence in the control loop of an operator that follows the procedures with no errors prevents the start-up of protection systems and contributes strongly to the mitigation of the accident. This does not stop the CVCS reaching alarm thresholds, but hinders the evolution of the reactor towards more severe and critical configurations.

The accident simulation was firstly carried out with no operator intervention (*no-operator control*) and then with operator intervention (*operator control*), in agreement with the control procedure and free of error. The results of these two cases are shown in Figures 4.10 - 4.14.

These figures contain the time dependent behaviour of the most important physical quantities (*mass flows, pressures,* and *VCT variables*) and human related variables (*stress function*) for the two cases and allow an immediate graphical and logical comparison.

CVCS Mass Flows

Figure 4.10 shows the mass flows in the *let-down line,* the *charge line* and *seal line* for both cases, i.e., with and without *operator control.*

- *let-down line mass flows*

In the case of *operator control*, the mass flow in the *let-down line* (Q_{ld-o}), i.e., the mass flow that is drawn out from the Primary Circuit for purification (Figure 4.7), decreases gradually up to 3200 seconds. The initial reduction of let-down flow derives from the progressive reduction of the pressure in the Primary Circuit that follows reactor shut-down. Up to 3200 seconds, the automatic system controls the let-down flow.

At that time the operator isolates the line in order to limit the rapid de-pressurisation and volume decrease in the Primary Circuit that follows the reactor shut-down, and, also, to avoid draining-out the Volume Control Tank. This action is carried out to support the Primary Circuit and to manage the CVCS (Figure 4.9).

In the case of *no-operator control,* as no action is carried out on the *led-down line* before 3200 seconds, no differences appear on the mass flows (Q_{ld}) before that time. However, after this time, as there is no human control, the let-down flow from the Primary Circuit does not stop (as in the case of *operator control*) but continues until 5000 seconds, when the protection system intervenes on a signal of low-level in the Pressuriser and isolates automatically the let-down line.

The Pressuriser is the major component in the Primary Circuit, located between the reactor vessel and the steam generators, and has the specific role of maintaining the pressure in the Primary Circuit, compensating for small changes in the water volume and density.

Consequently, in the case of *no-operator control,* some primary water is unduly spilled and a greater risk of draining-out the VCT occurs.

- *charge line mass flows*

In the case of *operator control*, the mass flow in the *charge line* ($Q_{ch\text{-}o}$), i.e., the mass flow that is sent back to the Primary Circuit (Figure 4.7), shows an initial increase due to the automatic regulation system that fully opens valve V39 in order to compensate for the volume decrease in the Primary Circuit. This generates an unbalance between let-down and charge flows in favour of the charge flow.

This automatic regulation stops at 1200 seconds, when the first *operator control* action takes place and the system is automatically repositioned to its nominal configuration. In particular, valve V39 returns to a partially closed position and this reduces strongly the charge flow.

The position of V39 is not changed by the operator for quite some time. However, the charge mass flow increases because of the progressive decrease of the pressure in the Primary Circuit.

At 7000 seconds, i.e., approximately two hours into the accident, the operator connects the Refuelling Water Storage Tank (RWST) to the charge line, by starting pump P04, to compensate for the depleted supply of make-up water.

This causes a small decrease of charge flow, which is recovered by the operator at 8200 seconds, with the start of the second pump P03, which doubles the feed flow from the RWST.

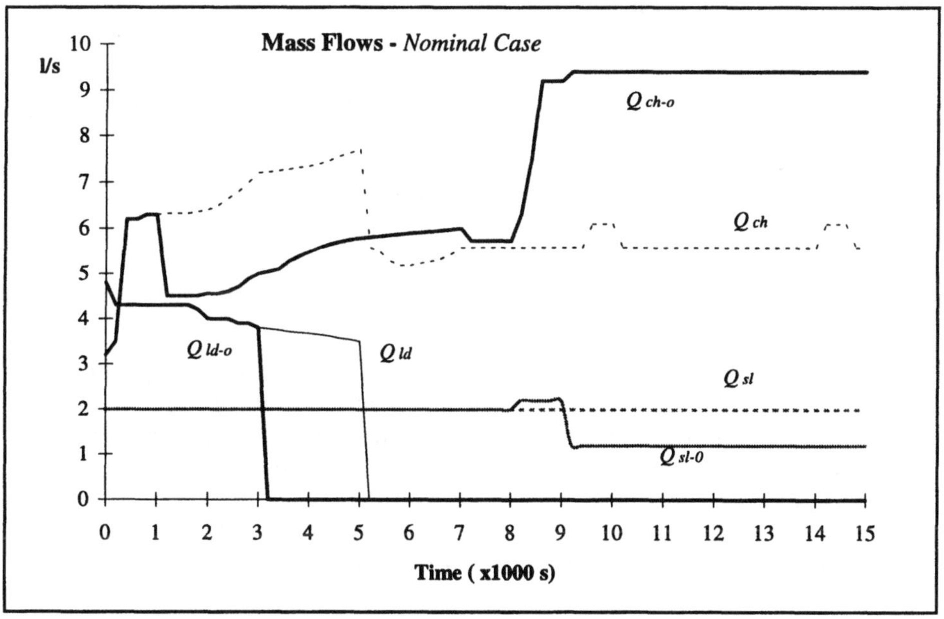

Figure 4.10. CVCS - Mass flows - Nominal Case

Later on, at 9200 seconds, the operator fully opens V39, further increasing the charge flow to its maximum. Then the mass flow $Q_{ch\text{-}o}$ remains constant for the rest of the accident evolution.

With *no-operator control*, the automatic regulation acts until 5000 seconds, increasing constantly the charge mass flow (Q_{ch}) and contributing in this way to the compensation for the volume decrease in the Primary Circuit.

However, at 5000 seconds, the activation of the protection system, resets all valves at their nominal initial state, partially shutting also V39 and reducing drastically the charge flow.

The rest of the time dependent behaviour of the charge flow is maintained almost constant, except from two small diversions, at around 9500 and 14000 seconds, due, as we will see in the following, to a repeated automatic switch (flip-flop) between the VCT and RWST, as the source of feed-water for the charge line.

In this way, the sequence with *no-operator control*, although initially quite successful in maintaining the charge flow at high level, is not able to preserve a fully satisfactory condition, like the human controlled sequence.

- *seal line mass flows*

The mass flow in the *seal line* is the mass flow that is sent to the seals of the primary pumps to safeguard from leaks from the Primary Circuit.

The *seal line* mass flows, in the case of *operator control* ($Q_{sl\text{-}0}$) and *no-operator control* (Q_{sl}), show quite substantial similarities.

In the case of *operator control*, when the RWST is connected to the feed system at 8000 seconds there is a small increase of the seal flow.

Later on at 9000 seconds, when the operator isolates the return line from the seals, according to procedure, the seal flow is then further the reduced. The seal flow, which must always be maintained in order preserve the isolation of the Primary Circuit pumps, is then fixed to a constant level represented by the water that can be sent to the labyrinths and pressure reducers.

In the case of *no-operator control*, as there is no operator intervention, the feed flow to the seals remains practically constant all throughout the sequence.

<u>Pressures</u>

The time dependent response of the pressure in the Seals and Primary Circuit are shown in Figure 4.11, for both cases of *operator control* and *no-operator control*.

The importance of these pressures for the control of the CVCS lies in the fact that the *seals pressure* ensures the isolation of the primary circuit, and thus safeguards from the leakage of highly radioactive primary water.

It is thus very important that the pressure in the seals of the primary coolant circulation pumping system always remains above the pressure in the Primary Circuit.

The importance of this function of the CVCS is further enhanced by the inclusion of the seals pressure amongst the variables affecting the stress function of the operator (Eq. 4.8).

The Primary Circuit pressure (p_{pr}), in reality has two slightly different behaviours in the two cases. But, as the differences are minor and do not affect the overall safety considerations of this study, we have only shown here the p_{pr} for the case *operator control.* In both cases, the Primary Circuit pressure remains below that of the seal pressure and thus within safety margins.

In the case of *operator control,* however, at 1200 seconds the automatic alignment of the system configuration with the reset of all components to their nominal initial state, causes a steep reduction of the $p_{sl\text{-}o}$.

While with *no-operator control* (p_{sl}) this adjustment occurs only at 5000 seconds, following the intervention of the protection system, as mentioned earlier.

At 7000 seconds and 8000 seconds, with the activation by the operator of the RWST and the start-up of both pumps P04 and P03, the pressure in the *seal line* settles at quite high values, with a very good safety margin.

The final increase in pressure $p_{sl\text{-}o}$. occurs when the operator isolates the return line from the seals, at 9000 seconds.

In the case of *no-operator control,* the seals pressure remains constant after 7000 seconds, except for the two deviations, at around 9500 and 14000 seconds, as occurs to the charge flow, due to the repeated switching between the VCT and the RWST as main source of feed-water for the charge line (see next sub-section).

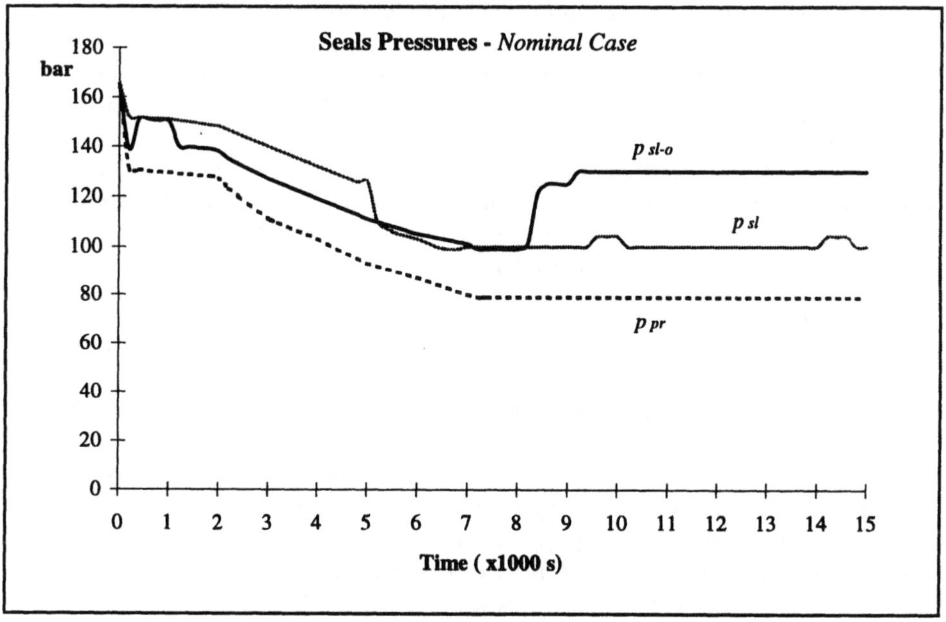

Figure 4.11. CVCS - Seals Pressures - Nominal Case

Volume Control Tank

The physical behaviour of the Volume Control Tank (VCT) is very important for the correct functioning of the whole CVCS system. Consequently, it is necessary that its critical variables are kept under control at all times of the transient.

In particular, the *outlet mass flow* (Q_{ovct}), the *make-up mass flow* (Q_{mkup}) (Figure 4.7) and the *level* (L_{vct}) give a clear representation of the behaviour of the VCT, for both cases of *operator control* (Figure 4.12) and *no-operator control* (Figure 4.13).

- *operator control*

The *level* is quite unstable during the first 1200 seconds as the automatic system acts on the regulation valve V68 in the attempt to maintain a constant water level in the VCT and generates large variations of *make-up flow*.

Subsequently, in the case of *operator control,* the operator intervention (Task 1) determines the progressive opening of V68 and the level increases for a while (Figure 4.12). However, after 2000 seconds the level starts decreasing because the *charge flow* is increasing, to compensate for the reduction of Primary Circuit volume, and the *let-down* mass flow decreases at first and is completely stopped at 3200 seconds by the intervention of the operator in Task 2 (Figure 4.10).

This effect is also shown by the behaviour of the *outlet mass flow* (Q_{ovct}), which increases for the first 1200 seconds under the actions of the automatic control, and then is reduced by the operator intervention at 1200 seconds. After 2000 seconds Q_{ovc} starts increasing, and this is reflected in the decrease of the VCT *level.*

At 7000 seconds, the operator isolates the make-up circuit, by shutting V68, and starts the RWST, as the source of make-up water is practically drained out (Task 3). The *make-up flow* drops to zero ($Q_{mkup} = 0$) and the VCT experiences an in-flow of water returning in the tank from the seals ($Q_{ovct} < 0$).

These two effects contribute to the increase of the VCT *level* until 9000 seconds, when the operator also isolates the return line from the seals. From this time on the VCT *level* remains constant, as there is no flow in or out of the tank. The charge flow is completely ensured by the RWST.

- *no-operator control*

As can be observed in Figure 4.13, the accident evolution is quite different in the case of *no-operator control.*

The dynamic behaviour of variables can be split in two well defined phases: before and after 5000 seconds.

Before 5000 seconds, the automatic control system tries, quite successfully, to maintain constant the VCT level by opening and closing the make-up line (V68). The VCT outlet mass flows (Q_{ovct}) increases constantly as the Primary Circuit pressure decreases and there is no reset of components to nominal states, as in the case of operator control at 1200 seconds with the partial closure of V39.

After 5000 seconds, the intervention of a protection system commences by starting the pumping system connected to the RWST and by closing the connection of the VCT with the make-up line ($Q_{mkup} = 0$).

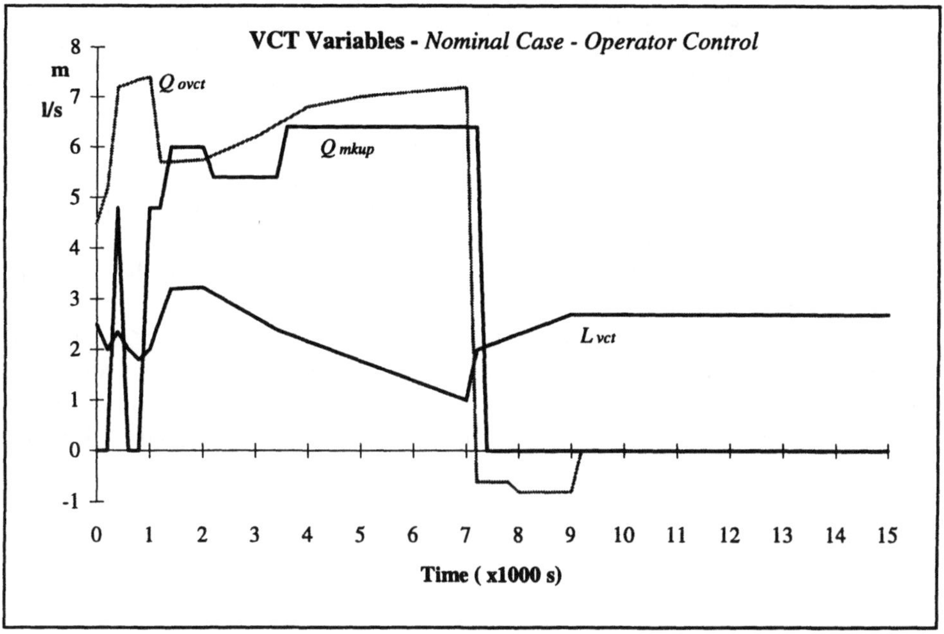

Figure 4.12. VCT Variables - *Operator Control* - Nominal Case

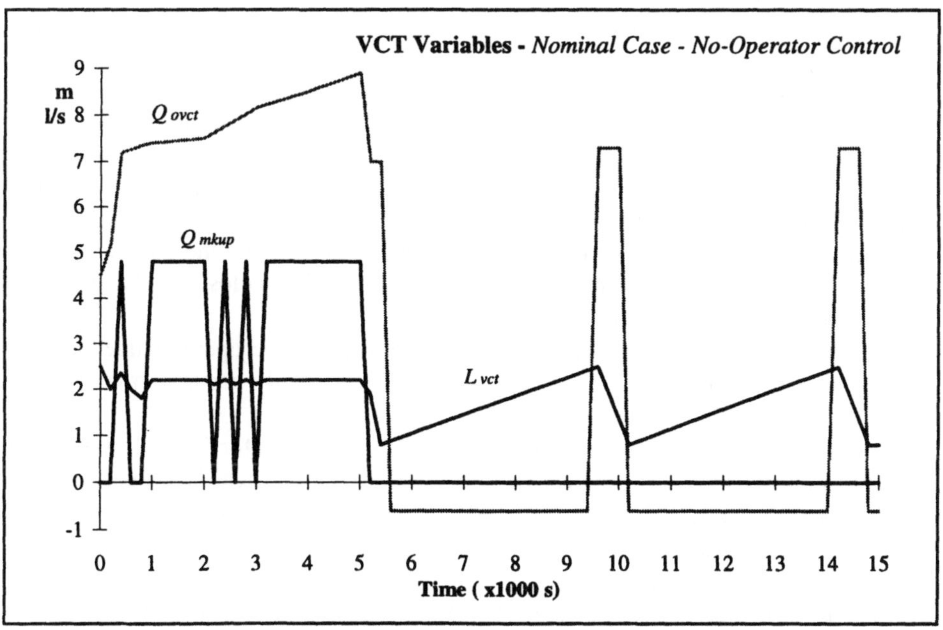

Figure 4.13. VCT Variables - *No-Operator Control* - Nominal Case

Similarly to the case of operator control, the outlet mass flows becomes negative, as the water fed into the system from the RWST returns from the pump *seals* and re-enters the VCT. The CVCS is controlled by the protection system, which is regulated mainly in accordance with the *level* of the water in the VCT.

This level increases as soon as there is no outflow from the VCT and some water returns from the seals. However, as soon as the *level* reaches the upper-threshold level of 4 meters, the protection system disconnects the RWST and re-installs the normal feeding to the charge line and seal line from the VCT.

This situation is encountered twice in the transient under study, namely, at around 9500 and 14000 seconds.

As the charge line is connected to the VCT, the outlet mass flow (Q_{ovct}) again becomes positive and quite sizeable, because the protection system opens all valves at 100%. This, in turn, causes a rapid decrease of the level to below the minimum threshold value (0.5 meters). The source of feed-water is then switched back to the RWST. The cycle then restarts until the upper-threshold value is reached again.

This situation does not occur in the case of the *operator control*, as the actions carried out by the operator are irreversible.

<u>Operator Stress</u>

The behaviour of the operator has been assumed to be free of errors for this nominal sequence.

However, the effect of the environment and plant can be measured in order to determine the level of stress of the operator, when all components respond according to design specifications.

The rate of change of the stress function versus time, calculated by Eqs. (4.8) - (4.13), is shown in Figure 4.14.

Initially the *stress* is negative, as there is a little effect due to time constraint (Eq. 4.9), the physical variables are still within normal boundaries, no alarms are active and the number of actions to be performed by the operator is small.

Thereafter, the stress function oscillates between positive and negative values until 7000 seconds, when the high number of active alarms and the operator activity become quite substantial. The operator has to manually change the source of feed-water from the VCT to the RWST and to isolate the VCT, by closing valve V68 and shutting down the make-up flow.

At around 10000 seconds, when the procedure reduces the operator intervention, the stress level drops quite substantially and becomes slightly negative, measuring a small level of boredom.

One interesting aspect of the stress condition of the operator is the low level of stress (boredom) shown at time zero and at 3000 seconds, when the values of F_{stress} fall below the minimum threshold (- 0.08) assigned by the Interaction Model for the generation of possible error conditions. In this case the error would be caused by lack of activity and distraction.

Similarly, the high workload generated by the intensive interaction with the plant control system from 7000 to 10000 seconds (2 - 2.5 hours inside the accident evolution) generates the pre-conditions for error making. Both stress thresholds (0.08 and 0.16) are reached in this period of the accident.

In this simulation it is also relevant the fact that we have considered all components in nominal behaviour. This implies that only a limited number of alarms are activated and that no extra workload is produced for the operator.

Only two alarms start before 2000 seconds, namely: the alarm of *low level* in the *charge-line*, (at t = *200 s*) due to the low pressure in the Primary Circuit; and the alarm of *high level* in the VCT (at $t \approx 1700$ *s*), which is immediately compensated by the operator control actions.

Later on in the accident, four more alarms appear: *low temperature* in the *charge line*; *low mass flow* in the *let-down line; high temperature* upstream of the regeneration and deionisation system; and *high temperature* downstream of one of the regeneration heat exchanger.

In summary, it results that, in some periods of the accident, the possibility of human error exists. It is also obvious that, in more complex conditions, as we will see in the following, the probability of error making would increase both in number and frequency.

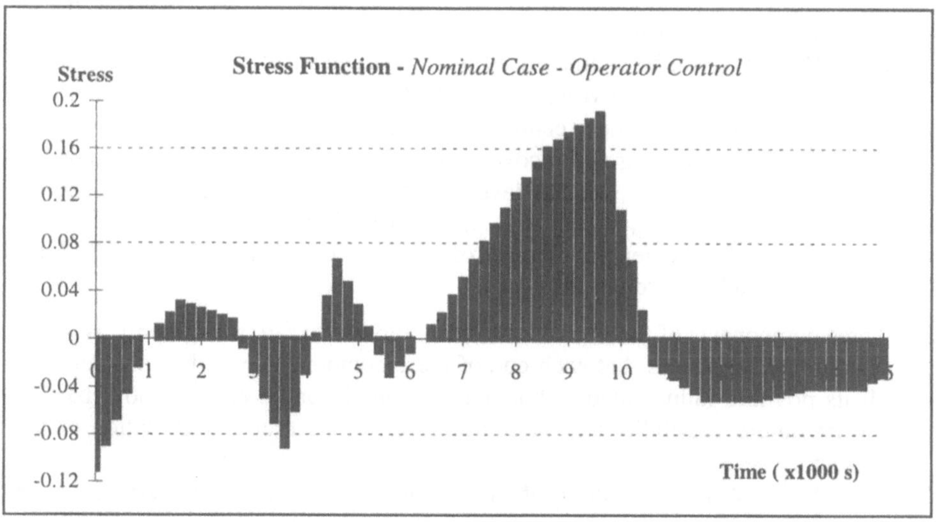

Figure 4.14. Operator Stress vs. Time. *Operator Control* - Nominal Case

Integral Case of CVCS Management

In the case when operator error and recovery are coupled with failures of hardware components, the situation can rapidly become very complex, as the combinations of possible errors, recoveries and failures that should be, in principle, analysed by the human-machine simulation are enormous.

In practice, the best way to reduce such complexity, is to identify and group together those components that may fail in combinations, and to specify selected erroneous human behaviours. In this way, the ability and experience of the analyst becomes very important, in defining crucial configurations and behaviours, and ultimately, saves a lot of computational and analytical work.

In this sample case, we have chosen the following combination of human behaviours and component failures:

1. The operator is represented according to the error modes and recovery that have been described previously, namely:
   a. Nominal behaviour and two error modes will be considered by means of the parameter *ERR*. Performance according to the control procedure will be related to *ERR=0*. Omission of one step in the performance of a task within the procedure will be identified by *ERR=1*; and omission of two steps in a task will be associated with *ERR=2*.
   b. Recovery will be characterised by the parameter *REC*. The value *REC=0* will imply no-recovery, while *REC=1* will activate recovery of all omissions during the execution of the procedure.
   c. Both error making and recovery depend on the environmental conditions and, in particular, on the level of stress. The two analytical functions applied are $F_{stress}(t)$ and $F_{recovery}(t)$. The threshold values of F_{stress} and $F_{recovery}$ for the generation of error and for recovery are: 0.08 and 0.16 for F_{stress} and 0.3 and 0.5 for $F_{recovery}$.

2. Three components of the CVCS have been allowed to fail, though only as initial condition. This means that each one of these components will be considered in all its possible failure modes, but only as initial conditions, and no repair or further failure possibility will be allowed during the transient evolution of the accident. These components are:
   a. Valve V01, which allows the isolation of the *let-down line*. V01 has been assumed to behave "nominally", according to design specifications, (parameter *V01 = 0*), or to be in a "failed state stuck open" (parameter *V01 = 1*).
   b. Valve V25, which separates the let-down flow into the flow that enters the VCT and the flow that is diverted to the Boron Recycle System (BRS). V25 has been assumed to behave "nominally", according to design specifications, (parameter *V25 = 0*), or to be in a "failed state stuck in the direction of VCT" (parameter *V25 = 1*).

c. Valve V73, which allows the isolation of one of the two feed lines from the Refuelling Water Storage Tank. V73 has been assumed to behave "nominally", according to design specifications, (parameter *V73 = 0*), or to be in a "failed state stuck closed" (parameter *V73 = 1*).

3. One component has been assumed to fail during the transient in relation to a specific physical condition. This is the regulation valve V68, which regulates the water make-up flow to the Volume Control Tank from the Boron Concentration Control System (BCCS). V68 has been assumed to always behave "nominally" (parameter *V68 = 0*) at the beginning of the accident and to fail closed (parameter *V68 = 1*), during the accident, when the amount of water in the BCCS reaches a minimum level, so that no more suction is possible.

The number of sequences generated by the Interaction Model (DYLAM) when combining all possible failed states of all components and operator error/recovery modes is quite high, and a further step of screening becomes necessary.

We will focus here only on the set of sequences generated from the initial condition "nominal" operator behaviour, i.e., *ERR=0, REC=0,* and "failed" component states, i.e., *V01 = 1; V25 = 1,* and *V73 = 1*. V68 is allowed to fail according to the time dependent physical evolution of the system.

In particular, 15 sequences are derived from this initial condition. The DYLAM diagram that describes these sequences is shown in Figure 4.15.

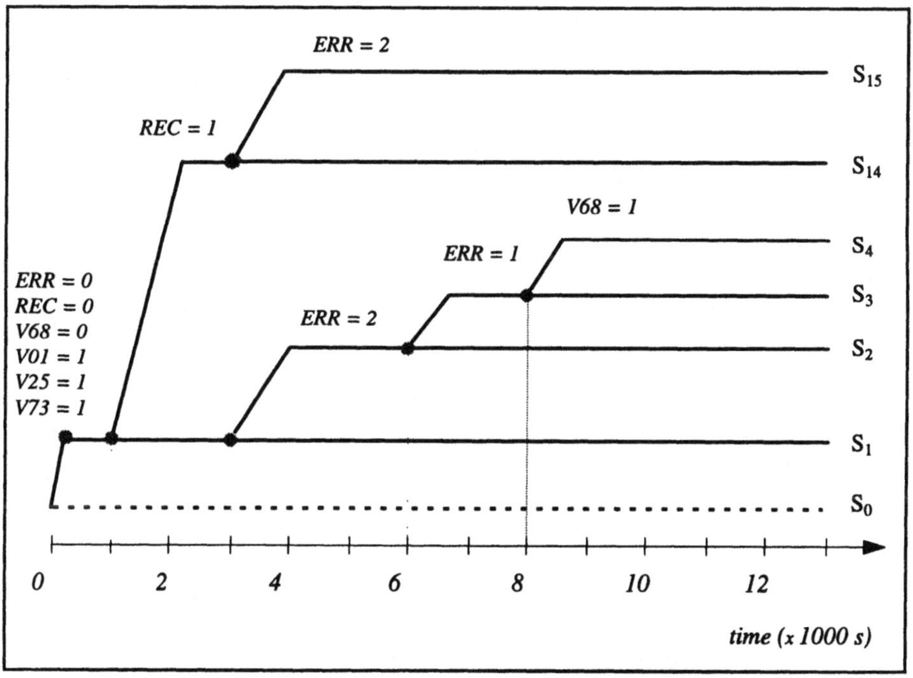

Figure 4.15. Generation of Sequences and descendants (DYLAM).

In this Figure we did not show the probabilities associated with each sequence, as this subject is not essential for the objectives of this sample case. Moreover, as we are focusing on the analysis of the operator behaviour, we will only discuss in detail three of these 15 sequences, as they are considered particularly important for the human-machine interaction aspects which they contain. These three sequences are: SEQ_1, SEQ_4, and SEQ_{15}.

Sequence *SEQ 1*

Sequence SEQ_1 is the generating (*"mother"*) sequence of the other two sequences that will be analysed later on, and is characterised by the three components of the CVCS in failed condition, i.e., *V01=1*, *V25=1* and *V73=1*, by error-free operator behaviour, i.e., *ERR=0*, and by the regulation valve on the BCCS performing according to design, i.e., *V68=0*.

The failures of the three valves do not affect the normal behaviour of the CVCS, i.e., the performance of the CVCS during normal operation, but they influence quite substantially the response of the system during accidents.

We will analyse firstly the behaviour of the physical variables and then we will consider the time dependent operator stress. The latter does not affect the accident evolution, as this sequence does not considers operator errors. However, it is interesting to notice how the calculated *stress function* varies according to the physical evolution of the system.

- *Physical Variables*

The results of this sequence are shown in Figures 4.16, 4.17 and 4.18, which show the mass flows in the CVCS (Q_{ch} = mass flow in the charge line, Q_{ld} = mass flow in the let-down line, and Q_{sl} = mass flow in the seal line), the physical variables characterising the VCT (Q_{ovct} = outlet mass flow, Q_{mkup} = make-up mass flow, and L_{vct} = water level in the VCT), and the pressures in the seal line (p_{sl} = pressure in the seals, and p_{pr} = pressure in the Primary Circuit).

The first effects of the malfunctions appear at around 1000 seconds, when the operator intervenes to control the let-down flow. The blockage of V25 in the direction VCT generates a rapid filling of the Volume Control Tank, causing the intervention of the protection systems already at 1600 seconds. This automatic intervention consists in the deviation of all let-down flow towards the Boron Recycle System (BRS) through V25.

However, this valve is stuck in the direction VCT, which is consequently filled-up to its maximum capacity (4 meters) at approximately 1800 seconds. This causes the opening of the relief and safety valves of the VCT (V26).

Therefore, the first goal of *CVCS Safeguard* is not completely fulfilled and only the opening of the relief valve V26 contributes to the control of the VCT level, at its maximum height.

The other physical quantities, in general, follow quite closely the nominal sequence (SEQ_0) for the first 7000 seconds of transient, except for the let-down flow (Q_{ld}), which is affected by the failure of valve V01.

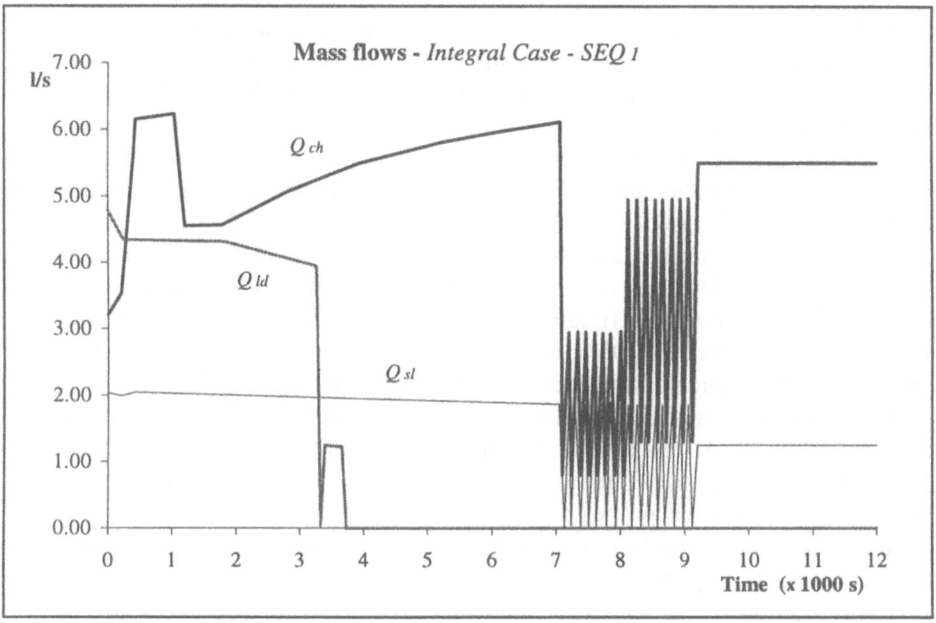

Figure 4.16. CVCS Mass flows - *Integral Case - Sequence* SEQ_1.

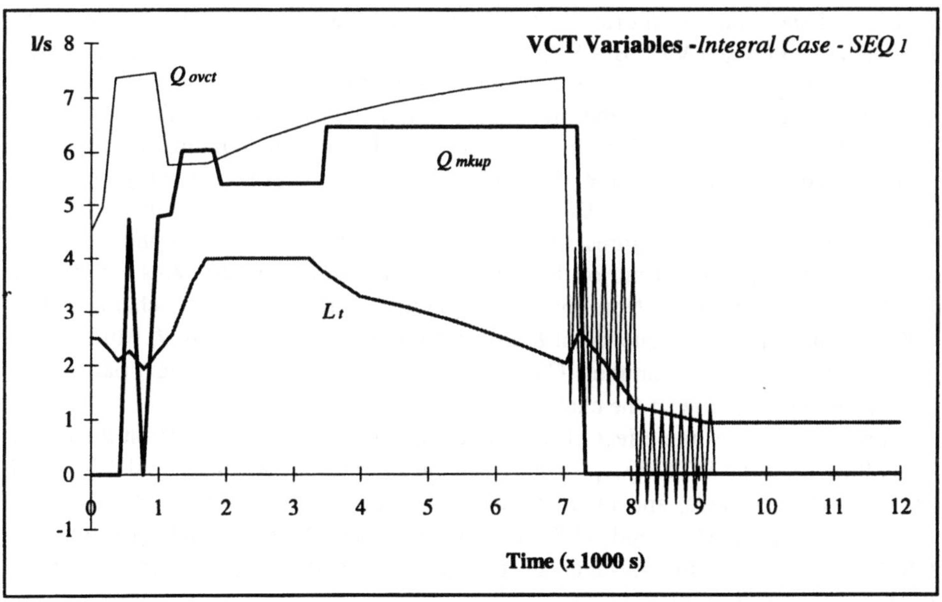

Figure 4.17. VCT Variables - *Integral Case - Sequence* SEQ_1.

At 3200 seconds, when the operator tries to isolate the let-down line, the failure of V01 in open position does not allow the successful performance of the operation. Moreover, as the complete sequence of actions of *Task 2-a* (Figure 4.9) requires that valves V09 and V10 are closed, and, as these two actions are carried out successfully, the pressure in the let-down line (not shown in the figures) starts rising very rapidly, causing the opening of a relief valve (V08) and the successive isolation of the line by the closure of valve V03.

This pressure oscillation and the opening of the relief valve affect (Figure 4.16) the let-down mass flow (Q_{ld}) around 3200 and 3700 seconds, when the let-down flow drops to zero, as V09 and V10 are closed, then rises to approximately 1 litre/second when the relief valve V08 opens, and, finally, returns to zero when the supplementary isolation of the let-down line is carried out by closing V03.

The critical effects of the failures of components are shown after 7000 seconds of transient (*Task 3*). At this time, the operator, according to procedure, connects the Refuelling Water Storage Tank to the charge line and isolates the make-up line, so the only source of water for the charge line is the RWST. However, the connection is hampered by the failed (closed) valve V73, and no water is fed to the reactor nor to the seals of the primary pumps.

The protection system intervenes immediately, at 7100 seconds on a signal of low mass flow in the seal line, by starting pump P03 and opening valve V72. These are in standby position and completely replace pump P04 and valve V73.

However, as the protection system is not designed to recognise valve failure, but is activated only on signals of low flow in the seal line, as soon as the seal feed-flow is re-established, the system returns to its design configuration, i.e., feed-water from RWST using P04 and V73. But V73 is failed closed, and, thus, the feed flow returns to zero. The cycle is then restarted by the response of the protection system.

This automatic switch on and off of P03, causes the oscillation of the mass flows in the charge line (Q_{ch}) and in the seal line (Q_{sl}), between maximum and minimum values (Figure 4.16), until valve V72 and pump P03 are kept permanently open and in operation, i.e., at time $t = 9200$. This operation is carried out by the automatic protection system that requires the permanent activation of the second pumping system (P03) only, as a result of the closure of the return from the seals carried out by the operator (*Task 5*). Note that the amplitude of the oscillations changes after 8000 seconds, as a consequence of the operator performance of *Task 4*, which has a very important influence on the VCT variables, as we will see in the next paragraph, and this affects the flow in the charge line.

Figure 4.17 shows the effect of the fluctuation between P04 and P03 on the outlet mass flow from the VCT (Q_{ovct}), which oscillates rapidly between 1.2 and 4 l/s from 7000 to 8000 seconds, and then between 1.1 and -0.8 (return of water inside the VCT) l/s, between 8000 and 9300 seconds. After this time, normal operations of CVCS for managing the accident are re-established and the system remains under control for the rest of the transient.

The change in amplitude and values of Q_{ovct} is a direct effect of the performance of *Task 4* by the operator and of the automatic control system. At 8000 seconds the operator starts permanently pump P03 and fully opens V39 to obtain maximum feed in the charge line. However, the automatic system continues to switch on and off pump P04, which is unable to provide flow because V73 is failed in a closed position.

Consequently, when P03 is the only functioning pump recognised by the automatic control system, then: a) P03 provides all the necessary flow; b) there is no need of flow from the VCT; and c) there is a return of water in the VCT from the seal line ($Q_{ovct} \approx - 0.8\ l/s$). Otherwise, when in conjunction with P03, also P04 should be operating, then: a) the flow provided by P03 is reduced; and b) the need of maximum charge flow leads to a request, by the automatic system, of further feed flow from the VCT, to back-up the flow of P03 ($Q_{ovct} \approx + 1.1\ l/s$).

This explains why the oscillations of outlet mass flow (Q_{ovct}) and charge flow (Q_{ch}) change in the intervals 7000 - 8000 seconds and 8000 - 9000 seconds.

A similar situation is shown by the time dependent responses of the pressure (Figure 4.18). In particular, it is important to notice that during the oscillations, from 7000 to 9300 seconds, the pressure in the seals drops to very low values. This stops the seal flow and could cause damage and leaks of contaminated water from the Primary Circuit.

The most important findings of this analysis is related to the design of the logic of protection system of CVCS. It results that the protection of the system connected to the RWST is activated only on a signal of low feed-flow in the seal line. Therefore, it does not recognise the impossibility to operate P04 and this causes the inappropriate performance of the system with the undesired oscillations of the pressures and mass flows. A review of the logic of the protection system is necessary and can be developed from the above analysis.

- *Operator Stress*

Even though the operator behaviour is error free, in this sequence, the time dependent stress function gives an estimate of the effect of failed components on the operator's cognitive performances.

We have already analysed the response of the stress function in the case of nominal behaviour (Figure 4.14), which showed that some peaks of boredom and stress in correspondence to moments of no activity, between 3000 and 4000 seconds ($F_{stress} = - 0.09$) and of high task performance between 7000 and 10000 seconds ($F_{stress} = + 0.18$).

In sequence SEQ_1 this trend is maintained (Figure 4.19). However, the incomplete isolation of the let-down line (due to the failed V01) combined with the opening of the relief valve V08 at around 3500 seconds and the fluctuating behaviour of the feed-water pumping system from the RWST (caused by the failure of V73), make the peak values of the *stress function* much greater than the nominal case, and also generate some form of instability in the operator's attitudes. In particular, a maximum value of the *stress function* is reached at 9000 seconds ($F_{stress} = + 0.49$), which is almost three times the value of the stress in nominal conditions. Similarly, the value of the *stress function* at around 5000 seconds oscillates between ± 0.22, i.e., 2.5 times the stress reached in the same time period in nominal conditions.

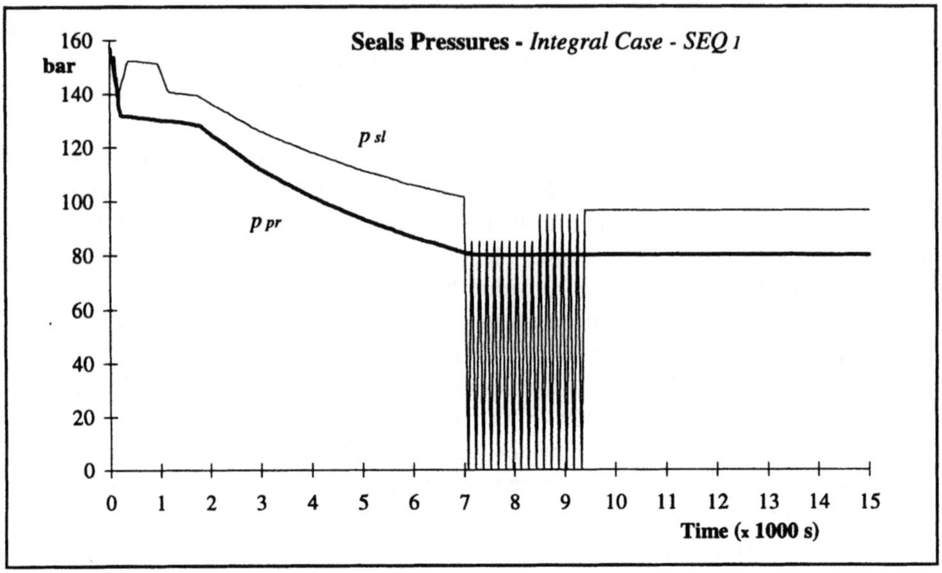

Figure 4.18. CVCS - Seals Pressures - *Integral Case - Sequence* SEQ_1.

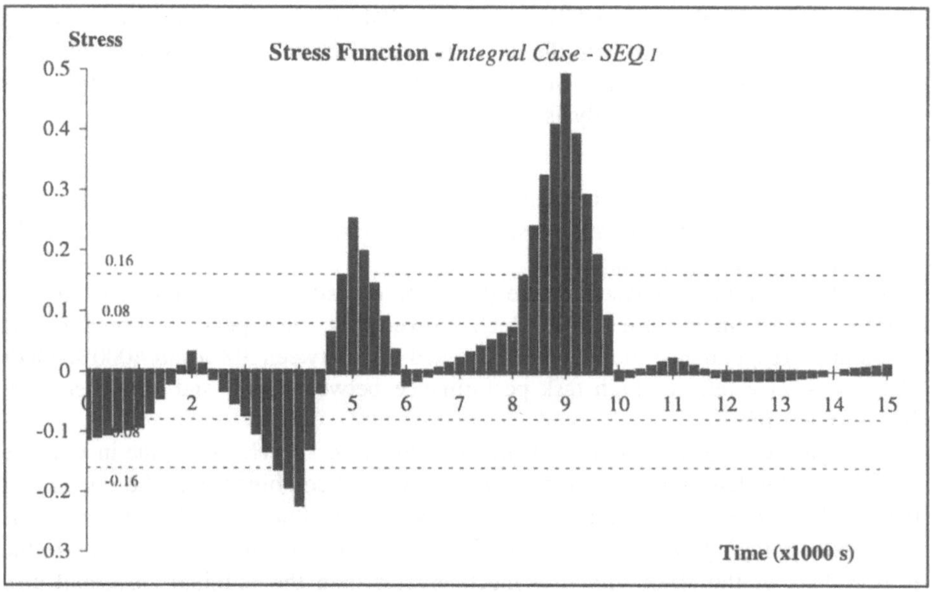

Figure 4.19. Operator Stress vs. Time - *Integral Case - Sequence* SEQ_1.

Towards the end of the transient, when the conditions of the CVCS become more stable, the stress function drops significantly to very small values, so as to simulate that the operator has completed his/her tasks successfully, has managed to control the system, even in the presence of failed or malfunctioning components, and there is no need of any further intensive interaction with the system.

A final consideration on the analysis of this sequence can be made. The operator has to perform an action within *Task 2,* initially considered of low importance by the task analysis and "tagged 4" (minimum value, Figure 4.9) for the simulation of the procedure, namely the closure of valve V03. This action compensates for the lack of response of V01 and follows the opening of the relief valve V08. Thus, in this case, this specific action becomes very important as it limits the amount of primary coolant which is released through the relief valve and contributes to a better management of the accident sequence.

By means of the analysis performed with the human-machine simulation architecture, we can further validate the quality of the task analysis and, in general, the working environment under study.

Sequence SEQ 4

As already explained, sequence SQ_4 originates from SEQ_1. It is one of the most complex sequences that may be studied, as it combines component failures and operator errors in correlation with each other (Figure 4.15).

This sequence shows the transition, at t = 3000 seconds, of the operator behaviour from an error-free state ($ERR=0$) to a state where two errors (ERR = 2) are made during the execution of each task performed in the procedure. At a later stage of the accident evolution (t = 5000 seconds), the error mode changes from state 2 to state 1 (ERR = 1), i.e., only one action is omitted for each task carried out. Finally, at 8000 seconds, there is a failure of valve V68 in failed closed position ($V68 = 1$), due to physical reasons, i.e., the depletion of make-up feed-water from the Boron Concentration Control System (BCCS).

It is important to note that the recovery has been inhibited in this sequence.

The results of the sequence are shown in Figures 4.20 - 4.23, where the physical and cognitive variables are presented. As SEQ_4 derives from SEQ_1, the two sequences are exactly equal up to 3000 seconds, when the first operator error occurs. At this time (Figures 4.23 or 4.19) the value of the stress ($t = 3100$ s, $F_{stress} \approx 0.1$) is just above the first threshold value (- 0.08) for state change and error generation ($ERR = 0 \Rightarrow ERR = 2$).

- *Physical Variables*

As in the previous case, it is noticeable how the level of the tank is firstly governed by the automatic behaviour of the make-up system (valve V68) which attempts to compensate for the level reduction due to the increase of the outlet flow (Q_{ovct}).

The malfunctioning of valve V25 results in the filling of the VCT with the level rapidly reaching the maximum height of 4 meters and the opening of the relief valve V26 (Figure 4.21).

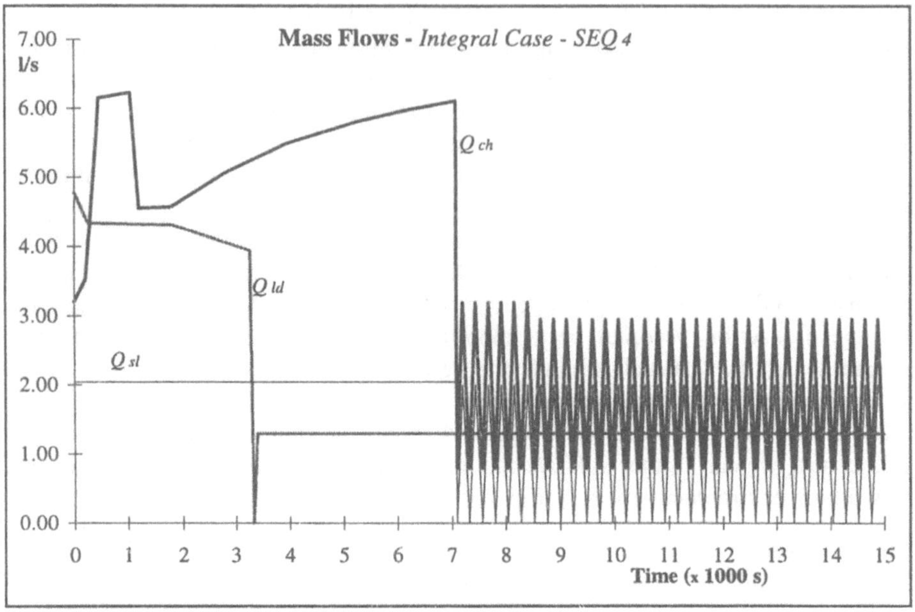

Figure 4.20. CVCS Mass flows - *Integral Case - Sequence* SEQ_4.

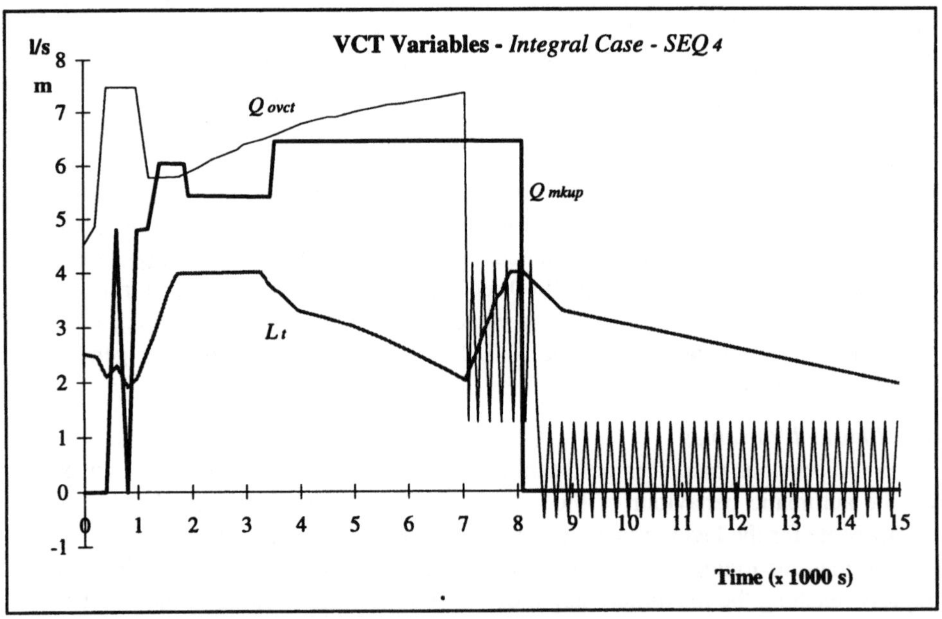

Figure 4.21. VCT Variables - *Integral Case - Sequence* SEQ_4.

From 3000 seconds, the operators begins to omit the performance of two actions in each task.

The first two omitted actions (*Task 2*) are the supplementary isolation of the let-down flow (closure V03) and the isolation of the excess let-down flow (closure V78).

The omission of the closure of V78 does not actually affect this accident, as the line is already isolated (normal stand-by conditions), and may be opened only by the protection system in specific cases. In this accident, these conditions are not met and thus the operator error does not affect the overall system performance.

The omission of the supplementary isolation of the let-down flow, instead, modifies the accident. Indeed, as the other actions of this task are completed (in particular the closure of V09 and V10), the let-down flow is actually stopped (Q_{ld} = 0, Figure 4.20). However, as valve V01 is failed open and V03 also being left open by the operator, the let-down flow of highly pressurised primary circuit water violently rises the pressure in the let-down pipe-network and causes the opening of the relief valve V08.

This implies the release of let-down flow to the Pressure Relief Tank (PRT) and further contributes to undesirable depletion of Primary Circuit water. This condition is not recovered and thus the let-down flow remains approximately at the level of 1.2 litres/second for the remaining time of the accident.

The operator performance improves after 5000 seconds, when only one action is omitted. However, three more actions are not carried out, one for each *Task 3, 4* and *5*. These errors, which affect quite strongly the behaviour of the system, are: a) the omission of the closure of the make-up flow at 7300 seconds (*Task 3*); b) the omission of the complete opening of valve V39, at 8000 seconds (*Task 4*); and c) the lack of isolation of the return flow from the seals, at 9000 seconds (*Task 5*).

In particular, the omission of the isolation of the make-up flow at 7300 seconds causes the rapid depletion of the source of feed-water from the Boron Concentration Control System (BCCS) (Figure 4.21) and the failure of valve V68 in closed position at 8000 seconds when the minimum level in the BCCS tank is reached and the make-up flows stops completely (Q_{mkup} = 0).

The lack of complete opening of V39 is not crucially important as, the oscillation of the feed-water from the RWST generates a great instability which absorbs the effect of charge flow.

The lack of isolation of the seal line at 9000 seconds causes the non-permanent activation of pump P03, as was done in SEQ_1, and the persistence of the oscillations of feed flow from the RWST. This effect is noticed in Figures 4.20 - 4.22, where the fluctuations of mass flows (Q_{ch} and Q_{ovct}) and pressures (p_{sl}) start at approximately 7000 seconds and continue for the whole transient.

In particular, the outlet flow from the VCT (Q_{ovct}) oscillates between, 1.2 and 4 l/s from 7000 to 8000 seconds as there is a need of feed-water from the VCT due to the inactive pump P04 and the on-off fluctuation of P03. Between 8000 seconds and the end of the transient analysis, P03 is put into permanent operation by the operator, and the oscillations of the Q_{ovct} now change for the same reasons already discussed in the case of SEQ_1. The Q_{ovct} fluctuates between - 0.8 and + 1.2 l/s, where the negative values imply that there is an inflow into the VCT coming from the return side of the seal line.

The dynamic behaviour of the level is more complex, as it depends on the combined effect of the make-up flow (Q_{mkup}) and the charge flow (Q_{ch}). At 7000 seconds the make-up flow is at maximum (Figure 4.21), but the charge flow (Figure 4.20) starts oscillating between 0.8 and 3 litres/seconds as a consequence of the failure of V73 and the already discussed fluctuating behaviour of the feed-water system connected to the RWST.

The net result of these combined effects is that:

- Between 7000 and 8000 seconds, the outlet flow from the VCT (Q_{ovct}) oscillates between 1 and 4 litres/second, because of the interaction between the make-up flow and the charge flow.
- Between 7000 and 8000 seconds, the level of the water in the VCT (L_t) increases up to its maximum (4 meters) and the relief valve V26 opens, as it did at the beginning of the transient, to control the level in the tank VCT. This effect lasts only for a short time, because the level starts to decrease as soon as there is no more feed-water from the make-up line (failure of V68).
- After 8000 seconds the lack of make-up flow and the persisting oscillation on the feed-water line from the RWST, cause the level to decrease, as the outflow from the VCT oscillates for the remaining time of the accident.

The pressure behaviour (Figure 4.22) in this SEQ_4 is particularly critical, starting from 7000 seconds, as the strong oscillations of the system, result in a loss of seal pressure (p_{sl}) and consequent loss of Primary Circuit water.

In this sequence, the omissions of the operator, the failures of components and the problem of inappropriate design of the protection system, concerning the RWST feed-water line, all combine to produce a complex "safety critical sequence".

The analysis of this sequence shows the importance and usefulness of applying human-machine simulation for evaluating design and for identifying flaws, which can only be discovered by extensive analyses on many different combinations of events and working conditions, independently of their probability of occurrence.

- *Operator Stress*

The performance of the system with failures, malfunctions and erroneous operator behaviour result in a very unstable and fluctuating dynamic variation of the stress.

The peak values of stress and boredom are much greater than those shown in SEQ_1 specially towards the end of the sequence, when the situation seems out of control (Figure 4.23).

The negative values of the stress functions are also connected with the fact that the omission of an action is equivalent to inactivity and thus, according to the model, the lack of action induces boredom. This is, however, followed by the increased criticality of certain physical quantities, e.g., level and pressure, which cause a return of stress. In any case, the overall behaviour of the stress function confirms that many safety issues exist with respect to the design of the protection and safety measures.

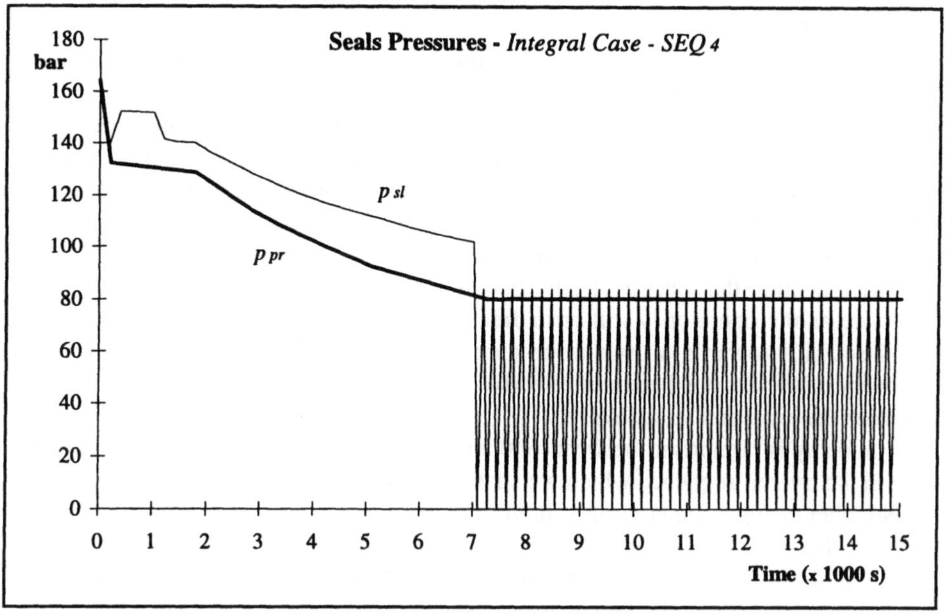

Figure 4.22. CVCS - Seals Pressures - *Integral Case* - *Sequence SEQ_4*.

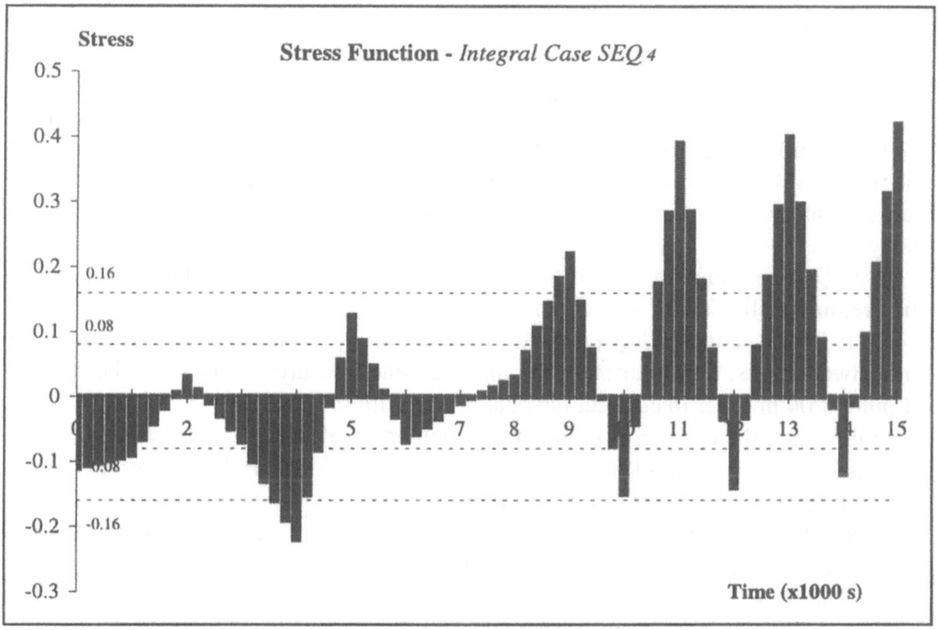

Figure 4.23. Operator Stress vs. Time - *Integral Case* - *Sequence SEQ_4*.

<u>Sequence SEQ 15</u>

Sequence 15 has been selected as it contains the consideration of recovery within the overall management of the system.

Like SEQ_4, also SEQ_{15} originates from SEQ_1 and shows the same behaviour up to the first event of operator "state change", which already occurs at 1000 seconds. Note that at this time (Figure 4.19) the value of the negative stress (boredom) is just above the threshold value (at $t = 1000$ s, $F_{stress} \approx -0.09$) and the variable associated with recovery changes from 0 to 1 ($REC = 1$).

The transition that occurs at this time, however, is one of recovery which is not effective as the operator is still error free. The next operator state change occurs at 3000 seconds towards a "two-error-state" ($ERR = 2$). In this case the complete recovery of the actions becomes effective, but there still is a delay to be considered between the error and its recovery!

The delay is comprised between 600 and 1000 seconds, and this plays a relevant role in the overall sequence which we shall discuss.

No other changes are envisaged in this sequence.

Note that an important difference exists between sequence SEQ_{15} and sequence SEQ_4. In SEQ_4 two errors are made starting from 3000 seconds, and only one error is made after 5000 seconds. While in SEQ_{15} two errors are made, and immediately recovered, starting from 3000 seconds. This "error-followed-by-recovery" condition is kept for the remaining time of the transient.

The results of this sequence are shown in Figures 4.24 - 4.27.

- *Physical Variables*

In practice, sequence SEQ_{15} shows some differences from SEQ_1 only after 3000 seconds, when two omissions per task are made and then recovered.

The first interesting event occurs in relation to the let-down flow (Q_{ld}) which shows a similar behaviour to the let-down flow of SEQ_1. However, the release from the relief valve V08 lasts longer (Figure 4.24), as the initial omission of the supplementary isolation of the let-down flow (closure V03) is recovered with a delay.

In SEQ_{15} the situation is well recovered, and no Primary Circuit water is lost to the Pressure Relief Tank, as occurs in the case of SEQ_4.

Up to 7000 seconds, SEQ_{15} is quite similar to SEQ_1. At that moment, the operator omits two actions: the isolation of the make-up line (closure of V68) and the start up of pump P04 in order to connect the charge line with the RWST.

In this case, the two errors compensate each other, at least for some time, because there is still water (up to 8000 seconds) in the Boron Concentration Control System, and this ensures feed-water to the charge line.

At 8000 seconds, the operator recovers from the errors and firstly starts P04 and then shuts V68. The first action starts the process of oscillation of the charge flow (Q_{ch}), seal flow (Q_{sl}), and make-up flow (Q_{mkup}) already described earlier. The second recovery is irrelevant, as the valve V68 has failed closed because the level in the Boron Concentration Control System has reached its minimum value. The make-up-flow is now stopped.

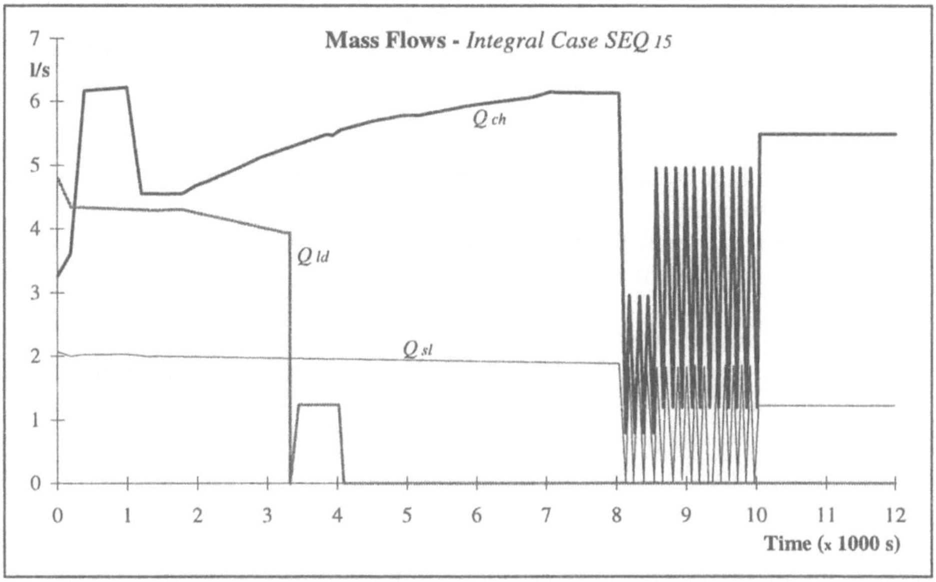

Figure 4.24. CVCS Mass flows - *Integral Case - Sequence SEQ_{15}*.

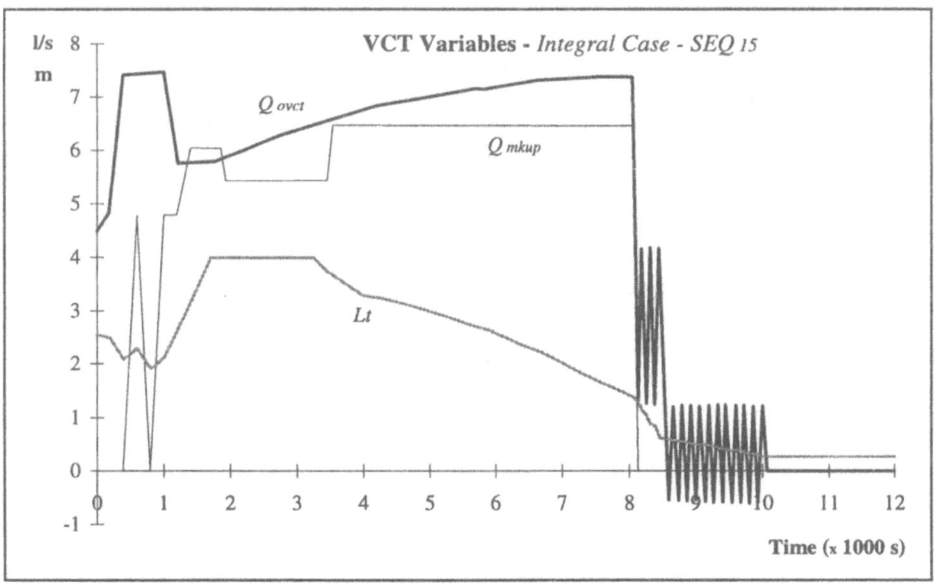

Figure 4.25. VCT Variables - *Integral Case - Sequence SEQ_{15}*.

The sequence proceed with the operator omitting to perform the actions foreseen in *Task 4*, at 8000 seconds, and recovering from the errors at 8600 seconds.

This causes the fluctuations of the mass flows to change their amplitude and maximum and minimum values in a similar manner as SEQ_1, in analogous conditions.

The same error and recovery event occur with respect to *Task 5*, at 9200 seconds, i.e., omission of the isolation of the return line from the seals and recovery from the error at 9800 seconds.

This causes the oscillations of the mass flows to continue approximately until 10000 seconds, when the situation returns under the complete control of the operator.

The seal pressure (Figure 4.26) shows strong oscillations, as in the other sequences, starting at 8000 seconds, in accordance with other physical quantities and operator actions.

As in SEQ_4 the seal pressure (p_{sl}) goes below the Primary Circuit pressure, but in this case (as in SEQ_1) only for the limited period of time, i.e., between 8000 and 10000 seconds.

- *Operator Stress*

The time dependent behaviour of the stress function shows that the recovery of errors helps in reducing the level of operator stress, reached during the performance of the procedure (Figure 4.27).

The stress peak reached at 5000 seconds, due to the first omissions made at 3000 seconds and subsequently recovered, is smaller than the equivalent peak reached at the same time in SEQ_1.

For the maximum stress level following the omissions of actions at 7000, 8000 and 9000 seconds, similar results have been found.

In the final period of the accident, the stress level drops to very low values, as the operator has managed to reach a stable CVCS condition and the situation is under complete control.

This situation is very similar to SEQ_1, with the same delay shown in the performance of control tasks, and is very different to SEQ_4, where the lack of control was also reflected in the very unstable behaviour of the stress function.

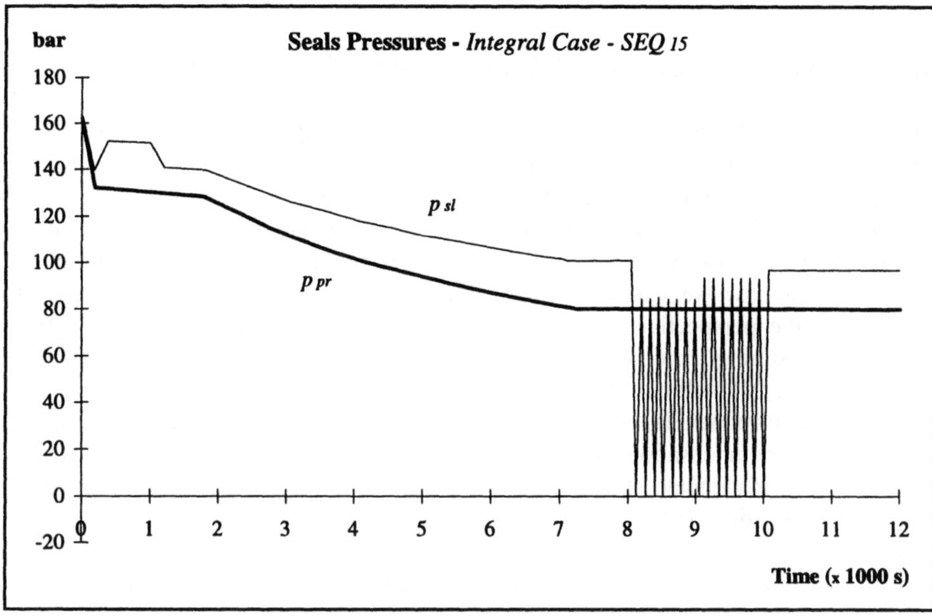

Figure 4.26. CVCS - Seals Pressures - *Integral Case - Sequence SEQ_{15}.*

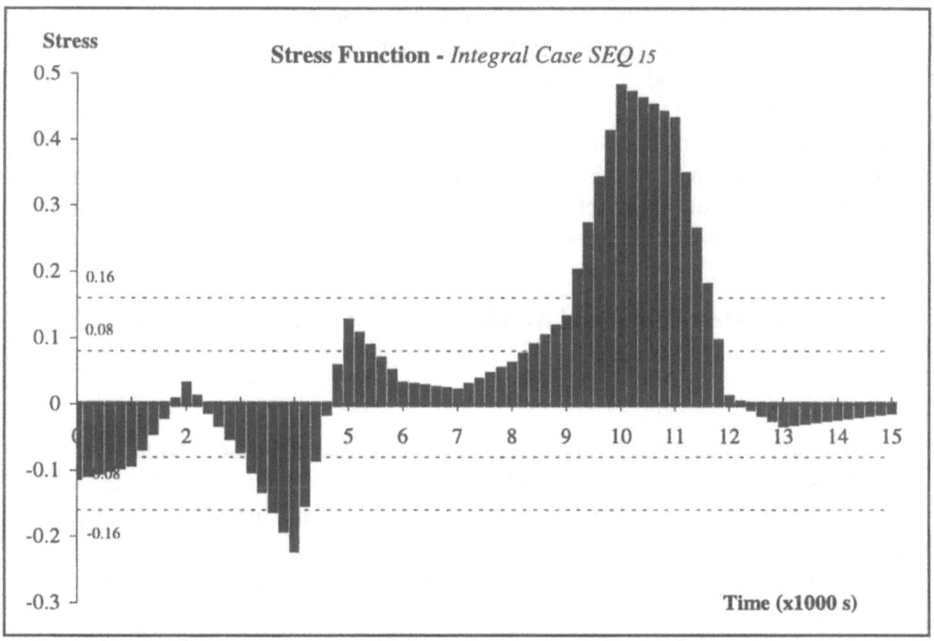

Figure 4.27. Operator Stress vs. Time - *Integral Case - Sequence SEQ_{15}.*

4.5 The Auxiliary Feed-Water System Case Study

The Auxiliary Feed-Water System (AFWS) has already been taken into consideration in the previous Chapter, as the plant model for the sample case based on the cognitive simulation COSIMO (§ 3.5.1).

However, while in the previous case the focus was mainly on the development of a simulation of cognitive functions and processes relating to a precise theoretical model, in this case we will be concerned with a safety and criticality assessment study from the perspective of the overall human-machine interaction and system management, during accident conditions.

In particular, we will concentrate on the accident management procedures, and on display and control panels, with the aim of validating their ability to support accident diagnosis and compensation, for a number of different situations. These situations will include inadequate functioning of interfaces and failure of hardware components, in addition to human errors and inappropriate behaviour.

Consequently, special attention will be dedicated to the description of the AFWS and its control and display systems and to the procedures that are to be followed in the case of failures or malfunctions during operation. Moreover, a human behaviour model, different than COSIMO, will be developed, which is more suited to the goals of the specific case study.

The AFWS sample case bears many similarities with the previous study dedicated to the analysis of the Chemical and Volume Control System. Obviously, the reference Framework for Human-Machine Interaction simulation is applied.

The DYLAM method is applied as Interaction Model for the evaluation of component failures and human errors.

A different model of operator behaviour and a different algorithm for modelling errors will be developed. In this way the reader can become acquainted with different alternative modes of approaching the problem of human modelling, within a predefined Human-Machine Interaction architecture.

4.5.1 Plant Description and Accident Scenario - Case Study AFWS

Plant Description

Although the Auxiliary Feed-Water System has already been described in brief in the § 3.5.1, we will now review the whole system in a much more detail fashion, covering *functions, structures, operability conditions* and *protection and control mechanisms.*

The Auxiliary Feed-Water System is an auxiliary, stand-by system, which operates on the secondary side of a Nuclear Power Plant, either when an emergency reactor shut-down (trip) occurs, or in some special "normal transient" conditions, such as during the reactor start-up process, or during the shut-down phases for the transition form the short term cooling (hot-shut-down) to the long term cooling (cold-shut-down).

In emergency conditions, the AFWS system is expected to feed-water to the secondary system of the reactor, in order to extract the heat generated in the nuclear core after the shut-down of the reactor (the decay heat).

This process lasts for several hours, until the long-term shut-down system takes over or reactor operations are restarted.

Functional and Structural Description

The AFWS is structured in two independent *main-feed-lines*, *TA* and *TB*, connected to the same *feed-tank*, *BA* (Figure 4.28).

Each *main-feed-line* provides feed-water to two *steam generators* and is connected to a pumping system, made-up of a *motor-driven-pump*, linked both to the grid and to the emergency Diesel power, and of a *turbine-driven-pump*, which is feed by the steam lines coming from the same *steam generators* of the reactor fed by that main-feed-line.

The injection of feed-water into the steam generators is controlled by an automatic, as well as a manual, regulation system, by which the isolation of each individual injection line may be obtained.

The system has been designed to grant the maximum redundancy and flexibility of functions between main-feed-lines and components. As an example, on each main-feed-line, the two pumps feed water to both steam generators connected to that main-feed-line.

The feed-tank is kept at constant pressure by nitrogen gas injection, and the make-up flow is activated either by a minimum tank level signal and/or by the start-up signal of the AFWS.

The *Steam Generators* are the focal point of system management and are the most delicate components of the system, because of their complex thermal behaviour and control problem.

Operability Conditions

During normal operations of the nuclear power plant, the AFWS rests in stand-by condition and its availability is constantly monitored by the plant control system.

The pumps are all in stand-by position, ready to operate at full power either in response to an automatic signal coming from the reactor protection system, or to a manual start-up signal activated from the control room or from an emergency panel.

Valves V031-V038 are in fully open position, while valves V151-V154 are in closed position and are automatically open with the AFWS start-up signal.

The AFWS starts operating in all accident conditions which result in the loss of main feed-water on the secondary side of the steam generators. This means that the AFWS is called to operate in many circumstances, which, although rare, cover a wide spectrum of possibilities that may be considered in a safety assessment.

In particular, the Auxiliary Feed-Water System operates every time there is a loss of electrical power, or an unexpected isolation of the main feeding systems, or a loss of coolant accident.

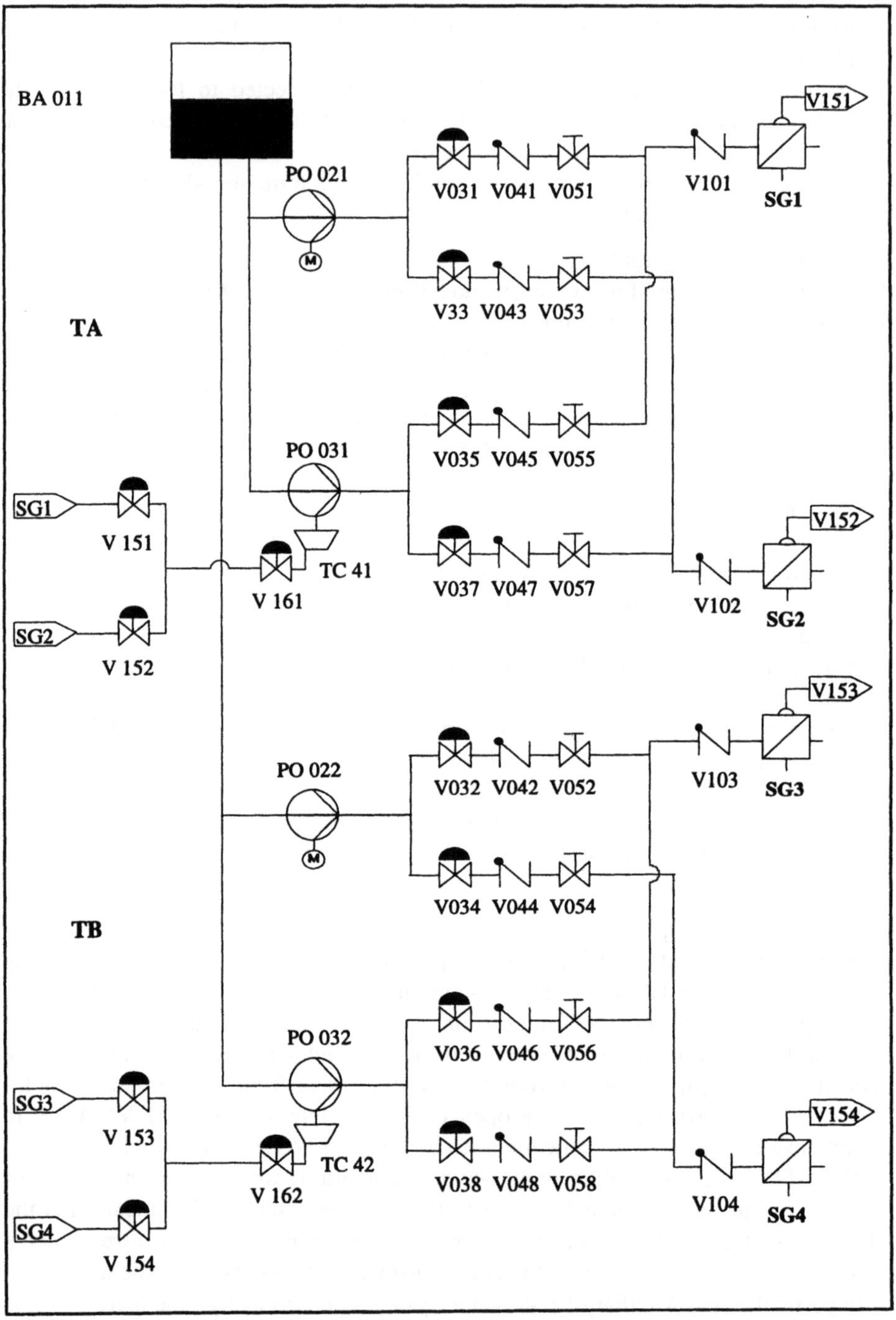

Figure 4.28 Schema of a typical Auxiliary Feed-Water System.

In all these cases the reactor is instantaneously shut-down by the protection system which also send the start-up signal to the AFWS. Apart from the reactor shut-down, other signal that cause the automatic start-up of the AFWS are: the low level in the steam generator, secondary side; the loss of coolant in the secondary circuit; the loss of electrical power.

Once the AFWS has started operating at full power, it is assumed than no control action is carried out by the operator for 30 minutes. Then some control actions are required in order to adjust the water flow to the progressive decay of the heat exchanged in the steam generators and to maintain the water level within acceptable boundaries.

The AFWS is expected to operate for a sufficient time as to allow either reactor start-up in case of rapid recovery from the accident conditions or further reactor cooling until the "Long Term Cooling System" takes over, in the case of a more serious accident that requires the total shut-down of the plant.

The maximum time of operation of the AFWS is of approximately 7.5 hours.

Control and Protection Systems

The control and protection system is carried out by collecting information through the indicators' panels and alarms and by regulating the plant performance through the actuators' panels.

The set of *indicators* and *alarms* play a crucial role in the human-machine interaction process and affect the human decision making, as they are the main source of information for the operator about plant performance.

Control and emergency procedures are developed in strict correlation with the information display system and control panels available to the operators. The indicators and alarms of the AFWS and Steam Generators are shown in Figures 4.29 and 4.30.

The *indicators* used for the management of the AFWS focus on the feed flows and on the important physical variables of the steam generators. In particular:

- The information on the AFWS is obtained from 9 numerical and 18 logical indicators. The numerical indicators are associated with the water level in the *feed-tank*, *BA*, the 4 mass flows and the 4 pressures downstream of the pumping system towards each steam generator (Figure 4.29).
- The information on the Steam Generators is obtained from 5 indicators: the water level, the pressure, the temperature, the void fraction and the outlet steam mass flow in each steam generator (Figure 4.30).

The *alarms* on the AFWS concern the *low level* in the feed-tank, the *oil pressure* in the feed pumps and the *void fraction* of the steam that feeds the turbines of the turbine-driven pumps (Figure 4. 29).

The *alarms* on the Steam Generators are related to the *high* and *low pressure* and *high* and *low water level,* and to *pressure change rate* during the accident (Figure 4.30).

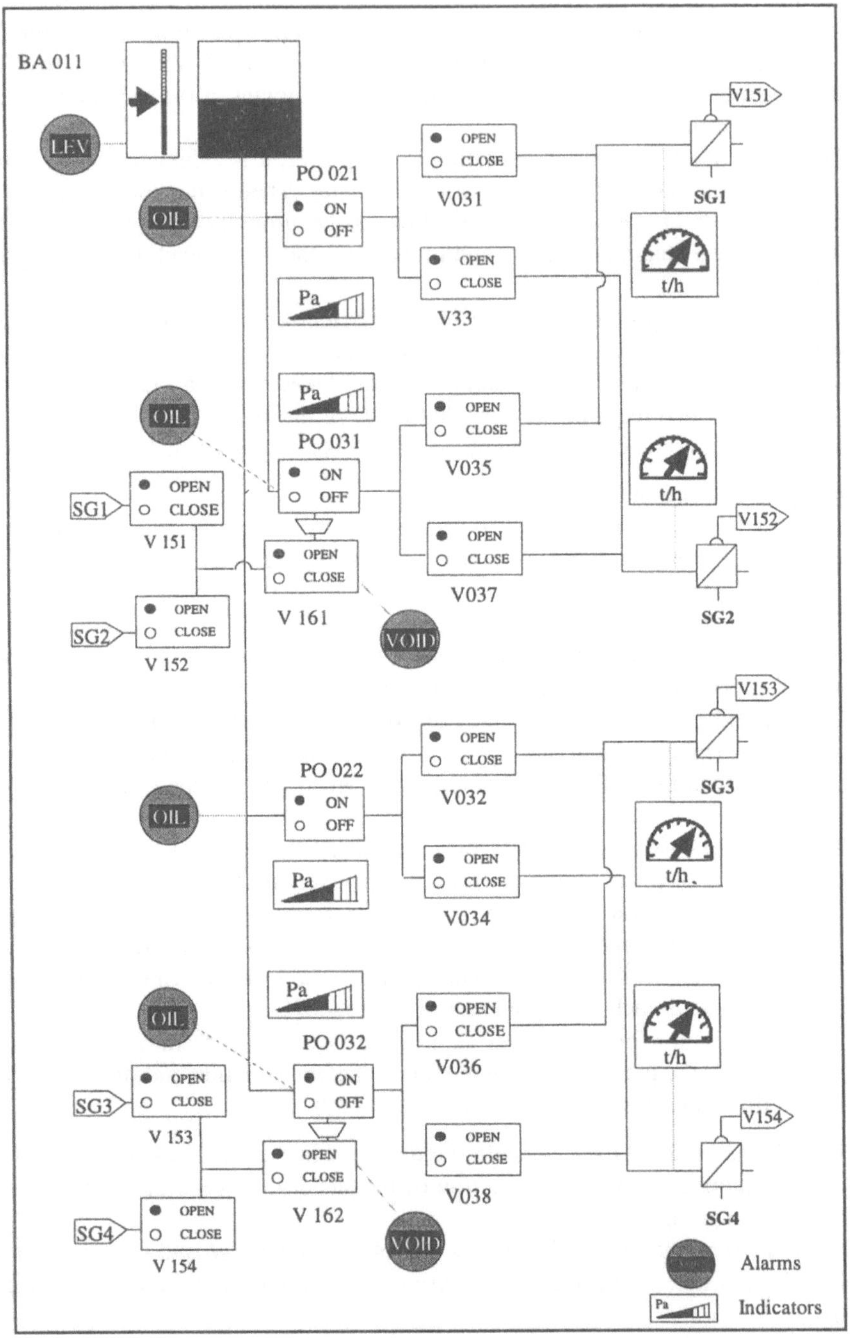

Figure 4.29. Indicators and Alarms of the Auxiliary Feed-Water System.

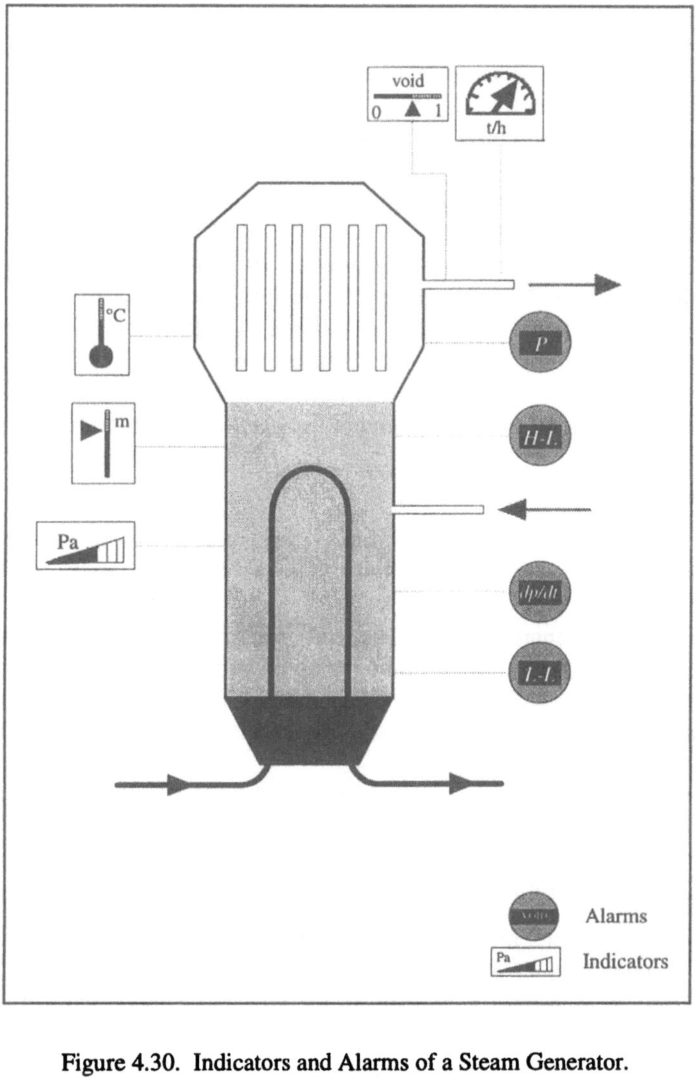

Figure 4.30. Indicators and Alarms of a Steam Generator.

The *actuators panel* is utilised to control and regulate mass flows and to set main components to specific positions in case of need.

The simulated AFWS actuators' panel contains 22 actuators by which all regulation components can be controlled (Figure 4.31). More precisely, 4 manual switches (*on/off*) are individually controlling each of the pumps; 8 manual switches (*open/close*) allow the positioning of the valves on the injection lines to fully open or closed positions (V031-V038); 6 other similar switches control the steam feed lines to the turbine driven pumps (V151-V154, and V161-V162).

The most important actuators are the 4 automatic regulators.

These govern the feed sections to each steam generator and act directly on the regulation valves (V031-V038) by adjusting opening or closure, so as to obtain the mass flows set by the regulators.

Normally, only the four automatic regulators, or controllers, are actually utilised for regulating the mass flows, including the isolation of a line or steam generator, which is obviously obtained by setting the desired mass flow to zero.

All other actuators are only used for setting fixed values, i.e., fully opened or closed valve, or pumps running or shut, and are utilised in the case of erroneous initial setting or failure or malfunctioning of the automatic regulation system.

Accident Scenario

Given the detailed representation of many components and instrumentation, the number of accident scenarios involving the AFWS, that may be taken into consideration, is very high.

In general, these scenarios cover "classical" initiating events, which can be further complicated by malfunctioning of components, in order to create special contextual and workload conditions which could affect operator behaviour.

Three main "classical" accident initiating events are considered in this sample case study, namely:

1. The *Loss of Off-site Power* (*LOP*) accident.
2. The *Loss Of Feed-water* (*LOF*) accident, due either to a leakage or a failure of a piping line on the secondary side of a steam generator.
3. The *Loss Of Steam-flow* (*LOS*) accident, due to turbine line failure.

In all these accidents, after the initiating event, the reactor is shut-down by the protection systems, and the automatic start-up of the AFWS is activated.

The operator is expected: a) to identify the type of accident and initiating event; b) to follow the dynamic sequence of events; and c) to manage the AFWS, in order to ensure gradual reactor cooling and de-pressurisation until the Long Term Cooling System takes over.

The goals and tasks, and the most appropriate behaviour of the operator will be discussed in detail in the forthcoming sections related to the Task Analysis and to the development of the operator model.

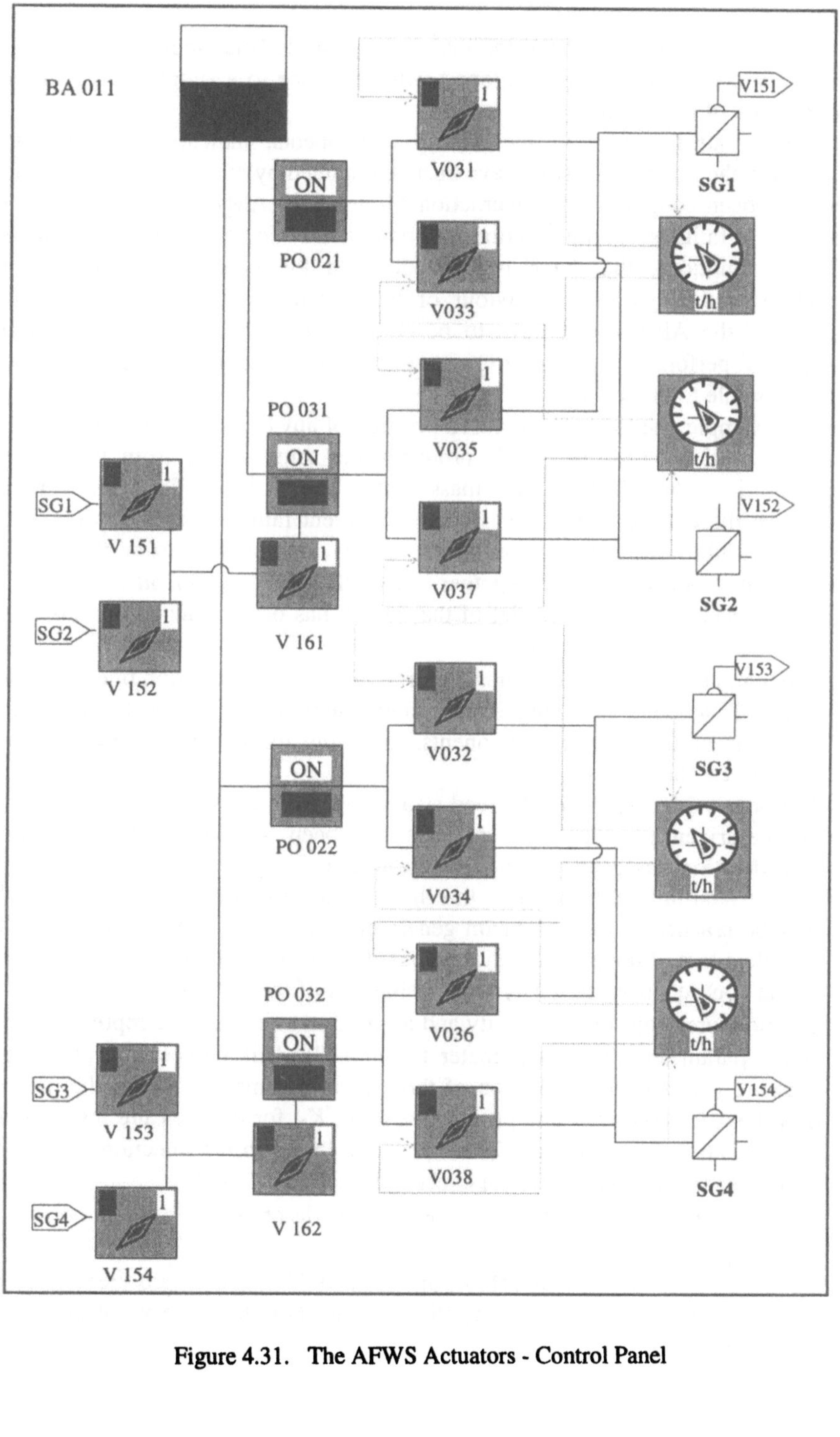

Figure 4.31. The AFWS Actuators - Control Panel

4.5.2 Machine Model and Simulation - Case Study AFWS

The numerical simulation of the thermal and hydraulic behaviour of the Auxiliary Feed-Water System has been developed with reference to a simplified geometrical configuration of the system.

However, all functions of the AFWS components, indicators and actuators, discussed in the previous section, have been maintained by means of a logical model, which has been coupled to the Interaction Model (DYLAM). This has enabled the simulation of failures and malfunctions of components and control instrumentation.

The assumptions, that have been made with the objective to simplify the simulation of the hydraulic behaviour of the system, do not affect the physical response of the AFWS, and have to be coupled with the logical models of the components' performance, in order to obtain an overall system behaviour which is as near as possible to the reality.

As an example, the failure in a closed position of any of the valves on an injection lines, e.g., V031, V041, or V051, produces the same effect, from the physical viewpoint, i.e., the blockage the mass flow. Therefore, while from a logical perspective these three valves may demand different failure rates and probabilities and may exhibit different failure states, modes and causes, from the physical viewpoint they can be "lumped" together in a single *super-component.*

In substance, the physical model of the AFWS has been obtained by making the following assumptions (Figure 4.32):

- The regulation, check and manual control valves on the injection lines have been "lumped" together in a single super-component, which is able to simulate all functions of these three components, including the one way flow limitation, typical of a check-valve.
- The physical behaviour of the feed-steam line, from the steam generators to the turbine-driven pumping systems, has not been simulated. However, some consideration has been included in the physical model of the possibility of a turbine fatal failure, based on a threshold value of minimum void fraction, that may be reached in case of steam generator malfunction. This event causes the complete loss of the pumping system based on the turbine-driven pumps and has strong consequences on the dynamic behaviour of the AFWS.
- The pressure drops due to gravity and all friction effects can be represented by a single parameter ("lump parameter technique"). This reduces the problem of pressure losses to the definition of three main parameters: K_1 for the pressure drop between feed-tank and pumping system; K_2 for the pressure losses within the regulation sections; and K_3 for pressure losses in the injection lines to the steam generators, i.e., across the check valves V101 - V104.
- The characteristic behaviour of pumps can be described by a linear correlation of the type: $Q = \alpha \Delta p + \beta$.
- The hydraulic and thermal behaviour of the AFWS and Steam Generators are loosely coupled, under the hypothesis that they have very different time constants.

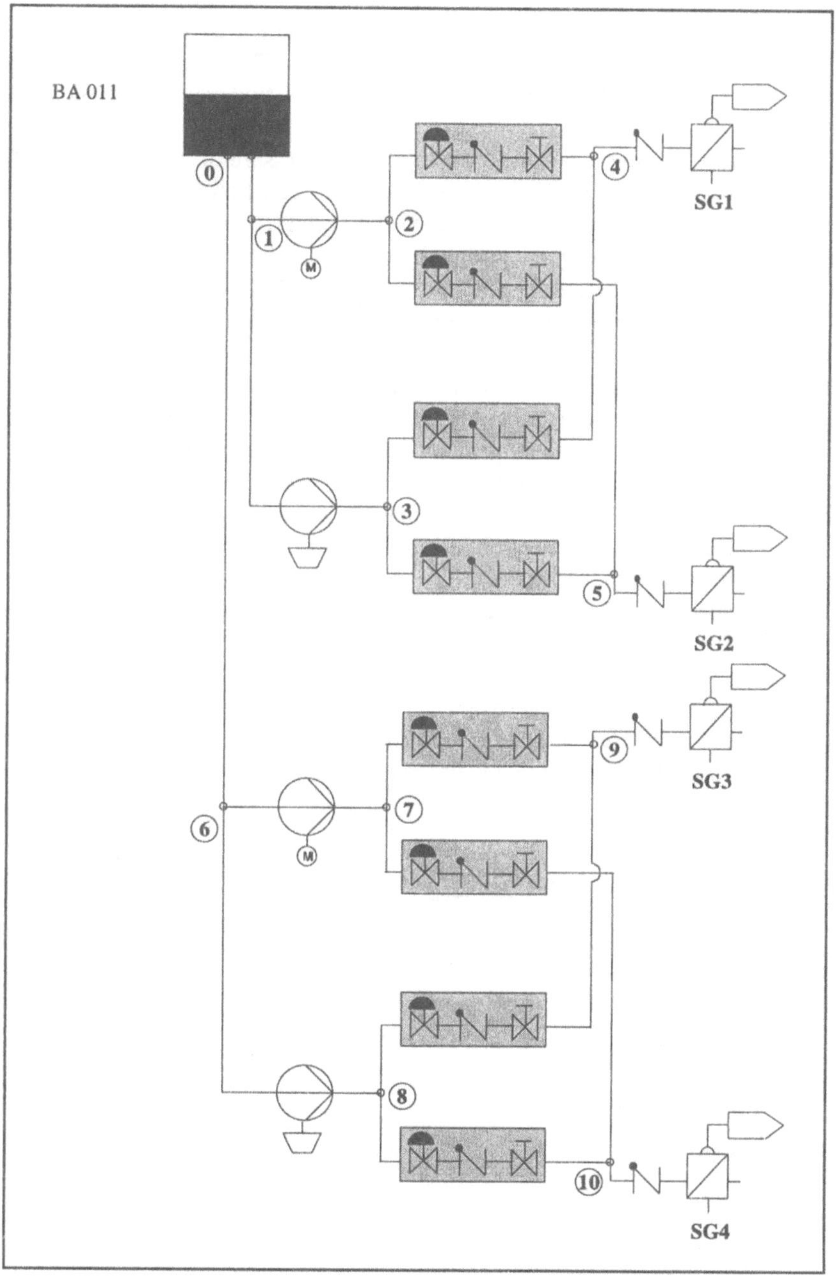

Figure 4.32. Simplified Scheme of the AFWS

Hydraulic Behaviour of the AFWS

The hydraulic behaviour of the AFWS has been obtained by defining, firstly, the boundary conditions that affect the hydraulic network. These have been coupled to the simplifying assumptions, previously discussed, so as to develop a simple physical model.

The boundary conditions consist of the hydraulic pressure of the water in the *feed-tank BA* and the pressures in the 4 steam generators. The first boundary condition is simply governed by the hydraulic head of the water in the tank, which changes relatively slowly and can be quite straightforwardly obtained by a simple calculation. While the pressures in the steam generator are much more complex to evaluate, as they depend on the thermal response of the whole system, and have much faster transient behaviour than the hydraulic head of the water in the tank.

The first problem tackled here is the calculation of the hydraulic equilibrium within the AFWS system, leaving the question of the coupling of the dynamic behaviour of the AFWS and steam generator for a forthcoming discussion.

Therefore, the AFWS has initially been structured in 10 important nodes and relative sections (Figure 4.33). Assuming that the pressures in the steam generators are known, the conservation equations for mass flows in all nodes of the system and for pressure drops within all sections have been developed and have been coupled with the equation that defines the pressure head in the tank, at the suction point (*node 0*).

The system of equations that has been developed is as follows. Here, for brevity, only the equations for the main-feed-line *TA* and for the water pressure at the suction point of the feed-tank have been reported. The equations for *TB* are omitted, as they are exactly symmetrical to those of *TA*, for their corresponding nodes and steam generators:

$$p_0(t+\Delta t) = \left\{ h(t) + \left[Q_{in}(t) - Q_{out}(t) \right] \frac{\Delta t}{B} \right\} \rho g + p_N \tag{4.17}$$

$$Q_{01} = Q_{12} + Q_{13} \tag{4.18}$$

$$Q_{12} = Q_{24} + Q_{25} \tag{4.19}$$

$$Q_{13} = Q_{34} + Q_{35} \tag{4.20}$$

$$Q_{4SG_1} = Q_{24} + Q_{34} \tag{4.21}$$

$$Q_{5SG_2} = Q_{25} + Q_{35} \tag{4.22}$$

$$Q_{01} = K_1 A_{01} \sqrt{|p_0 - p_1|} \tag{4.23}$$

$$Q_{12} = a(p_1 - p_2) + b \tag{4.24}$$

$$Q_{13} = a(p_1 - p_3) + b \tag{4.25}$$

$$Q_{24} = K_2 A_{24}(t)\sqrt{|p_2 - p_4|}\,H(p_2, p_4) \tag{4.26}$$

$$Q_{25} = K_2 A_{25}(t)\sqrt{|p_2 - p_5|}\,H(p_2, p_5) \tag{4.27}$$

$$Q_{34} = K_2 A_{34}(t)\sqrt{|p_3 - p_4|}\,H(p_3, p_4) \tag{4.28}$$

$$Q_{35} = K_2 A_{35}(t)\sqrt{|p_3 - p_5|}\,H(p_3, p_5) \tag{4.29}$$

$$Q_{4SG_1} = K_3 A_{4SG_1}(t)\sqrt{|p_4 - p_{SG_1}|}\,H(p_4, p_{SG_1}) \tag{4.30}$$

$$Q_{5SG_2} = K_3 A_{5SG_2}(t)\sqrt{|p_5 - p_{SG_2}|}\,H(p_5, p_{SG_2}) \tag{4.31}$$

....

....

where:

Q_{ij}	Mass flow from *node i* to *j* or from *node i* to *steam generator* SG_k
Q_{in}	Make up flow in the feed-tank BA011 = constant
Q_{out}	Mass Flow out of Tank BA011 = $Q_{01} + Q_{06}$
p_i	Pressure in *node i*
p_{SGi}	Pressure in *Steam Generator i*
K_i	Pressure loss coefficient
Δt	Time interval of hydraulic simulation of AFWS
h	Water level in Tank BA011
B	Tank base surface
p_N	Nitrogen pressure in tank = constant
A_{ij}	*Valve setting parameter*: $A_{ij} = 0$, valve closed, $A_{ij} = 1$, valve fully open, $0 \le A_{ij} \le 1$, valve partially open by operator or automatic regulation
$H(p_i, p_j)$	Step function for the simulation of the effect of one way check valves:

$$\begin{cases} H(p_i, p_j) = 1 & p_i \ge p_j \\ H(p_i, p_j) = 0 & p_i < p_j \end{cases} \tag{4.32}$$

The system of equations (4.17) - (4.31) for *main-feed-line TA* and the equivalent system of equations for *main-feed-line TB* are systems of non-linear algebraic equations which describe the hydraulic equilibrium within the AFWS at a specific instant. These systems of equations have to be solved simultaneously at each time instant t of the simulation. The solution is obtained by the application of the generalised iterative method of Cross (Douglas, Gasiorek, and Swaffield, 1983), and generates the mass flows in all sections and pressures in all nodes of the AFWS system at times $t_i = t_{i-1} + \Delta t_{AFWS}$.

The boundary conditions, which are updated at each time interval of the hydraulic calculation, Δt_{AFWS}, are obtained from the calculation of:

- The operator actions, which define the *setting* of the automatic and manual regulations of the actuators, i.e., the parameters A_{ij}, in the systems of equations; and
- The pressures in the steam generators P_{SGi}, which are derived from the calculation of the thermal dynamics within the SGs.

Thermal Behaviour of the AFWS and Dynamic Solution Algorithm

Thermal Behaviour of the AFWS

The thermal behaviour of the AFWS is related to the response of the Steam Generators to the heat transfer from the Primary Circuit of the Reactor and to the feed-water flow from the AFWS.

The model utilised for the simulation of the physical behaviour of a Steam Generator is the same as the model already described in Chapter 2 (§ 2.5.2). For completeness, we will only show the solving equations and main parameters:

$$\frac{dh}{dt} = \frac{1}{V_{tot}\left(\frac{\partial v}{\partial p} + v\frac{\partial v}{\partial h}\right)}\left[v\frac{\partial v}{\partial p}Q_{eff} + v^3(w_{out} - w_{in})\right] \tag{4.33}$$

$$\frac{dp}{dt} = \frac{1}{V_{tot}\left(\frac{\partial v}{\partial p} + v\frac{\partial v}{\partial h}\right)}\left[-v\frac{\partial v}{\partial h}Q_{eff} + v^2(w_{out} - w_{in})\right] \tag{4.34}$$

where:

h	enthalpy	t	time	p	pressure
v	specific volume	w	mass flow	x	mass vapour quality
M	total mass of water	T	temperature	Q	exchanged power
V	volume	z	water level	c	specific heat

and

$$Q_{eff} = Q + w_{in}(h_{in} - h) - w_{out}(h_{out} - h) \tag{4.35}$$

and:

$$x = \frac{h - h_{liq}}{h_{vap} - h_{liq}}$$

$$v = v_{liq} + x(v_{vap} - v_{liq})$$

$$\frac{\partial v}{\partial p} = \frac{v_{vap} - v_{liq}}{h_{vap} - h_{liq}}$$

$$\frac{\partial v}{\partial p} = \frac{v[T(p)] - v[T(p - 0.5)]}{0.5}$$

The boundary condition defined by the exchanged power in the steam generator between the Primary and Secondary Circuit is assumed to be known and is described by an exponential equation of the type:

$$Q = Q_0 e^{-\alpha t} \tag{4.36}$$

which follows the decay heat generated in the reactor after shut-down.

The equations for the state variables enthalpy and specific volume of the water in liquid and vapour conditions, are evaluated with reference to equilibrium temperature and pressure:

$$h_{liq},\ h_{vap},\ v_{vap},\ T(p) \quad \textit{from tables or polynomials (Meyer - Pittroff)}$$

or

$$h_{liq} = c_p T$$

$$h_{vap} = c_p T + \lambda$$

$$v_{vap} = v_{vap}(T)$$

$$T = T(p) = \textit{equilibrium liquid - vapour}$$

$$h_{in}(t) = \; constant \quad or \quad = h_{in}(t - \Delta t)$$

$$h_{out}(t) = h_{out}(t - \Delta t) \approx h(t - \Delta t)$$

$$w_{out}(t) = f\left[p(t - \Delta t)\right] \quad or \quad = \; w_{out}(t - \Delta t)$$

$$w_{in}(t) = w_{in} \quad from\ operator\ model$$

Dynamic Solution Algorithm

The coupling of the simulation of the hydraulic equilibrium within the AFWS with the simulation of the pressure behaviour within the Steam Generators (SGs) leads to the description of the dynamic behaviour of the integrated system.

As already mentioned, the technique utilised is based on the fact that the two systems (AFWS and SG) have rather different time constants. The time constant for the hydraulic behaviour of AFWS is much greater than that of the thermal response of the Steam Generators. In practice, this means that the time step utilised to solve the equations of the steam generators is much smaller that the time step that is used for solving the dynamic behaviour of the AFWS: $\Delta t_{SG} \leq 0.001\ \Delta t_{AFWS}$.

Therefore, the two simulations have been coupled with an explicit method with no other iterations in addition to those already utilised for the individual solutions of the systems of equations describing AFWS and SG.

The numerical solution and the integration of the calculations of the AFWS and SG are obtained by the following logical flow of calculations and assumptions (Figure 4.33):

1. At each macro-time step of the AFWS, Δt_{AFWS}, the mass flows to the Steam Generators (Q_{SGi}) are kept constant, while equations (4.33) - (4.36) solve the problem of the dynamic variation of pressures and enthalpies in the Steam Generators, using their own micro-time step, Δt_{SG}.
2. The simulation of the SGs performs a whole cycle of calculations for a sufficient number of Δt_{SG}, to cover a Δt_{AFWS}. Then the new values of the pressures reached in the Steam Generators are utilised to update the mass flows in the sections and pressures in the nodes the AFWS, by the system of equations 4.17, 4.18 etc.. These calculations define the new updated mass flows to each steam generator, and a new cycle is then started.

This loop allows to calculate the response of the physical system, and has to be coupled and synchronised with the simulation of operator behaviour in order to define the accident evolution.

Therefore, while the flow-chart of Figure 3.33 shows the solution of the thermal and hydraulic coupling of the Auxiliary Feed-Water System with the Steam Generators, the complete interaction chart can be developed only once the human behaviour model is defined.

The simulation of the AWFS has been developed in a computer program in standard language FORTRAN 77.

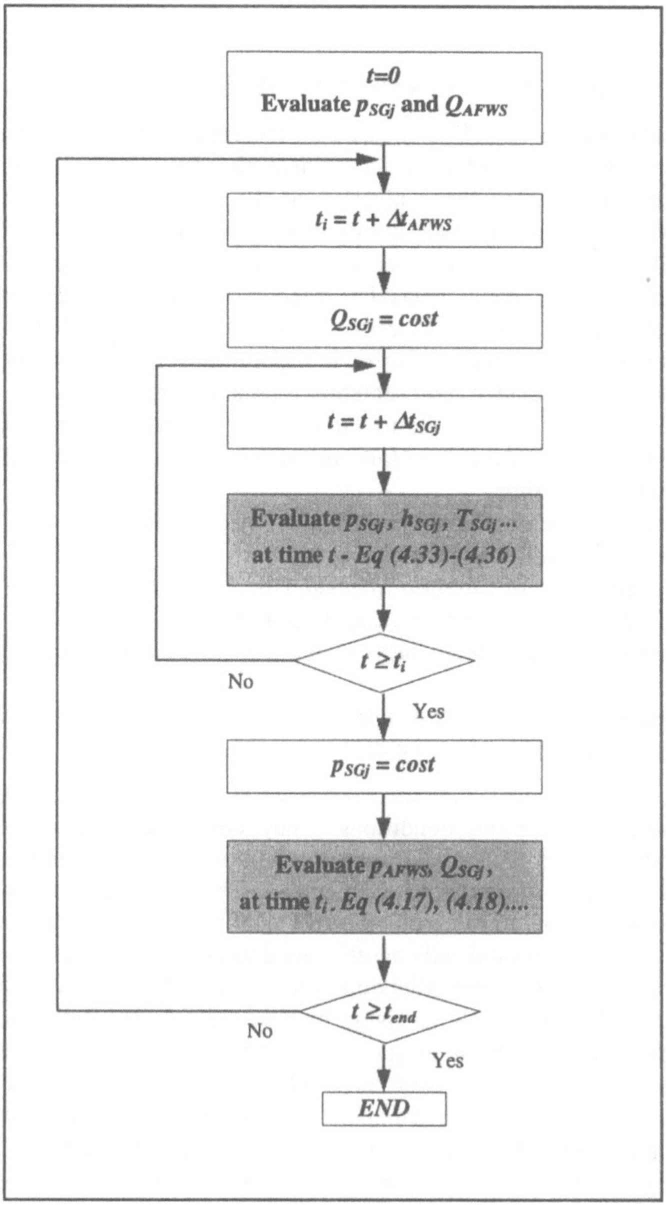

Figure 4.33. Flow diagram of the solution algorithm for the AFWS simulation.

4.5.3 Human Model and Simulation - Case Study AFWS

Definition of Problem Boundaries and Aim of the Simulation

The case study of the AFWS is focused on the application of the Human-Machine Interaction simulation architecture to the analysis of the control procedures in the case of accidents, and to the evaluation of the operator response to component failures and malfunctioning of the information display system.

These objectives can be achieved once a detailed model of state variables and physical behaviour of components has been developed, in combination with the response of control panels. Moreover, the control panels have been further subdivided in *indicators* and *actuators*, in order to differentiate between the various alternatives that this detailed representation is able to offer.

With this variety, it is possible to simulate the system physical behaviour due to hardware component failure, or to false information display, or even to both failures occurring at the same time.

In particular, some combinations of human-machine interaction will be selected, so as to enhance the following topics:

1. The development of *operator strategies and cognitive processes*, (rather than behavioural performances), in case of accident.
2. The analysis of the effectiveness of the *control instrumentation* for diagnostic and planning processes.
3. The analysis of *control and emergency procedures* for managing the AFWS during crucial phases of an accident.

The accidents that will be considered have already been described, at least for what concerns their initial conditions. They cover the area of "design basis" accidents, and, thus, no abnormal transient conditions will be studied.

This is reckoned sufficient for the objectives of the case study, even if the method is obviously applicable to more complicated configurations and accidents, provided that the appropriate physical and cognitive models are considered.

The simulation of operator behaviour will concentrate on the cognitive functions and processes that govern the decision making and on the context conditions that may influence human behaviour, specially in relation to the control display panels. The question of "false alarms" and "false information" will be a major issue, typical of plants with many redundancies and comprehensive information displays.

Human errors and flaws will be considered as systematic predefined events, as well as, logical consequences of external factors, such as hardware components malfunctions and information display failures.

The interaction model will be based on the DYLAM approach, that offers the possibility to combine probabilities of failures with logical configurations of human behaviour and system performance, within a dynamic simulation architecture.

The specific application of the human-machine interaction architecture for human reliability assessment will not be developed in detailed (Cacciabue, 1994).

Cognitive Task Analysis and Field Study of Working Context

The Task Analysis for the case study of the Auxiliary Feed-Water System has been performed with respect to a specific design of a real AFWS, which is very similar to that shown in Figure 4.28 - 4.32 and for which the control and emergency procedures are well known.

Unfortunately, no field experiments could be carried out and, thus, the hypotheses and assumptions made during the critical review of procedures and interfaces could not be sustained by, or adapted to, the results of empirical analysis.

This aspect will not impinge on the methodological approach nor will modify the demonstration purposes that we intend to pursue with this case study.

However, the reader must be aware that, in a real working context, the step of empirical analysis and field study lies at the foundation of the whole human-machine simulation approach. Therefore, this important element of the Task Analysis can not be simply omitted or performed at a shallow level. On the contrary it should always be carried out in depth, and should be part of the background notions and culture of any human factors specialist.

Performance of Procedures

As already mentioned, the AFWS is automatically activated as a consequence of well defined initiating events and physical criteria.

These criteria are based on the following signals arriving in the control room:

- Low level in the Steam Generators.
- *Loss Of Feed-water* (*LOF*) to a Steam Generator on the secondary side or *Loss Of Steam-flow* (*LOS*) from a steam generator outlet.
- *Loss of Off-site Power* (*LOP*).

The AFWS can also be started manually, even if this is only a back-up system which is not usually considered.

The objectives of the emergency management procedures differs slightly in relation to the type of accident under control. In principle, four main goals can be identified, covering the entire spectrum of possible human interactions. These four goals can be summarised as follows:

1. *Identification* of the type of accident. It is important to distinguish between two major categories of accidents, namely: a) failure of a pipeline with loss of coolant; and b) loss of electrical power, maintaining the complete integrity of the mechanical system.
2. *Identification* and *isolation* of the sub-system where the event has occurred, including the connected steam generator. This goal applies only in case of failure of a pipeline or, more in general, of a large leakage of coolant.
3. *Initial Regulation* of mass flow to remove the residual heat, generated inside the reactor as consequence of the nuclear decay following shut-down, and to sustain the process of transfer from hot-shut-down to cold-shut-down conditions. This activity aims at safeguarding the integrity and the cooling of the reactor core and reactor containment buildings.
4. *Gradual regulation* to complete the de-pressurisation of the secondary side.

In order to reach these objectives, the operator has to perform a series of checks, inspections and actions during the whole transient period of the accident.

We will now focus on the tasks which involve the activity of a *single operator* controlling the Auxiliary Feed-Water System, even if the analysis of the accident management procedures has clearly shown that:

a) Several other control and emergency systems have to be managed during the accident, in order to protect the whole reactor.
b) Teamwork has to be carried out, with operators co-operating directly, by visual and aural contact, as well as collaborating remotely, by simple aural interaction.

The whole period of operation of the AFWS is rather extended, as that the system has to sustain the passage from hot-shut-down to cold-shut-down, and should cover approximately seven hours of operations. This is approximately the duration of operation of most emergency and protection systems which are utilised to control an accident in a nuclear power plant.

In the case under study here, however, the vast majority of decisions, actions and interventions are condensed in the first half hour of the accident, and consequently, we will only concentrate on this relatively short time interval.

Focusing on the specific tasks related to the control procedures of the AFWS, the following activities are foreseen (Figure 4.34):

1. *Identification*
   The task of *Identification* implies that the operator differentiates between the major types of accident, namely *Loss of Feed-water (LOF)*, *Loss of Steam-flow (LOS)* and *Loss of Off-site Power (LOP)*. This task can be performed by verifying values and trends of certain physical variables, namely: mass flows, pressures and levels. The identification process demands a certain period of evaluation and validation of the hypotheses that are being formulated, as the system evolves dynamically. A number of (mental) iterations are expected.
   A. The loss of coolant accident may occur either on the cold-leg (feed-line, *LOF*), or on the hot-leg (steam line, *LOS*) of the secondary side circuit of the reactor cooling system. The Primary Circuit is assumed to remain intact. The effects of such a type of accident, in terms of size and rate of change of pressures, mass flows and levels, obviously depend on the dimension and the location of the leakage. In any case, in a rather short time period, i.e., *five minutes*, these physical quantities reach values which are outside the norms, and thus enable the identification of the line and the corresponding Steam Generator where the failure has happened. Both of these need to be isolated before taking any corrective action and regulation.
   B. The *Loss of Off-site Power* does not lead to very relevant differences between the state variables of the feed lines and steam lines of the system. On the contrary, all quantities start a gradual change from their normal values in a uniform way. This rate of change, in the case of LOP, is much smaller than in the case of LOF/LOS accidents.

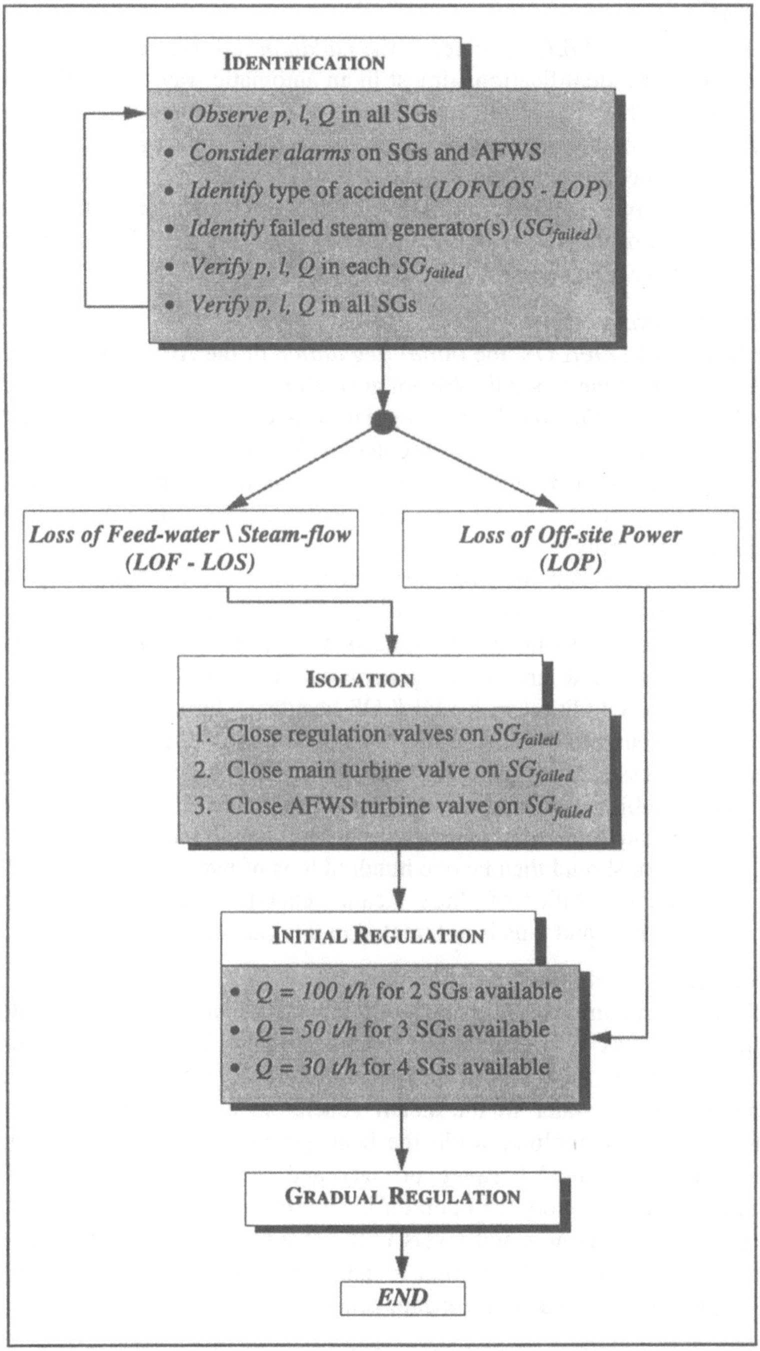

Figure 4.34. Flow diagram of the AFWS simulation.

2. *Isolation*

   In the case of LOF/LOS accidents, the *Isolation* task is straightforwardly carried out, following identification, almost in an automatic way. The following actions are performed:

   2.1 Closure of the regulation valves on the feed-line of the steam generator where the failure has occurred.

   2.2 Closure of the main turbine valve of the same steam generator.

   2.3 Closure of the valves that feed steam to the turbine driven pumping system of the AFWS (Figure 4.28).

3. *Initial Regulation*

   In the case of LOF/LOS, the *Initial Regulation* of the AFWS is started following the *isolation* of the failed line/steam generator.

   In the case of LOP, the Initial Regulation is carried out immediately after the *identification* task. This initial regulation depends on the number of available steam generators and hardware components, as well as on the type of accident. In particular, with reference to the available SGs, the following regulation requirements apply:

   A. In the case that all four steam generators are available, typical system configuration encountered in a LOP situation, an initial mass flow of thirty tons of water per hour (*30 t/h*) needs to be fed to each steam generator.

   B. In the case that three steam generators are available, which is the "normal" expected condition in a LOF/LOS accident, the initial mass flow to each steam generator should be fifty tons of water per hour (*50 t/h*).

   C. In the case that only two steam generators are available, i.e., when the unavailability or failure of some hardware components has further complicated the accident scenario, the initial mass flow to each steam generator should then be one hundred tons of water per hour (*100* t/h).

   The unavailability of three steam generators is considered outside the design basis and thus is not treated in this case study.

4. *Progressive Regulation*

   Following the initial phase of the accident, which last approximately thirty minutes, the task of the operator is to adjust in time, by a *progressive regulation*, the mass flow to each steam generator. This task has the objective of maintaining the water in the steam generators at a constant level, and thus ensuring reactor cooling, while the heat generated in the nuclear core (decay heat), and exchanged between primary and secondary circuit, decreases. To perform this regulation, the operator has to maintain under constant observation temperatures, pressures and levels in the steam generators and must reduce the mass flows appropriately in order to maintain a balance between the water flows and the heat exchanged in the steam generators.

All these observations, regulations and control actions, in general, are performed by the operator using the information display and control panels (Figure 4.29 - 4.31). In particular, all regulations and isolations of feed-flow to steam generators are performed by acting on the automatic flow regulation system (Figure 4. 31), which allows the setting of the desired mass flow (zero setting meaning isolation) to each steam generator.

The automatic system adjusts the opening of the interested regulation valves, so as to respond to the regulator settings. The other actuators available to the operator for controlling the mass flows are utilised only if the automatic system does not respond. They do not allow a refined adjustment of the mass flows and are operated only locally and manually.

As already mentioned, we will concentrate on the first phase of the procedure, i.e., the first 30 minutes of the accident, when the operator has the tasks to observe the evolution of the accident, to formulate a first identification of the initiating event and type of accident, and to carry out the isolation, if necessary, and the initial regulation. The rest of the control procedure is not within the scope of the study.

The reasons for such a choice is that the second period of AFWS control is not very demanding in terms of mental activity and that the main objective of this case study is to focus on cognitive processes, particularly related to the diagnostic function, rather than the actual performance of control actions.

Factors Affecting Behaviour

The objectives of this sample case are to perform a "standard" analysis of accident evolution in the case of human errors of omission in the performance of tasks, and to study the effects of failure of components and control instrumentation on the cognitive and decisional processes of the operator.

The variety of situations and scenarios that may be generated is due to the combination of three main elements: the *initiating event;* the *failure of components;* and the *human errors.*

- Three types of *initiating event* are considered, namely *LOF*, *LOS* and *LOP*. They can be further detailed, thus increasing the variability of case studies, by defining location and dimension of failures.
- *Failure of components* cover many possible events, including time dependent failures, hardware failures, malfunctions of actuators and indicators, etc.
- *Human errors* can focus on inappropriate *identification* or *planning* in the early stages of the accident and/or on errors of omission during the performance of control tasks.

This produces an enormous variability which needs to be reduced by making a series of assumptions resulting in a selected number of sequences.

Following a first set of assumptions a matrix of accident configurations can be devised:

1. *Hardware components* and *control instrumentation* are separated, even if they belong to the generic category of components, and exhibit similar reliability performances. The reason for such distinction is that we are particularly interested in studying the response of operators to faulty information of display panels, rather than to failures of hardware components.

2. *Recovery* from errors has not been considered as a random or probabilistic event, but has been associated with the reasoning process of the operator. Therefore, recovery becomes a consequence of external factors, such as alarms, and relevant information, which may induce a revision of previously performed activities and a correction of errors.

Forms of Erroneous Behaviour and Main Case Studies

Six types of case studies have been defined and are formatted as follows (Table 4.6):

- The type of initiating event is indicated by a *letter*: **A** refers to the initiating event *LOF*; **B** to the initiating event *LOS*; and **C** to the initiating event *LOP*.
- The type of failure of either human, or hardware component, or control instrumentation is indicated by a *number*: **0** identifies nominal component (no failure) and error-free human behaviour; **1** identifies human errors, but nominal behaviour of hardware components and control instrumentation; **2** means failure of hardware components but nominal human behaviour and control instrumentation; and **3** means failure of control instrumentation (actuators and indicators) but nominal human behaviour and hardware components.

Table 4.6. Matrix of Case Studies for the AFWS analysis

Case Study	**Operator Behaviour**	**Hardware Components**	**Actuators & Indicators**	**Initiating Event**
Case 0A	*A*	*A*	*A*	*LOF*
Case 1A	*F*	*A*	*A*	*LOF*
Case 2A	*A*	*F*	*A*	*LOF*
Case 3A	*A*	*A*	*F*	*LOF*
Case 0B	*A*	*F*	*A*	*LOS*
Case 0C	*A*	*F*	*A*	*LOP*

F = Faulty performance
A = Appropriate "nominal" performance

Table 4.6 contains the whole spectrum of accidents that will be studied with the human-machine simulation for the AFWS. The initiating event on which the most emphasis has been placed is the Loss Of Feed-water on the secondary side. This is because the LOF accident presents the most complex and demanding activity in the first 30 minutes of transient and may evolve in many different directions according to the occurrence of component failures.

More precisely, the six cases that will be analysed are characterised by the following conditions and behaviours:

Case 0A

Case 0A is the nominal configuration, where operator and components (hardware and information display) behave and respond according to design specifications and procedure. This case allows the evaluation of accident management. The choice to analyse the *LOF* accident is related to the relative complexity of the diagnostic phase and initial regulation.

Case 1A, 2A

Case 1A is the configuration, where all components (hardware and information display) respond according to design specifications and procedure, but the operator may make some errors. The type of scenario is, like in the previous case, the *LOF* accident. The errors that can be studied have been restricted to errors of omissions and/or inappropriate planning and performance of control actions.

In particular, we will consider the lack of observation of mass flow, pressure and level on the steam generator where the failure has occurred, combined with an error of planning control actions.

For example, if the *LOF* event has occurred on the SG_1 feed-line, we will study the error of observation of the variables of SG_1 and the systematic exchange between SG_2 with SG_1 at the level of control actions. This means that all procedural actions carried out on SG_1 are performed on SG_2 and vice versa.

Case 2A considers nominal human behaviour and nominal control instrumentation coupled with possible failures of hardware components.

Case 3A

Case 3A exhibits nominal operator behaviour, and hardware components, but considers faults in the information display panels. The selected accident scenario is again the *LOF*. In particular, two possible cases of false alarms will be analysed.

In one case the alarm will refer to the steam generator on the same main feed-line as the steam generator where the failure occurred. For example, initiating event *LOF* on SG_1 and false alarm on SG_2.

The second case will consider the event of false alarm on a steam generator completely independent from the steam generator where the event occurred. For example, initiating event *LOF* on SG_1 and false alarm on SG_3.

Case 0B and 0C

Case 0B, and *Case 0C* are similar scenarios for what concerns the nominal behaviour of the operator, hardware components and control instrumentation. However, they differ for the initiating event.

Case 0B is the case of *Loss Of Steam-flow* (*LOS*) due to a failure on the secondary side hot leg of a steam generator. *Case 0C* is the loss of off-site power with no failures of components or human errors.

The failures of hardware components that could be considered are mainly related to the inappropriate functioning of the pumps or main control valves on the injection lines of the steam generators. However, these failures will not be considered with the intention to reduce complexity and variety of scenarios.

A final consideration on Table 4.6 is that the 6 cases described in the matrix, except for *Case 0A*, represent in reality *families* of accidents in that we have only limited the combinations of initiating events, human erroneous behaviours and malfunctions of components/control instrumentation.

However, we have also to account for the variety due to the number and types of errors and malfunctions that can be logically considered and the number of components and indicators that may fail.

Theoretical Model of Operator Behaviour and Numerical Algorithms

Basic Concepts and Model of Cognition

The model of operator behaviour for the AFWS case study has been developed in such a way as to represent the observations, decisional processes and control actions carried out by an experienced operator, as foreseen by the designer of the control procedures.

Specific attention has been dedicated to cognitive processes, in particular the diagnostic phase, and it has been assumed that the operator is familiar with the procedures contained in the operating manuals and tries to follow instructions.

The events of errors will be either artificially included in the model, with no specific reference to working environment conditions (this modelling development has been discussed in the previous case study), or will be induced by the interpretation of faulty signals (false alarms, false indications) or malfunctioning information display panels.

The theoretical model chosen as reference operator model of the AFWS is COCOM (§ 3.2.6), limited to the "Competence" component of the modelling, which is substantially contained in the Very Simple Model of Cognition (Hollnagel, 1993a). The basic cognitive functions of *Perception, Interpretation, Planning* and *Execution* are considered, but the *Allocation of Resources* and the *Memory and Knowledge Base* are only superficially included in the model.

In practice, little attention will be paid to the "Control" component of the model, which selects the type of operator's competencies that are being applied (in terms of COCOM: "strategic, tactical opportunistic or scramble" control). This simplification is coherent with the assumptions made of paying little attention to the working environment, which, on the contrary, in COCOM terms affects the control mode.

In addition to the paradigm of the Very Simple Model of Cognition, we have included the concept of *salience*, already discussed in several sections of this book and utilised for the development of the simulation COSIMO (§ 3.5.3).

The *salience* allows the screening of information perceived by the operator and is particularly important in a modelling architecture like the one discussed for the AFWS, where many indicators (27 on the AFWS and 5 for each steam generator) and alarms (7 on the AFWS and 4 for each steam generator) are simulated.

The *salience* of a cue represents the content of information associated with that cue, as perceived by the operator, and is made of two parts: the *physical salience*, which is linked to the sensory relevance of a cue, e.g., physical appearance, sound, colour etc.; and *cognitive salience*, which is a measure of the importance associated to certain information by the operator, following expectancy, reasoning process, procedural requirements etc. As the concept of salience has already been widely described and discussed earlier, we will not dwell on it any further.

The information contained on the display panels and the physical quantities related to the performance of the AFWS and steam generators are transformed into measures and values of plant behaviour by a model based on *fuzzy logic* and *fuzzy functions*. In this way, the operator model is able to develop the cognitive functions of *Perception, Interpretation, Planning* and *Execution*.

All fuzzy measures are calculated in accordance to reference quantities derived from the task analysis and, specially, from the field studies.

Main Assumptions

The model starts by describing the operator behaviour from the instant of accident initiation and reactor shut-down ("scram").

The event of failure to "scram" the reactor is considered outside the design basis accident condition. Similarly, it is assumed that the operator perceives that the reactor has been shut-down as a consequence of an accidental event and that an emergency situation is under development.

As already mentioned, the most important aspects of the accident sequence are shown during the first 30 minutes of transient, when the operator has to identify the accident type, isolate the failed steam generator sub-system, if necessary, and perform the initial regulation. In many modern nuclear power plant, during the first five minutes of accident, no operator intervention is allowed. This is a design requirement which has the purpose to ensure that the operator team has a minimum time period for understanding the situation and selecting the most appropriate emergency or control procedure.

During the initial five minutes the system follows an automatic procedure, while the cognitive processes are developed and, in most cases, a decision is taken and the execution of control actions is carried out as soon as possible.

The erroneous behaviour of the operator, as already mentioned, will only be forced into the simulation by the input data and will not be correlated to working environment conditions. However, false alarms will be considered, as a contextual cause of inappropriate cognitive processes and behaviour.

Similarly, the recovery from errors or inappropriate behaviour will be considered only as a consequence of the cognitive processes connected to the information perceived and elaborated during the time evolution of the accident.

Model and Simulation of Salience and Perception

From the analysis of the tasks (Figure 4.34), the operator has initially to focus on, at least, 12 indicators, describing the state and trend of 3 physical variables for each steam generator.

These variables are particularly important for understanding the situation, for identifying the accident type, and for deciding the initial intervention and regulations. These variables are:

- Steam generator pressure in each (p_{SGi}).
- Steam generator water level (L_{SGi}).
- Steam generator inlet mass flow (from AFWS feed-line) (Q_{SGi}).

The first quantities are shown on the AFWS panel, while pressures and levels in each SG appear on specific display panels.

The operator is assumed to focus only on these 12 quantities (3 per steam generator), even if the display panels contain a very high quantity of other information. In practice, this other information becomes relevant only in the case of specific malfunctions of components, when a local and very particular indication, such as, for example, the state of a valve (*open* or *closed*) or a pump (*running* or *inoperative*), allows the immediate diagnosis of a failure.

Therefore, indications, other that the 12 initially considered, are taken into consideration only as a second order iteration process.

These 12 quantities are actually taken into account by the simulation of the operator through the concept of the *salience*.

From the beginning of the accident, the *salience* of the 12 variables related to the steam generators is calculated and normalised to *1*. The salience of each indication is dynamically updated with the accident evolution.

In order to consider the contribution of cognitive and physical salience to the overall *salience*, each indicators has been associated with three different measures: *cognitive salience*, *physical salience "trend"*, and *physical salience "value"*.

- *Cognitive Salience*

The *cognitive salience* of an indicator *i* (cs_i) has been calculated by considering the contribution of two parts, namely:

a) A linear function which starts at level *0.5* at time zero and drops progressively to zero in 300 seconds. This cognitive salience derives from the procedures, that require to keep under observation SGs main variables for the first 5 minutes. It vanishes as time passes and the understanding of the situation improves.

b) A stepwise function that increases the *cognitive salience* of all indicators relative to a certain steam generator, or AFWS main feed-line, when an alarm occurs on that steam generator or main feed-line.

$$cs_i(t) = cs_i(t-\Delta t) + 0.5(1 - t/300) + H(al_{SGi})0.5e^{-\lambda_i(t-t_{al})} \qquad (4.37)$$

where:

$H(al_{SGi})$ Step Function:

$H(al_{SGi}) = 1$ if an alarm has occurred, at time t_{al} on the SG or main feed line connected to the indicator *i*.

$H(al_{SGi}) = 0$ if no alarms occur in connection to the indicator *i*.

λ constant of decay of the influence of an alarm on cs_i.

- *Physical Salience "trend"*

The *physical salience* "trend" of an indicator *i* (pst_i) has been calculated by considering the relative rate of change of the variable associated to that indicator in an interval of 60 seconds.

This rate of change has then been coupled with a fuzzy rule that maximises the *physical salience* "trend" when the rate of change over 60 seconds is at least 30 %.

The following fuzzy rule and equation applies:

$$pst_i(t) = H_i \frac{1}{0.25}\left[\left|\frac{v_i(t) - v_i(t-\Delta t)}{v_i(t)}\right| - .05\right] + G_i \qquad (4.38)$$

where:

v_i Physical variable corresponding to the indicator *i*.

R_v $R_v = |[v_i(t) - v_i(t-\Delta t)] / v_i(t)|$, rate of change of the variable corresponding to the indicator *i*.

H_i, G_i Step Functions:

$H = 0$	if $R_v < 5\%$ or $R_v > 30\%$
$H = 1$	if $5\% \le R_v \le 30\%$
$G = 0$	if $R_v < 30\%$
$G = 1$	if $R_v > 30\%$

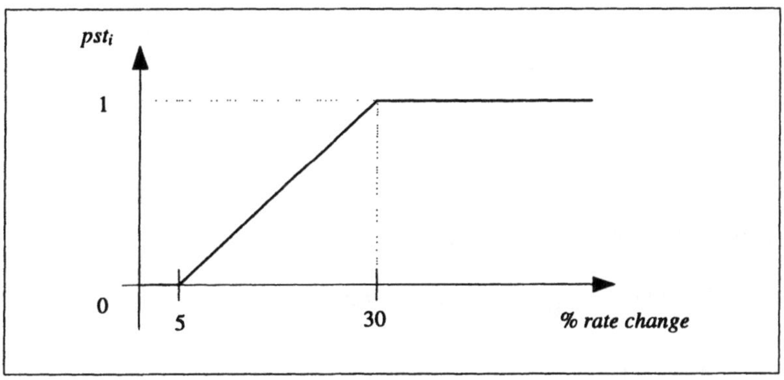

- *Physical Salience "value"*

The *physical salience* "*value*" of an indicator *i* (psv_i) is associated with the indicator readings relative to the 12 variables describing the behaviour of the 4 steam generators, i.e., pressure, level and inlet mass flow for each steam generator.

The fuzzy rules that are applied depend on the "normal" or "critical" values of each of the physical variables associated with the different indicators: in particular the following fuzzy functions apply.

1. Indicators of *Inlet Mass Flow to SGs* (Q_{SGi}):

$$
\begin{aligned}
&for\ Q_{SGi} < 50\ t/h,\ or\ Q_{SGi} > 160\ t/h \Rightarrow psv_{Q_{SGi}} = 1 \\
&for\ 90 \le Q_{SGi} \le 120\ t/h \qquad \Rightarrow psv_{Q_{SGi}} = 0 \\
&for\ 50 \le Q_{SGi} \le 90\ t/h \qquad \Rightarrow psv_{Q_{SGi}} = \left(90 - Q_{SGi}\right)/40 \\
&for\ 120 \le Q_{SGi} \le 160\ t/h \qquad \Rightarrow psv_{Q_{SGi}} = \left(Q_{SGi} - 120\right)/40
\end{aligned}
\tag{4.39}
$$

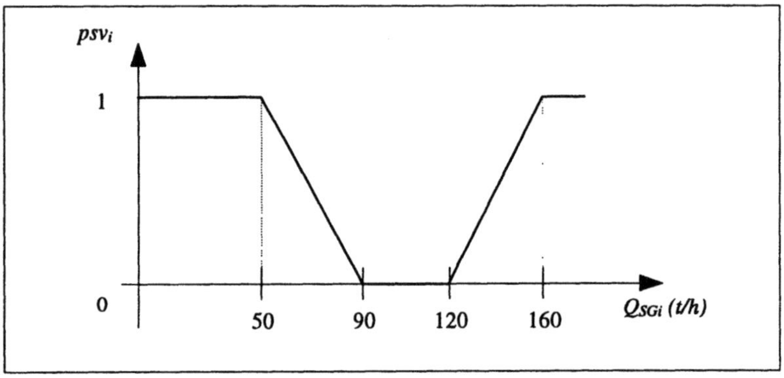

2. Indicators of *SG Pressures* (p_{SGi}):

$$
\begin{aligned}
&for\ p_{SGi} < 40\ bar,\ or\ p_{SGi} > 70\ bar \Rightarrow psv_{p_{SGi}} = 1 \\
&for\ 50 \le p_{SGi} \le 60\ bar \qquad \Rightarrow psv_{p_{SGi}} = 0 \\
&for\ 40 \le p_{SGi} \le 50\ bar \qquad \Rightarrow psv_{p_{SGi}} = \left(50 - p_{SGi}\right)/10 \\
&for\ 60 \le p_{SGi} \le 70\ bar \qquad \Rightarrow psv_{p_{SGi}} = \left(p_{SGi} - 60\right)/10
\end{aligned}
\tag{4.40}
$$

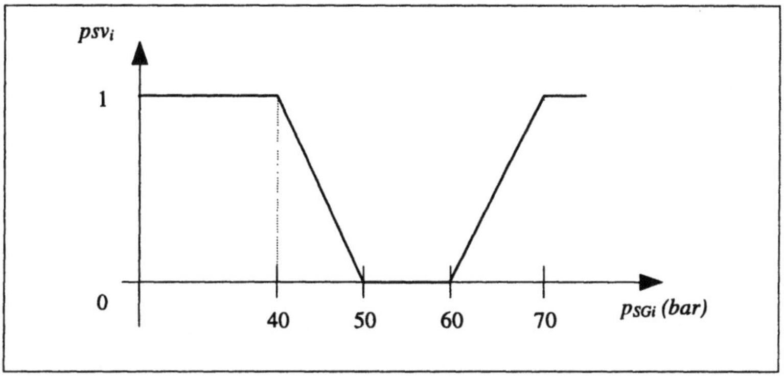

3. Indicators of *SG Levels* (L_{SGi}):

$$\left.\begin{array}{ll} for\ L_{SGi} < 6\ m,\ or\ L_{SGi} > 10\ m & \Rightarrow psv_{L_{SGi}} = 1 \\ for\ 7 \le L_{SGi} \le 9\ m & \Rightarrow psv_{L_{SGi}} = 0 \\ for\ 6 \le L_{SGi} \le 7\ m & \Rightarrow psv_{L_{SGi}} = \left(7 - L_{SGi}\right) \\ for\ 9 \le L_{SGi} \le 10\ m & \Rightarrow psv_{L_{SGi}} = \left(L_{SGi} - 9\right) \end{array}\right. \qquad (4.41)$$

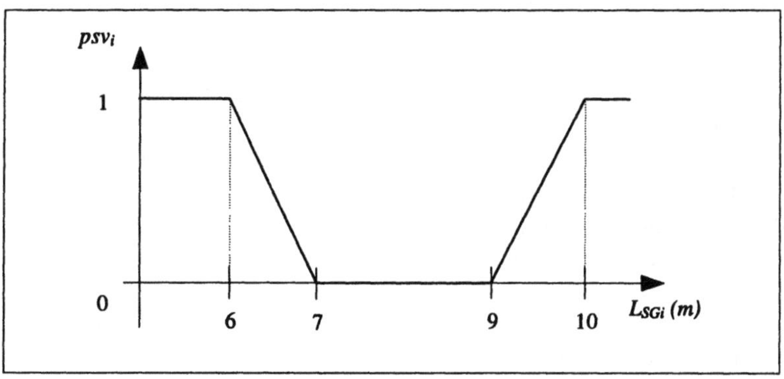

In summary, the overall time dependent *salience* associated to each indicator is calculated as the arithmetic mean of the contribution of the three different salience measures, i.e., the *cognitive salience*, the *physical salience "trend"*, and the *physical salience "value"*, individually calculated by means of the equations (4-37 - 4.41).

In this way, the normalised value of the *salience* can be obtained by the following equation:

$$sal_i(t) = \frac{cs_i(t) + pst_i(t) + psv_i(t)}{3} \tag{4.42}$$

By Equation (4.42), the *salience* associated to an indicator is always smaller or at maximum equal to *1:*

$$0 \leq sal_i(t) \leq 1 \tag{4.43a}$$

In order to differentiate between the information that is taken into consideration or is disregarded, a time dependent minimum threshold value of salience is defined, $sal_{min}(t)$, and the importance of each perceived variable is calculated in relation to sal_{min}, as the accident progresses. Note that $sal_{min}(t)$ is a function of the maximum value of the salience reached in the previous time step: $sal_{min}(t) = f[sal_{max}(t\text{-}\Delta t)]$.

The indicators whose salience falls below sal_{min} are ignored, i.e.:

$$< \textit{if}\ [sal_i(t) < sal_{min}(t)]\ \textit{then,}\ sal_i(t) = 0 > \tag{4.43b}$$

The salience of all the other indicators is utilised to identify the steam generator on which the operator focuses the attention and the initial diagnosis and planning.

<u>Model and Simulation of Interpretation and Planning</u>

On the basis of selected information the process of diagnosis, or interpretation, is carried out.

In practice, when the amount of perceived information focuses on a specific steam generator, as its physical parameters are rapidly diverging from expected normal values and trends, then the accident that is diagnosed is the *Loss Of Feed-water* (*LOF*) or the *Loss Of Steam-flow* (*LOS*). In both these cases, the initial isolation and regulation procedures are quite similar and to distinguish between the two types of accidents requires a more accurate reasoning on the mass flow in inlet and outlet of the steam generator.

In any case, the procedures for the management of the *LOF* and *LOS* accident for the first 30 minutes of the accident are the same.

On the other hand, when the information concerning the four steam generators shows similar values and trends for the physical variables (Q_{SGi}, p_{SGi}, and L_{SGi}) and no unusual behaviour is perceived, then the accident that is diagnosed is the *Loss of Off-site Power* (*LOP*). This diagnosis takes place approximately after 30 minutes of observation of accident evolution.

The numerical simulation adopted to represent the diagnostic process is based on the concept of the *diagnosticity* (D_{SGi}), which represents a measure of the relevance of the perceived information, relative to a specific accident type and scenario. More precisely, the *diagnosticity* is always associated with one specific steam generator, and represents the relevance of that steam generator for the scenario under development.

The *diagnosticity* of a steam generator is described by a normalised expression based on the *salience* of the indicators related to the steam generator versus the *salience* of all perceived indicators:

$$D_{SGi}(t) = \frac{\sum_{j=1}^{in_{SGi}} sal_{j,i}^2(t)}{\sum_{k=1}^{k=4} \sum_{j=1}^{in_{SGk}} sal_{j,k}^2(t)} \tag{4.44}$$

where in_{SGi} is the number of indicators perceived for the steam generator SG_i.

Applying Equation (4.44), the Steam Generators that reach a *diagnosticity* greater than 0.4 ($D_{SGi} \geq 0.4$) are candidates selected for diagnosis. The steam generator that exhibits the highest *diagnosticity* is finally chosen, and the initial *isolation* starts according to procedures.

When all steam generators remain at similar values of diagnosticity for some time, then the *LOP* scenario is selected.

The effect of the alarms is quite important in the whole dynamic process of initial diagnosis and revision based on the feedback of new information. Indeed, when an alarm occurs, the *cognitive salience* of all indicators associated to the steam generator where the alarm has appeared, is instantaneously increased of 0.5. The effect of this abrupt increase in salience, and thus in *diagnosticity*, is utilised to generate in a "natural" way the process of recovery from errors. In this way the recovery is not forced into the cognitive process in an artificial numerical way, but is calculated and bears some relation to certain dynamic events.

The effect of alarms on the cognitive salience, however, vanishes with time exponentially (time constant λ, Eq. 4.37). In this way it is possible to disregard the alarms that are redundant for the event under way and do not add new information.

A typical example of this type of alarm is the alarm associated to the low level in the feed-tank of the AFWS, which is activated a few minutes inside the accident and remains active for the whole period of the accident, without having a real impact on the information content nor on the operator's behaviour.

When the operator has performed a diagnosis and has defined the type of accident and available steam generators, the initial actions of isolation and regulation are selected according to the procedure and are straightforwardly performed.

These two steps of the cognitive model are not very deeply developed in the simulation, as they are considered a simple implementation of well known and prescribed plans, and as we are particularly interested in the study of the diagnostic process.

The diagram that describes the sequence of the dynamic process of perception, interpretation (diagnosis), planning and action is shown in Figure 4.35. In particular, the distinction between the first 300 seconds, when no actions are allowed, and the remaining period of the accident, up $\approx$ 2000 seconds, is clearly shown.

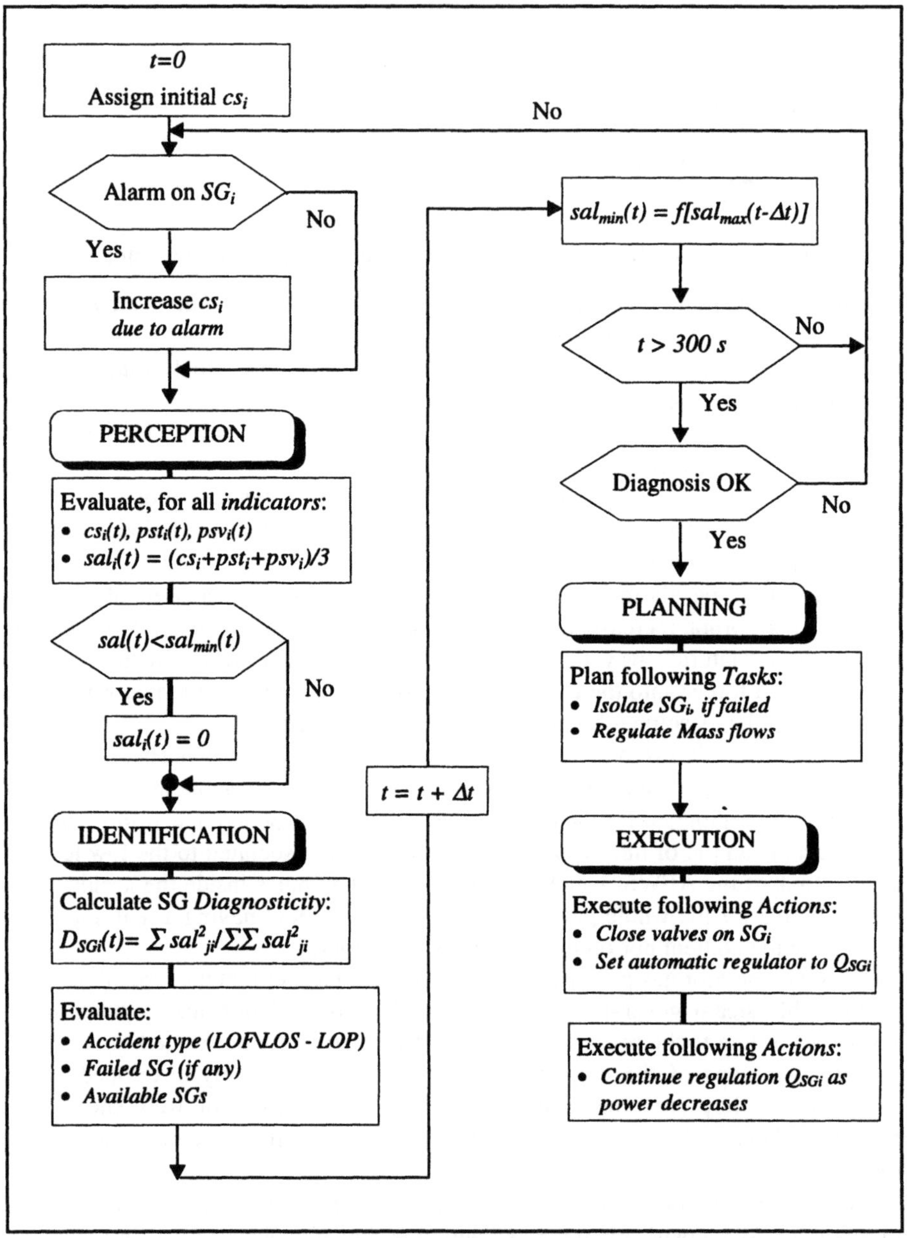

Figure 4.35. Flow diagram of Operator Simulation.

Programming Language and Environment

The simulation of the operator has been developed in combination with the model of the AFWS and Steam Generator.

As the numerical algorithms of the human model are very simple and do not require particularly difficult iterations and solution methods, any programming language and computer architecture is able to represent and contain the model.

Moreover, given that the models of the physical response of the AFWS, SGs and control panels have been developed in FORTRAN 77, like the DYLAM program, utilised for the Interaction Model, we have also developed the operator model in FORTRAN 77.

This has made the integration of the simulations very easy and has not reduced the completeness and accuracy of the computer program with respect to the simulation algorithms.

4.5.4 Interaction Model - Case Study AFWS

The Interaction Model for the case study of the Auxiliary Feed-Water System is based on the DYLAM method, as in the previous sample case of the Chemical and Volume Control System.

As we have already seen in the previous sections of this Chapter, DYLAM is able to combine logical and probabilistic change of states of components, in relation to time, and physical evolution of the simulated system. These peculiarities make DYLAM the ideal approach for modelling the Interaction Model in a Human-Machine simulation architecture.

In the case of the AFWS, the whole simulation sequence has been structured in such a way that the DYLAM algorithms are able to correlate human behaviour and machine responses in correspondence with the evolution of the accident (Figure 4.36).

In particular, at the beginning of the accident ($t=0$) DYLAM ascertains the initial configuration of the human-machine system in terms of “state” of hardware components, information displays, and human errors. After that, the first time step of the simulation is performed.

Firstly, the simulation of the AFWS and Steam Generators is carried out. The models of AFWS and SG calculate, with their internal algorithms and time progression scheme (Figure 4.33), the evolution of the physical quantities (mainly T_{SGi}, p_{SGi}, Q_{SGi}) for a time step of the overall simulation.

The human model, then updates its variables, in terms of perception, interpretation, planning and execution (mainly sal_i, D_{SGi}, and settings of the automatic regulators, Q_{SGi}) with respect to the new values of the physical quantities.

In this way the synchronism is obtained for what concerns the plant performances and human interactions.

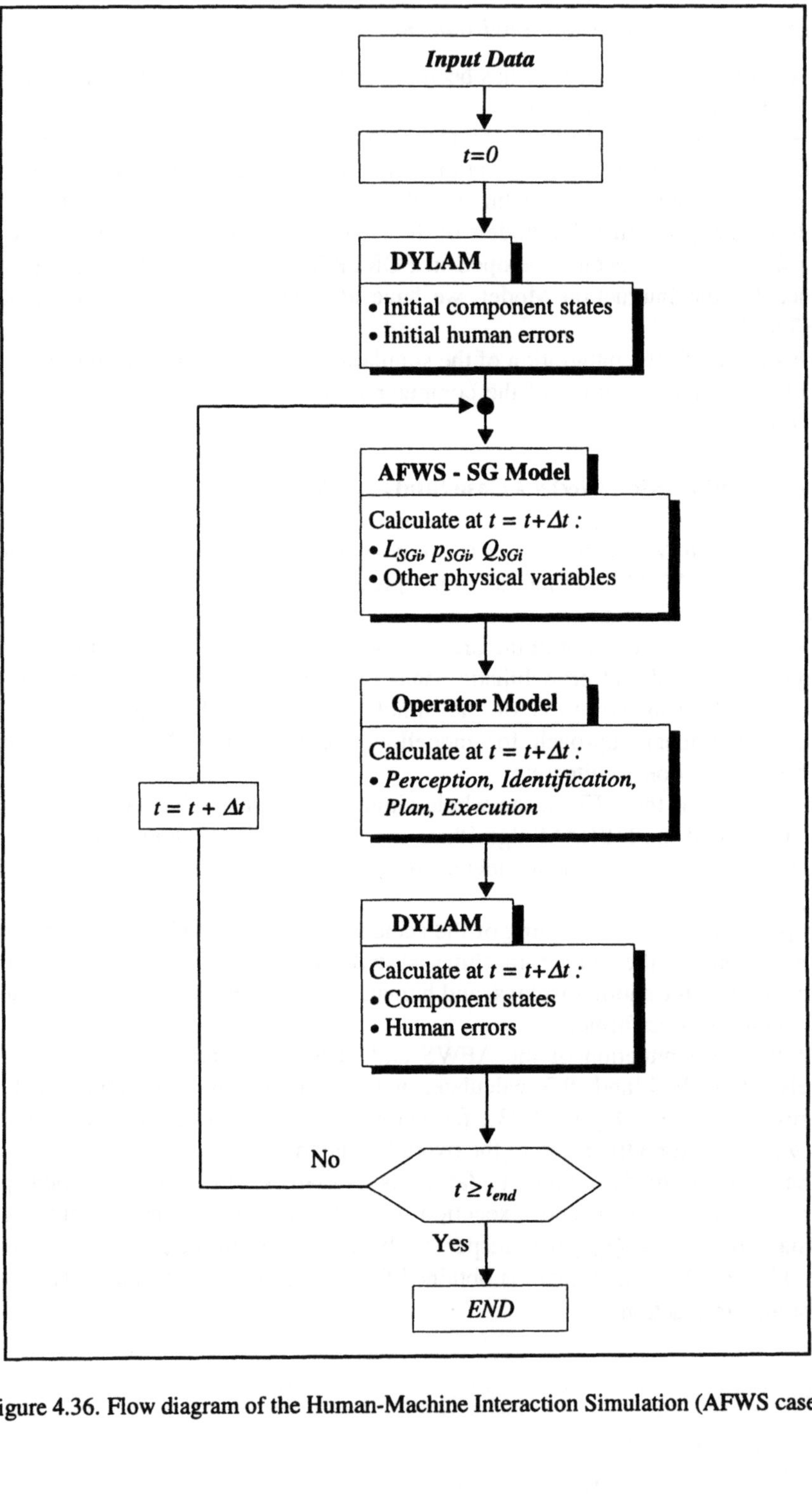

Figure 4.36. Flow diagram of the Human-Machine Interaction Simulation (AFWS case).

At this point the DYLAM method, in accordance with the updated physical and behavioural quantities, calculates the new logical and probabilistic state of the components, which will then affect the evolution of the accident during the following time step (Δt).

The cycle is then repeated until the end of the simulation is reached.

In this case, the use that has been made of DYLAM is quite reductive with respect to the potentiality of the method, as we will only apply the combinatorial framework offered by the method, while very little use will be made of the probabilistic model for evaluating accident frequencies due to component failures and operator errors.

We will discuss in the following two sections the data for the simulations of the Auxiliary Feed-water System and operator, so as to characterised in more detail the accident scenarios in combination with the matrix of *Case Studies* already selected and shown in Table 4.6.

Interaction Model and Machine Components

The physical characteristics of the AFWS and Steam Generators have been derived from a standard nuclear power plant system (Amendola, 1986; El Wakill, 1971).

The nominal pressure of the steam generators has been set at 5.6 MPa, and the total volume at 271 m^3. The opening pressure of the safety valves of the steam generators has been fixed at 8.2 MPa and the closing pressure (following opening) at 7.0 MPa. The pump performances have been derived from standard characteristic of pumps operating at the regimes and conditions typical of nuclear power plants. The AFWS feed-tank capacity has been assumed of 1140 m^3.

Concerning the reliability of components and display panels, all hardware components, actuators, indicators and alarms have been modelled following the Failure Mode and Effect (FMEA) analysis, according to the DYLAM requirements.

In particular, the following possible states have been considered:

1. Hardware Components (valves/pumps):
   a. *Nominal state.* Working according to design specifications.
   b. *Stuck.* Stuck in the position that the component had at the time of failure.
   c. *Failed open/on.* Valve failed open. Pump failed in working mode.
   d. *Failed closed/off.* Valve failed closed. Pump failed off.
2. Actuators:
   a. *Nominal state.* Working according to design specifications.
   b. *Failed.* Not responding to commands.
3. Indicators:
   a. *Nominal state.* Working according to design specifications.
   b. *Failed.* Indicator failed in a fixed position.
4. Alarms:
   a. *Nominal state.* Working according to design specifications.
   b. *Failed on.* Alarm erroneously activated.
   c. *Failed off.* Alarm not responding to critical values of physical variables.

As we have modelled a high number of components and display instruments, the possible combinations of failures and malfunctions is enormous.

In reality, with reference to the case studies described in Table 4.6, the only accident cases that will be studied in detail in the next section are: a) the nominal behaviour of all components for the initiating events *LOF* (*Case 0A*), *LOS* (*Case 0B*), and *LOP* (*Case 0C*); b) the possibility of failure of alarms on steam generators, for the initiating event *LOF* (*Case 3A*).

The study of failures of crucial components, such as regulation valves, check valves, or pumps, for the initiating events *LOF* (*Case 2A*), will not be pursued.

Interaction Model and Human Behaviour

The data for the simulation of the operator model are combined with the definition of the *salience*, associated with the physical quantities, and with the error modes for the DYLAM model.

The quantities that modify the salience of the information display have already been discussed, while the error modes that have been included in the operator simulation cover the possibility of inappropriate performance of the cognitive functions.

In particular, a number of possible forms of erroneous performance for each cognitive function have been theoretically identified. These are:

1. Perception:
   a. *Nominal state.* All information correctly perceived.
   b. *Failed-1.* Mass flows (Q_{SGi}) not perceived.
   c. *Failed-2.* Mass flows (Q_{SGi}) and pressures (p_{SGi}) not perceived.
   d. *Failed-3.* Mass flows (Q_{SGi}), pressures (p_{SGi}), levels (L_{SGi}) not perceived.
2. Identification
   a. *Nominal state.* Diagnosis correctly performed.
   b. *Error in diagnosis.* Wrong *SG* identified ($SG_1 \Leftrightarrow SG_2$, or $SG_3 \Leftrightarrow SG_4$).
3. Planning
   a. *Nominal state.* Plan correctly developed / understood.
   b. *Error in planning.* Wrong actuators selected ($SG_1 \Leftrightarrow SG_2$, or $SG_3 \Leftrightarrow SG_4$)
4. Execution
   a. *Nominal state.* Actions correctly carried out.
   b. *Error-1.* Wrong regulator setting ($\Delta Q_{SGi} = +30\%$).
   c. *Error-2.* Wrong regulator setting ($\Delta Q_{SGi} = -30\%$).
   d. *Error-3.* Wrong regulator setting ($\Delta Q_{SGi} = +50\%$).

As for the case of the components of the AFWS, not all possible failure modes have been considered in the test cases, because the variability introduced by their combination is very high.

A number of selected cases will be performed combining the accident scenarios defined in Table 4.6 with specific human error modes and component behaviours.

4.5.5 Results of Case Study AFWS

As a result of the discussion on the previous section the variety of the accident cases that will be discussed has been largely reduced.

The criterion adopted for such a reduction has been mainly based on the relevance that a particular combination of failures or human errors may have on the development of the accident.

The role of DYLAM for the generation of sequences and combination of failures, related to probabilistic criteria, has been minimised. DYLAM has been utilised as an Interaction Model only for what concerns the logical sequencing of events and for the management of data and time step.

The sample cases shown in Table 4.6 will be studied, except for the failures on hardware components. The detailed configuration of each accident will be defined in accordance to the specific initiating event.

In practice, three classes of Case Studies will be developed:

1. *LOF* - SG_1: Loss Of Feed-water on the inlet line of the secondary side of Steam Generator 1 (SG_1). Four specific cases of failures and errors will be considered:
   1.1. *Case 0A - LOF -* SG_1 *- Nominal Conditions*
   1.2. *Case 1A - LOF -* SG_1 *- Errors of Perception and Planning*
   1.3. *Case 3A - LOF -* SG_1*, False Alarm on* SG_2
   1.4. *Case 3A - LOF -* SG_1*, False Alarm on* SG_3

2. *LOS* - SG_1: Loss Of Steam flow on the outlet of the secondary side of Steam Generator 1 (SG_1). Only one specific case will be developed:
   2.1. Case 0B - LOS - SG_1 *- Nominal Conditions*

3. *LOP* : Loss of Off-site Power. Only one specific case will be developed:
   3.1. Case 0C - LOP - Nominal Conditions

Table 4.7 contains the detailed specifications of the accidents that will be studied.

The safety assessment study of the AFWS, with respect to human factors, could be carried out by an extended application of the method described above.

In this case, it would be necessary to utilise probabilities of human errors and the study should cover the whole spectrum of combinations of human errors and/or component failures.

A comparison could then be carried out between "classical" human reliability methods with the application of a dynamic method, which could show advantages and drawback of both types of approaches.

This exercise would be even more complicated than the current study and, it is outside the scope of the present book.

Finally, as already mentioned, even if the duration of operation of the AFWS is of several hours, we will focus only on the first 30 minutes of transient, as the majority of the interactions between operator and control system, specially at the level of cognitive processes, takes place in this initial period.

Table 4.7. Detailed Matrix of Case Studies for the AFWS analysis

Case Study - *Initiating Event*		**Operator Behaviour**	**Hardware Components**	**Actuators & Indicators**
LOF - SG_1	**Case 0A**	*Nominal*	*Nominal*	*Nominal*
	Case 1A	*Errors perc. and plan.*	*Nominal*	*Nominal*
	Case 3A	*Nominal*	*Nominal*	*False Alarm SG_2*
	Case 3A	*Nominal*	*Nominal*	*False Alarm SG_3*
LOS - SG_1	**Case 0B**	*Nominal*	*Nominal*	*Nominal*
LOP	**Case 0C**	*Nominal*	*Nominal*	*Nominal*

Case Study Loss Of Feed-water - AFWS

The class of accidents *LOF* is characterised by the initiating event of Loss of Feed-water on the inlet line of the secondary side of a steam generator.

The same initial accident configuration and phenomena will be adopted for all case studies related to the *LOF* condition. These can be summarised as follows:

- The set of data that characterise the AFWS and SG systems and describe the initiating event are shown in Table 4.8.
  In particular the failure is assumed to occur in the feed-line of Steam Generator 1 (SG_1), with a surface area of 10 cm^2.
- The failure is located downstream of the inlet mass-flow sensor of SG_1 of the AFWS (Figure 4.37).
- Following the above condition, the indicated feed-flow to SG_1 on the display panel is substantially different from the actual flow of coolant entering or escaping from the steam generator. This flow can be positive (entering SG_1) or negative (escaping from SG_1) is relation to:
  a) the hydraulic balance derived from the dynamic pressure change inside SG_1,
  b) the pressures downstream of the feed-pumps of the AFWS,
  c) the dimension of the failure (10 cm^2), and
  d) the position of the control components.
- The operator receives from the display panel an erroneous signal concerning the feed-flow to SG_1.

Table 4.8. Typical Input Data for the models AFWS - SG and Operator

Description	Units	Data
Dimension of inlet-leg break	m^2	1.10^{-3}
Dimension of outlet-leg break	m^2	1.10^{-2}
Failed Steam Generator	-	1
Accident type. 1 = Inlet, 2 = Outlet, 0 = No failure	-	1
Δt_{AFWS} for AFWS simulation	s	60
Δt_{SG} for SG simulation	s	10^{-3}
Δt_{Oper} for Operator simulation	s	60
Switch for Operator Model = 0; External user = 1	-	0
Exchanged power Primary-Secondary circuit	W	$9.37\ 10^8$
Steam Generator nominal pressure	MPa	5.63
Inlet Mass flow at time = 0	Kg/s	489
Liquid mass	Kg	56031
Total SG volume	m^3	271
Inlet flow enthalpy	J/Kg	$1.1\ 10^5$
Opening pressure of safety valves	MPa	8.2
Closing pressure of safety valves	MPa	7.0
Coefficient for the evaluating steam flow to atmosphere	-	$5.52\ 10^{-6}$
Coefficient for the evaluating steam flow to turbine	-	2

Figure 4.37. Loss of Feed-water flow on the secondary side of SG_1.

Case 0A - LOF - SG_1 - *Nominal Conditions*

In the "nominal" case of Loss Of Feed-water on the secondary side of Steam Generator 1, all components respond according to design specifications and no human error is considered.

During the first 5 minutes, the operator can not intervene on the control panel and is limited to collect and verify information and develop diagnosis.

This period is very important for the cognitive processes as it determines the initial diagnosis and intervention. The variable that governs the simulation of these cognitive processes is the *salience.*

Table 4.9 shows the time dependent values of the *salience*, obtained by applying Eq. (4.2), for the main physical variables (Q_{SGi}, p_{SGi}, and L_{SGi}).

Table 4.9. Case 0A - LOF -SG1. Salience of physical variables

Salience	t = 60 s	t = 120 s	t = 180 s	t = 240 s	t = 300 s
sal_{QSG1}	0.4	0.3	0.2	0	0
sal_{LSG1}	0.4	0.3	0.2	0	0
sal_{pSG1}	0.4	0.479	0.586	0.69	0.797
sal_{QSG2}	0.4	0.32	0.31	0.31	0.31
sal_{LSG2}	0.4	0.3	0.2	0	0
sal_{pSG2}	0.4	0.3	0.2	0	0
sal_{QSG3}	0.4	0.3	0.2	0	0
sal_{LSG3}	0.4	0.3	0.2	0	0
sal_{pSG3}	0.4	0.3	0.2	0	0
sal_{QSG4}	0.4	0.3	0.2	0	0
sal_{LSG4}	0.4	0.3	0.2	0	0
sal_{pSG4}	0.4	0.3	0.2	0	0

The salience associated to each physical variable of a steam generator is used to calculate the *diagnosticity* of that Steam Generator (Eq. 4.44), which defines the degree of confidence of the operator in diagnosing an event related to that steam generator.

The time dependent values of the *diagnosticity* for the four steam generators, is shown in Figure 4.38.

It is quite obvious that the steam generator affected by the accident is identified very shortly after the initiating event. Indeed, after 4 minutes of the accident, the diagnosticity of SG_1 is already approaching 1 (D_{SG1} = 0.83), while the diagnosticity of the other SGs are very small and decreasing.

It is however interesting to note that, as steam generators SG_1 and SG_2 are fed by the same main feed-line (*TA*) (Figure 4.28), the feed-flow to SG_2 is affected by the strong pressure transient on the inlet line of SG_1, where the failure has occurred. For this reason, the salience of some of the quantities of SG_2, and, thus, the *diagnosticity* of SG_2, are not rapidly levelling to zero, as occurs in the case of SG_3 and SG_4.

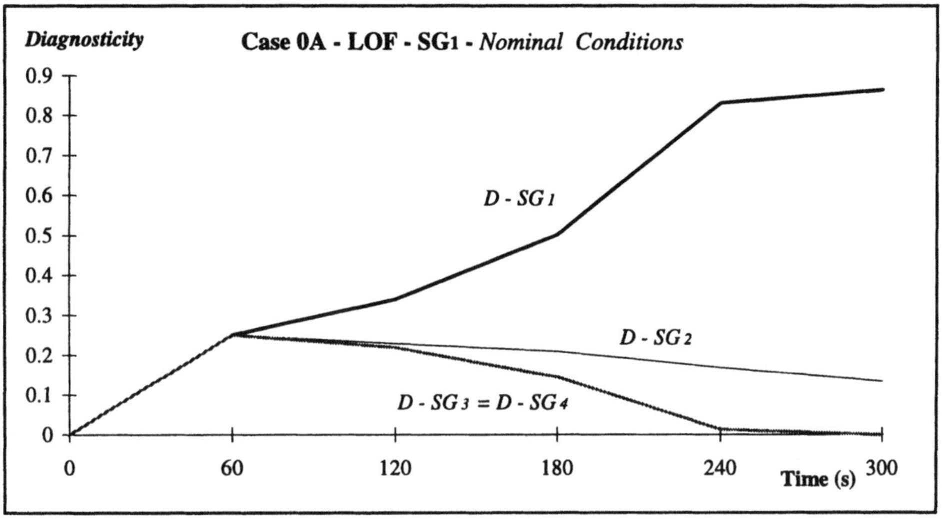

Figure 4.38. Case 0A - LOF -SG_1. Diagnosticity (D) of SG_1 - SG_4

The effects of human-machine interaction are shown in Figure 4.39, which contains the time dependent behaviour of the most significant quantities of SG_1 for the first 30 minutes of accident. The equivalent quantities for the other steam generators do not exhibit any particular behaviour. It is important to remember that Q_{SG1} represents the actual flow that either enters or escapes from the steam generator SG_1 (Figure 4.37), while the operator perceives and decides on the basis of the information of the display panel, which shows the flow provided by the AFWS.

The following comments and observations can be made.

- During the first 300 seconds, a certain amount of water is lost through the break in the feed-line. Moreover, the effect of loss of feed-water causes an increase of the pressure (p_{SG1}), which contributes further to the decrease of the flow of water that actually enters the steam generator. Indeed, at approximately 300 seconds, without operator actions, the pressure in the steam generator is such that the net flow of feed-water becomes negative, i.e., the water coming from the AFWS is completely lost through the pipe break and, in addition, some water actually starts escaping from the steam generator ($Q_{SG1} < 0$, at $t \approx 300\ s$).
  The outflow from SG_1 also oscillates between a maximum and a minimum, in frequency with the change in pressure (p_{SG1}). At approximately 1900 seconds the steam generator is dried-out, as the water level has dropped below the inlet nozzle and, thus, no more outflow of water from SG_1 is possible.

- At 300 seconds, the operator can act on the failed steam generator by isolating and performing the initial regulation. This means that the valves and the automatic regulators of feed-flow to the different steam generators are positioned, according to the procedure, at the following settings (Figure 4.34):
  - *Q_{SG1} = 0 t/h (Isolation)*
  - *Closure of main turbine valve and AFWS turbine valve relative to SG_1*
  - *Q_{SG2} = Q_{SG3} = Q_{SG4} = 50 t/h = 13.9 Kg/s*
- The mass flow escaping from SG_1 suddenly increases, as consequence of the isolation, which results in the abrupt loss of the pressure head due to the AFWS pumping system. A large amount of water is now escaping from the steam generator and the level begins to rapidly decrease.
- The lack of water inside the steam generator causes the increase of the pressure (p_{SG1}) and the cyclic opening and closing of the safety valves, between *8.2* and *7 MPa*. This process occurs at 780 seconds (and is repeated at *t=1350 s*), when the pressure inside SG_1 firstly reaches 8.2 MPa and, then, drops to 7 MPa, following the opening of the safety valve.
- The level in SG_1 is affected by the net inlet mass flow. It remains almost constant during the first 5 minutes and then rapidly drops to zero as large amount of water are lost trough the pipe break.

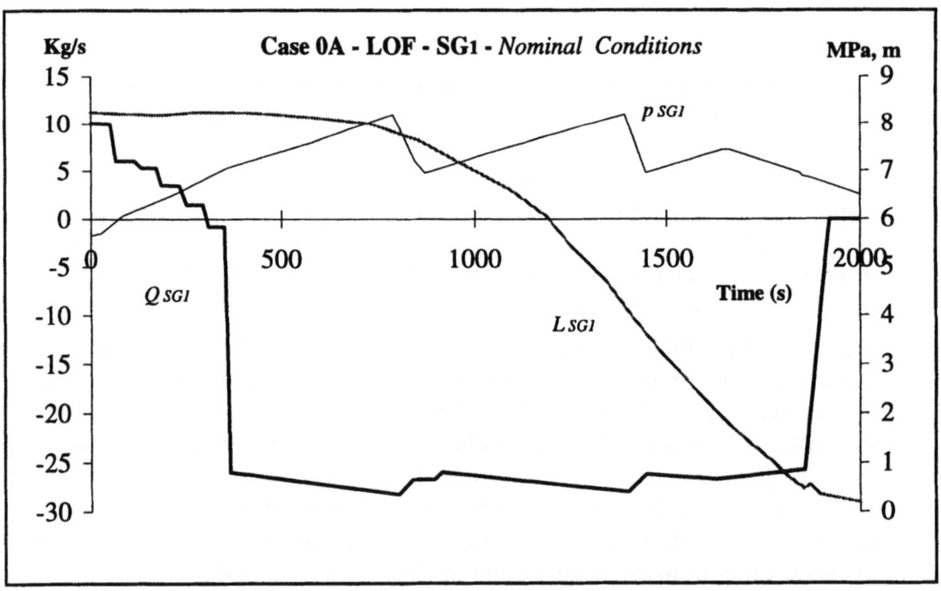

Figure 4.39. Case 0A - LOF -SG_1. Physical variables of SG_1. Mass Flow, Q_{SG1} (Kg/s); Pressure, p_{SG1} (MPa); Level, L_{SG1} (m).

Case 1A - LOF - SG_1 - *Errors of Perception and Planning*

The class of accidents containing human errors covers all possible combinations of inappropriate perception, identification, planning and execution, as described in the previous section.

We have already made some assumptions concerning the specific mode of errors for each cognitive function. However, the combination of these errors leads to a high number of possible scenarios and sequences.

As we intend to show the potential scope of application of the method, rather than performing the safety assessment of the AFWS, we will limit ourselves to the study of one specific combination of errors.

In particular, given the initial condition of loss of feed-water flow on the secondary side of SG_1, the combination of the two following human errors has been selected:

Error 1. Lack of perception of all indicators of SG_1.

Error 2. Systematic planning error in selecting the actuators of SG_1 and SG_2.

These two errors imply that:

a) When performing the diagnosis, the operator does not take into consideration the salience of the physical variables relative to SG_1; and
b) The control actions planned on SG_1 are actually carried out on SG_2 and vice-versa ($SG_1 \Leftrightarrow SG_2$).

This combination of errors is very singular, as it matches inappropriate perception with shallow or incorrect planning, which, by pure chance, compensate each other and result in the performance of the correct procedure. This situation is very dangerous, as it, usually, lasts for short periods of the accident and may resolve only local problems, leaving unresolved and suspended the critical issues of the accident.

The results of the analysis of this case study are presented in Figures 4.40 - 4.42, which show time dependent evolution of mass flows, pressures and levels in each of the four steam generators.

- Remembering that the initiating event and the state of components are the same as the scenario previously discussed, i.e., the *LOF* - SG_1 accident, and that the operator cannot intervene on the control panels during the first 5 minutes, we can conclude that the simulation of the initial 5 minutes of accident is exactly the same as in the previous case.
  However, the operator cognitive processes during this period differ quite substantially from those of the previous (nominal) case.
  As the operator does not perceive the signals of the indicators relative to SG_1, the initiating event is not properly recognised.
  The operator perceives the data relative to the other SGs, and in particular those of SG_2, which are altered by the failure on SG_1 (SG_1 and SG_2 are fed by the same main feed-line, as shown in Figure 4.28).
  Consequently, the operator's attention and diagnosis concentrate on SG_2.

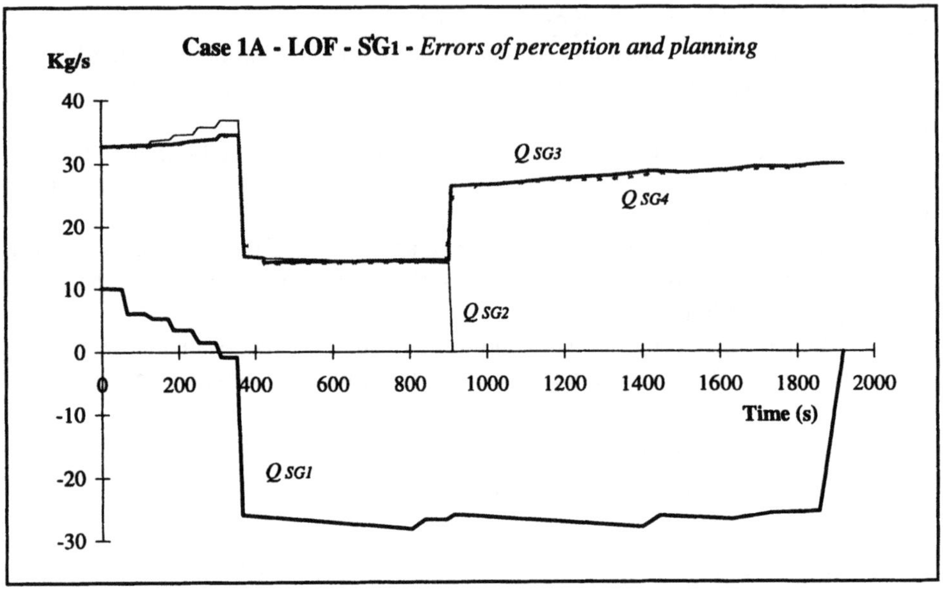

Figure 4.40. Case 1A - LOF -SG1. Mass Flows, Q_{SG1} - Q_{SG4} (Kg/s).

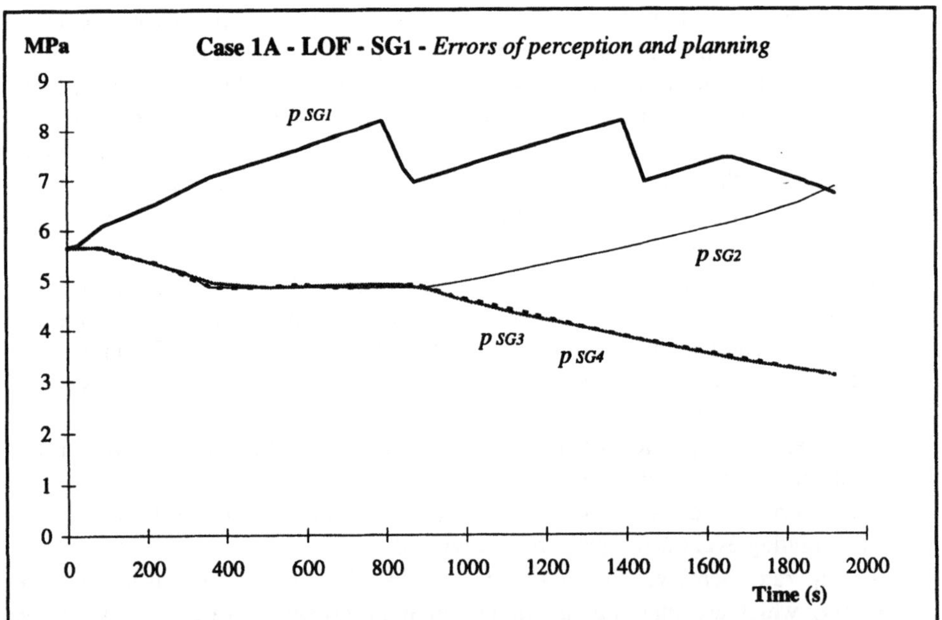

Figure 4.41. Case 1A - LOF -SG1. Pressures, p_{SG1} - p_{SG4} (MPa).

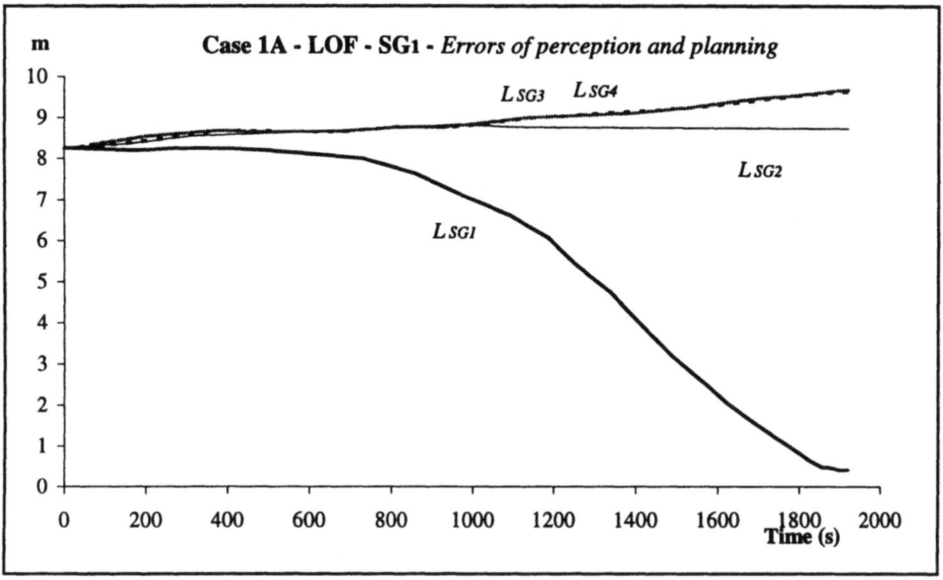

Figure 4.42. Case 1A - LOF -SG1. Levels, L_{SG1} - L_{SG4} (m).

- At the moment of actually implementing the initial isolation of SG_2 ($t = 300\ s$), following the diagnosis, the second human error occurs: the control actions planned on SG_2 (isolation) are actually carried out on SG_1.
  The operator implements the procedure and actually regulates at *13.9 Kg/s (50 t/h)* the mass flow in SG_3, SG_4, and, SG_2. The regulation on SG_2 is due to the systematic error of exchanging the actuators of SG_1 and SG_2.
  In this way, the management of the accident in the first minutes is actually performed according to procedure and it is appropriate for the scenario under development, even though two errors have been made: the two errors have neutralised each other and everything seems under control.
- However, the basic error remains. Later on ($t = 780\ s$), when an alarm on SG_1 appears, due to high pressure and opening of the safety valves, the attention of the operator is focused on SG_1 and a new diagnosis is performed. The decision to isolate SG_1 is taken, but, as the error of planning persists between actions concerning SG_1 and SG_2, the operator actually isolates, erroneously, SG_2.
- At this time ($t = 900\ s$), only two steam generators remain available for cooling the reactor, namely SG_3 and SG_4.
  The procedure then demands that, with only two SGs available, the feed-flow to each of them is set at 100 *t/h* (*27.8 Kg/s*). This is why at 900 seconds, in correspondence with the isolation of SG_2 ($Q_{SG2} = 0$, Figure 4.40), the mass flow to SG_3 and SG_4 is raised to 27.8 Kg/s.

- From this time on, given that also SG_2 is isolated and that the heat transfer from the primary circuit continues, the pressure in this steam generator (p_{SG2}) starts to increase constantly, like in SG_1 (Figure 4.41).
- The behaviour of the water level (Figure 4.42) shows a similar time dependence as in the previous case. The failure in the pipeline of SG_1 causes the rapid depletion of the water inventory and a decrease of the level (L_{SG1}), while the levels in the other steam generators remain quite constant. In SG_2, the level (L_{SG2}) starts decreasing slightly after the isolation, as a direct consequence of the vaporisation and the increase of pressure.

This case study shows quite substantially the potentiality of the proposed architecture of Human-Machine Interaction, which, as we will also see in the forthcoming examples, consists of the capacity to represent and analyse the accident in a dynamic context and to combine the human behaviour and cognitive processes with the system performances and responses.

Finally, the inclusion of probabilities in this type of analysis could be performed by means of the DYLAM algorithms, by which we could extend the potential application of the proposed method to human reliability and probabilistic safety assessment studies (Cacciabue, 1997).

Case 3A - LOF - SG_1 - False Alarm SG_2

The case study of Loss Of Feed-water on Steam Generator 1, in the event of a false alarm on Steam Generator 2, aims at evaluating the consistency of the emergency procedures in the case of a failure (the false alarm) that may generate erroneous interpretations and expectations by the operator.

The table of the salience values associated with the variables of the four steam generators shows very clearly the effect that this particular false alarm plays on the decision of the operator about the failed steam generator ($t \leq 300$ s), and on the subsequent phases of the accident ($300 \leq t \leq 840$ s) (Table 4.10).

Indeed, during the first 5 minutes of the accident, the salience of the variables relative to SG_2, which are already affected by the failure on the feed-line of SG_1, is further increased by the false alarm (cognitive salience): the values of the overall salience of the mass flow (Q_{SG2}), level (L_{SG2}) and pressure (p_{SG2}) of SG_2 are all well above the minimum.

Consequently, the operator becomes immediately convinced that the problem lies with SG_2 and "looses interest" in controlling the behaviour of the physical quantities of the other steam generators, in particular SG_1, which presents quite substantial deviations from "normal" emergency conditions. From the simulation viewpoint, the *salience* of all variables, except those of SG_2, becomes smaller than the minimum salience level, and thus is set to zero, according to Eq. (4.43b).

The possibility to review the initial diagnosis is completely ruled out by the simulation, which, on the contrary, forces the attention of the operator into a sort of a "tunnel vision", focusing the observation only onto the physical variables of SG_2.

Table 4.10. Case 3A - LOF - SG_1. False Alarm SG_2. Salience of physical variables.

Salience	t = 60 s	t = 180 s	t = 300 s	t = 780 s	t = 840 s
sal_{QSG1}	0.4	0	0	1	1
sal_{LSG1}	0.4	0	0	0.5	0.5
sal_{pSG1}	0.4	0	0	1	1
sal_{QSG2}	0.9	0.81	0.314	1	1
sal_{LSG2}	0.9	0.7	0	0	0
sal_{pSG2}	0.9	0.7	0	0	0
sal_{QSG3}	0.4	0	0	0	0
sal_{LSG3}	0.4	0	0	0	0
sal_{pSG3}	0.4	0	0	0	0
sal_{QSG4}	0.4	0	0	0	0
sal_{LSG4}	0.4	0	0	0	0
sal_{pSG4}	0.4	0	0	0	0

The identification of SG_2 as the failed steam generator and the decision to isolate it as soon as possible, i.e., after the first five minutes of accident, are already reached after 1 minute, as is shown in Figure 4.43. At $t = 60\ s$, the *diagnosticity* of SG_2 ($D_{SG2} = 0.628$) is greater than the minimum threshold ($D_{Min} = 0.4$) for diagnosing failure (according to input data). This is consistent with the fact that the simulation does not consider the salience of the other physical variables and every observation converges towards SG_2.

Only later on ($t = 780\ s$), after the occurrence of the (pressure) alarm, this time true, on SG_1, the operator simulation considers again the salience of the variables of SG_1.

Naturally, at this point, the already abnormal behaviour of these variables, due to the failure on SG_1 inlet line, coupled with the boost in cognitive salience due to the alarm, result in very relevant values of overall salience ($sal_{QSG1} = 1$, $sal_{pSG1} = 1$, $sal_{LSG1} = 0.5$). Remembering that the step increase in cognitive salience due to an alarm is 0.5, it is clear that SG_1 variables (especially the mass flow and the pressure) have not been previously noticed because the earlier diagnosis had completely focused the attention on SG_2.

Similar considerations could be made by observing the trend of the salience associated with the variables of SG_2: the overall salience decreases as the relevance of the alarm progressively disappears and the actual values of SG_2 physical quantities are only slightly different than the expectations (SG_2 is affected by the failure on SG_1).

In substance, at approximately 800 seconds, a totally new diagnosis is performed. This time, SG_1 is identified as the failed steam generator, and a new isolation and regulation become necessary ($D_{SG1} = 0.7$, Figure 4.43).

However, as the alarm on SG_2 remains active, the *diagnosticity* of SG_2 does not completely vanish, even if it is greatly reduced ($D_{SG2} = 0.3$).

The operator decides to isolate SG_1, but, as some uncertainty remains on the state of SG_2, this steam generator is left isolated. The regulation that follows is based on the only two available steam generators, namely SG_3 and SG_4.

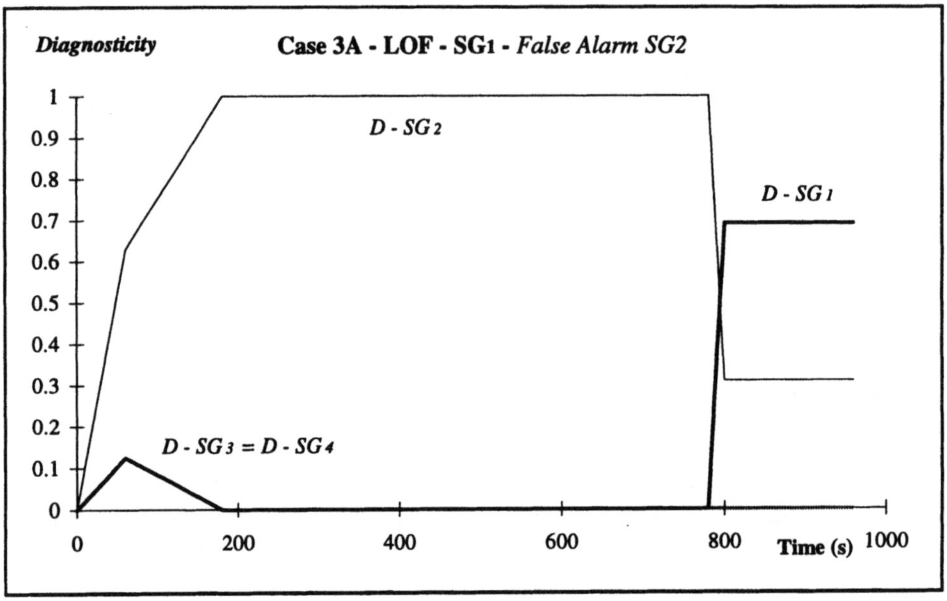

Figure 4.43. Case 3A - LOF - SG_1. False Alarm SG_2.. Diagnosticity (*D*) of SG_1 - SG_4

The effect of the Human-Machine Interaction is reflected in the dynamic behaviour of the physical variables of the four steam generators, specially in terms of mass flows and pressures.

It is important to remember that the mass flow, relative to the failed steam generator (Q_{SG1}), represents the actual flow entering or escaping from the steam generator, and not the mass flow indicated in the control room, as the failure is located downstream of the mass flow sensors (Figure 4.37).

In particular, Figures 4.44 and 4.45 show the mass flows (Q_{SGi}) and pressures (p_{SGi}) in the four steam generators. The following considerations can be made:

- The inlet flow to SG_1 is strongly affected by the failure on the inlet line, and in the first 5 minutes of transient, as the accident is the same of the two case studies previously discussed, the mass flow behaves in exactly the same way, and becomes slightly negative just before the end of these first five minute of the accident.

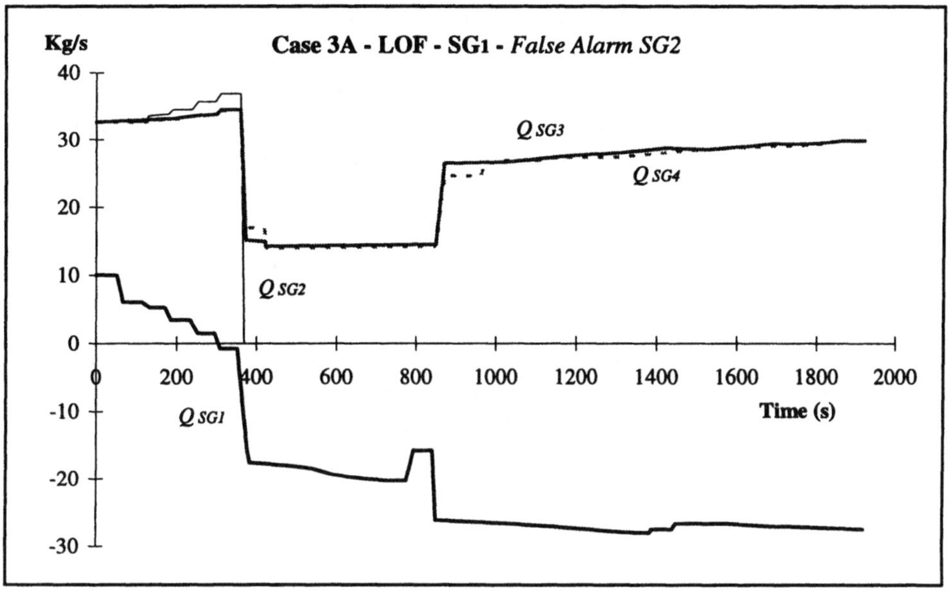

Figure 4.44. Case 3A - LOF - SG_1. False Alarm SG_2. Mass Flows, Q_{SG1} - Q_{SG4} (Kg/s).

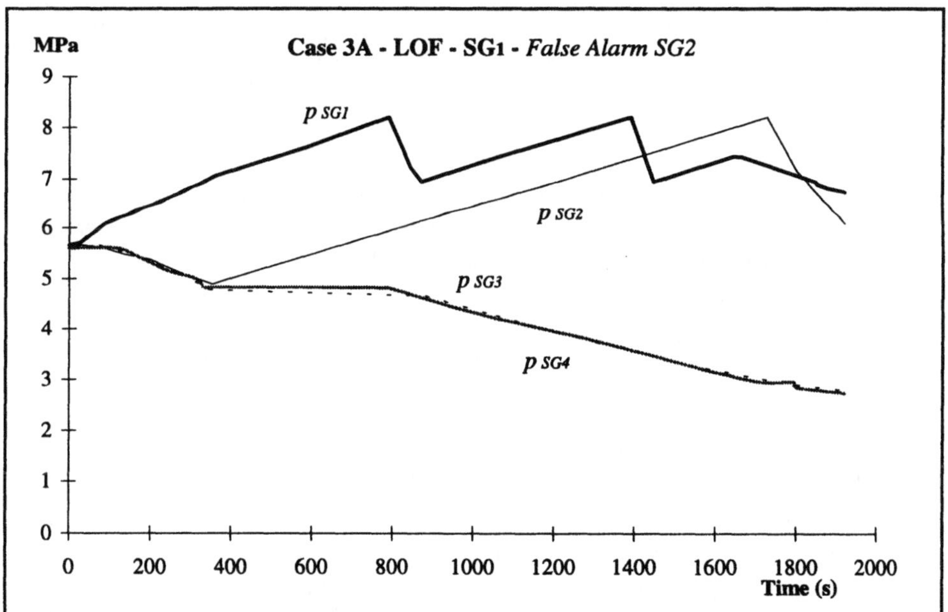

Figure 4.45. Case 3A - LOF - SG_1. False Alarm SG_2. Pressures, p_{SG1} - p_{SG4} (MPa).

- The operator, following the first diagnosis, isolates SG_2 ($Q_{SG2} = 0$) as soon as possible ($t = 360\ s$), and regulates the mass flow to the other steam generators at $50\ t/h$ ($13.9\ Kg/s$) ($Q_{SG1} = Q_{SG3} = Q_{SG4} = 13.9\ Kg/s$).
- The mass flow escaping SG_1 becomes even greater, as the pressure of the pumping system of AFWS is reduced by the regulation. However, the amount of water that escapes SG_1 is actually less that in the previous cases, because the operator has not completely cut off the flow that is sent to SG_1, and has, thus, maintained the pressure head.
- The behaviour of the mass flow from SG_1 is then affected by the pressure transient and the opening of the safety valves at approximately 800 and 1350 seconds, as clearly shown in Figure 4.45. In particular, at 800 seconds, when the pressure drops as a consequence of the first opening of the safety valves, a greater reduction of the water leakage than in the previous case studies is observed. This is because the AFWS is still pumping water to SG_1.
- At 840 seconds, when the operator also isolates SG_1, and completely cuts off the water flow and pressure head from AFWS to SG_1, the amount of water that escapes SG_1 is increase, as in the previous case studies. This slightly different dynamic behaviour of pressure and mass flow and the reduced amount of water that escapes from SG_1, results in a steam generator with some water inventory at 1800 seconds of simulation, while, in the previous cases, steam generator dry-out was attained.
- The behaviour of the physical quantities of SG_2 are strongly affected by the initial diagnosis (Figure 4.44). The mass flow to SG_2 is completely cut off by the isolation manoeuvre and is never re-established, as the doubts derived from the false alarm remain, as shown by the values of the *diagnosticity* of SG_2 (D_{SG2}). No recovery action on SG_2 is actually performed.
- The pressure in SG_2 starts increasing as soon as the isolation is performed (Figure 4.45). The derivative of the pressure in time is such that at approximately 1800 seconds the safety valves on SG_2 open ($p_{SG2} \geq 8.2\ MPa$) to keep the steam generator pressure under control.
- The mass flows to SG_3 and SG_4 are initially regulated at $13.9\ Kg/s$, but are later ($t = 840$) raised to $27.8\ Kg/s$ when the operator isolates SG_1 and does not re-open the circuit to SG_2, remaining with only two steam generators to control the accident transient (Figure 4.44).
- Finally, the pressures in SG_3 and SG_4 follow the expected behaviour in this phase of the accident, when the regulation of mass flow is slightly too high and thus the pressures starts to slowly decrease in time.

The study of Human-Machine Interaction in the case of this false alarm, shows that the designs of procedures, interfaces, control and hardware systems need to be developed in strict iteration with each other. This avoids mishaps and very expensive corrections, which become inevitable when the different designs are performed independently from each other and in sequence.

Unfortunately, this principle of design integration has rarely been followed in the past and has been the primary cause of very severe accidents and catastrophes that, in some cases, have been fatal for a whole plant.

Case 3A - LOF - SG_1. False Alarm SG_3

The results obtained with the simulation of the Loss of Feed-water accident on SG_1, in the presence of a false alarm on SG_3, are presented in Table 4.11 and Figures 4.46 - 4.48.

This case study differs from the previous one, as the false alarm is associated to a steam generator which does not have any connection to SG_1, where the failure has occurred. This difference is crucial for the management of the accident.

The measure of the salience of the indicators due to the false alarm and to the physical phenomena following the initiating event, shows how, firstly, the operator focuses on the variables of SG_3. Yet, these variables do not present anomalies with respect to the expected values, as the accident occurred on the feed-line of SG_1, which is completely independent of SG_3.

The dynamic evolution of the salience of the physical variables of SG_3, and, thus, of the *diagnosticity* (D_{SG3}) (Figure 4.46), simulate the fact that the operator is, initially, "tempted" to diagnose an accident on SG_3 and focuses only on the variables of that steam generator, disregarding the other indicators.

Table 4.11. Case 3A - LOF - SG_1. False Alarm SG_3. Salience of physical variables

Salience	t = 60 s	t = 180 s	t = 300 s	t = 780 s	t = 840 s
sal_{QSG1}	0.4	0	0	1	1
sal_{LSG1}	0.4	0	0	0.5	0.5
sal_{pSG1}	0.4	0	0	1	1
sal_{QSG2}	0.4	0	0	0	0
sal_{LSG2}	0.4	0	0	0	0
sal_{pSG2}	0.4	0	0	0	0
sal_{QSG3}	0.9	0.7	0	0	0
sal_{LSG3}	0.9	0.7	0	0	0
sal_{pSG3}	0.9	0.7	0	0	0
sal_{QSG4}	0.4	0	0	0	0
sal_{LSG4}	0.4	0	0	0	0
sal_{pSG4}	0.4	0	0	0	0

However, as time passes the variables of SG_3 remain at a "normal" level, and thus their salience disappears. Consequently, the *diagnosticity* of SG_3 drops to zero before the end of the five minutes period.

As soon as the operator can perform some control actions, i.e., after 5 minutes, the following conditions are found: a) no steam generator is diagnosed as failed; b) no isolation action takes place; and c) the cognitive processes of search for further information (perception) and identification of the failed steam generator continues (Figure 4.35).

At approximately 800 seconds, the appearance of the alarm on SG_1, due to the increasing pressure and opening of the safety valves, attracts the operator's attention towards that steam generator and favours the consequent immediate (correct) diagnosis of the event ($D_{SG1} = 1$, at $t = 800$ s).

The isolation and regulation actions that follow are in perfect accordance with the procedures and bring the accident under control.

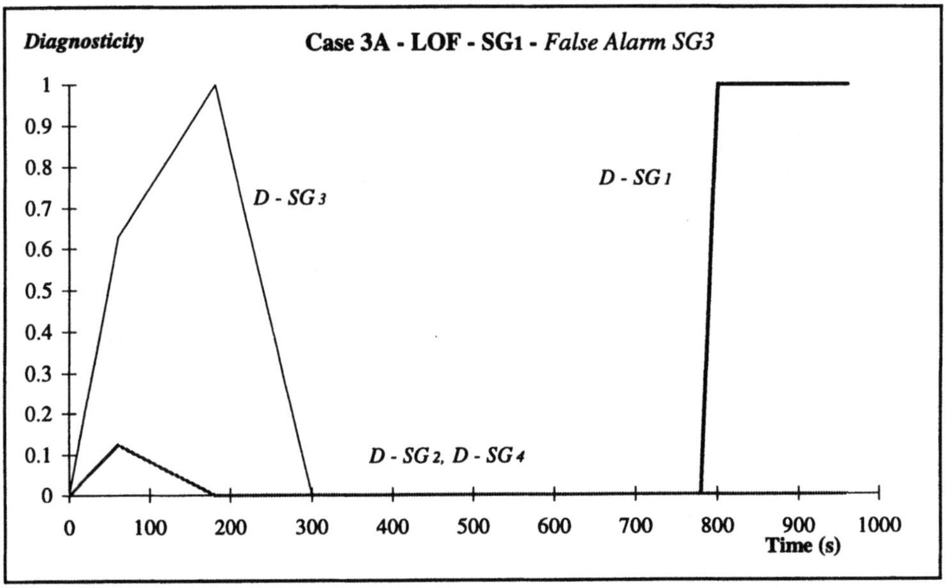

Figure 4.46. Case 3A - LOF - SG_1. False Alarm SG_3. Diagnosticity (D) of SG_1 - SG_4

The interaction of the operator with the AFWS can be further analysed by considering the time dependent behaviour of the main physical variables of the steam generators.

In particular, from Figures 4.47 and 4.48, the following considerations are made:

- The mass flow in SG_1 presents a constant negative rate (decrease) and becomes negative at approximately 300 seconds (Figure 4.47). This loss of water from SG_1, through the break in the inlet "leg" of the secondary side, continues and increases as the pressure in SG_1 grows. No operator control action is carried out during this period, as a consequence of the uncertainty of the diagnosis.

- At approximately 800 seconds, the safety valves of SG_1 open automatically [$p_{SG1}(t=800) > 8.2\ MPa$] and the pressure in SG_1 diminishes for sometime (Figures 4.47 and 4.48). The mass flow that escapes from the steam generator is slightly reduced.
- As soon as the operator performs the correct diagnosis and isolates SG_1 the mass of water lost from SG_1 increases suddenly, because the feedback pressure of the AFWS pumping systems is lost and the outflow from SG_1 encounters little resistance. In this case, as in the previous one, the reduced amount of water lost through the break, during the first 800 seconds, helps in reaching the 1900 seconds without attaining the dry-out condition in SG_1 (Figure 4.47)
- The mass flows in the other steam generators (Q_{SG2}, Q_{SG3}, Q_{SG4}) are not regulated at 300 seconds, as the uncertainty of the initiating event does not lead to the successful identification of the failed SG. When the correct diagnosis if performed, the regulation of Q_{SG2}, Q_{SG3}, and Q_{SG4} can be carried out.
  In this case, as we still have 3 steam generators available for controlling the system and the regulation of the mass flow is set at *13.9 Kg/s* per steam generator ($Q_{SG2} = Q_{SG3} = Q_{SG4} = 50\ t/h$) (Figure 4.47).
- In practice, in this case the AFWS has been over-cooled during the first phases of the accident and then becomes more stable, after the isolation of the failed steam generator, and the regulation of the flow to the other steam generators.

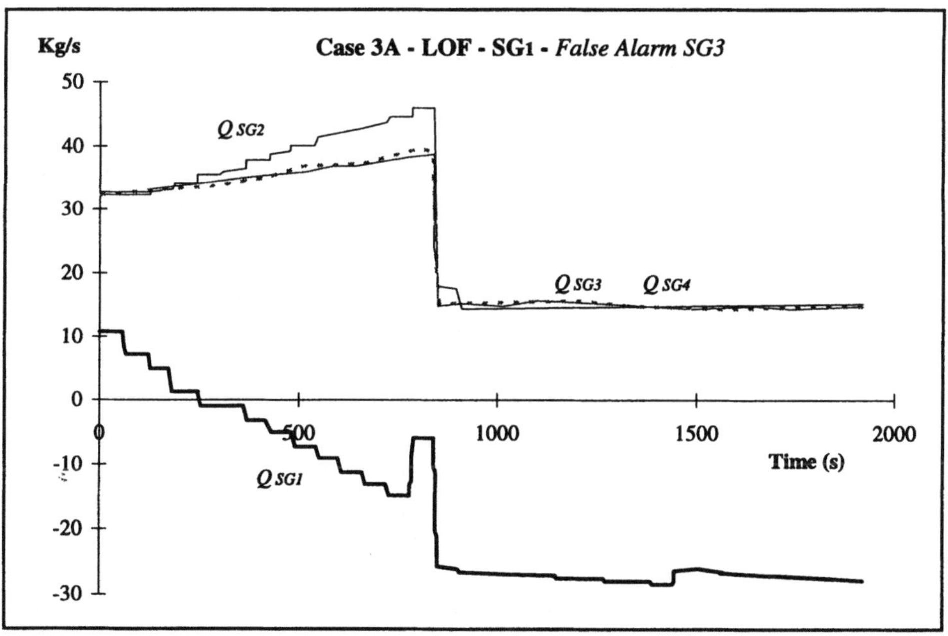

Figure 4.47. Case 3A - LOF - SG1. False Alarm SG3. Mass Flows, Q_{SG1} - Q_{SG4} (Kg/s).

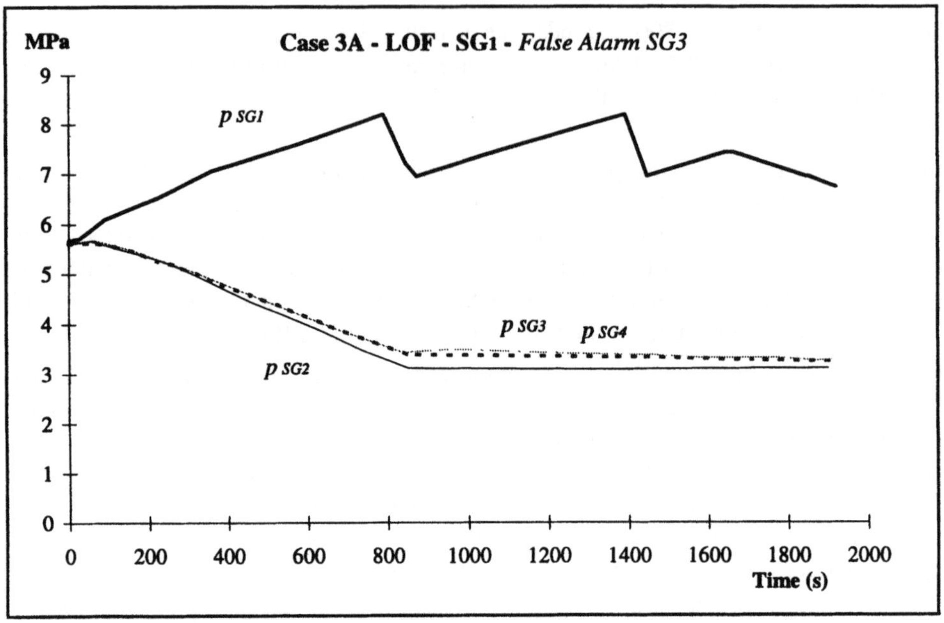

Figure 4.48. Case 3A - LOF - SG_1. False Alarm SG_3. Pressures, p_{SG1} - p_{SG4} (MPa).

This case study has shown the robustness of the control and emergency procedures, in the presence of a false alarm on a steam generator (SG_3) not connected to the same line as the steam generator where the failure has actually occurred.

The comparison with the previous case, where the false alarm on SG_2 had not been well managed, leads to the conclusion that the control procedures seem to have failed considering the hydraulic problems derived from the special design of the AFWS.

In this case, a more stringent procedure for the identification of the failed steam generator could successfully solve the problem of the false alarm on SG_2, and it would not be necessary to wait for the occurrence of the second alarm to recover the situation.

Case Study Loss Of Steam-flow - AFWS

The case study of Loss Of Steam-flow (*LOS*) (*Case 0B - LOS - SG_1*) accident has been simulated by assuming that that the accident occurs on the outlet ("hot-leg") of the secondary side of Steam Generator 1 and that the dimension of the break in the pipeline is 100 cm^2.

The initiating conditions of this class of accidents differ from the previous one by the location of the failure. This is a crucial difference, as the water that is lost through the break is much less than before, even if the surface area of the break is 10 times larger that in the case of the *LOF* accident. This is due to the fact that steam is actually released instead of water, and the pressure transient is less rapid and completely different.

As in the case of the *LOF* accident, the steam-line break is located upstream of the steam flow indicator and isolation valve. Therefore, it is not possible to stop the outflow of steam through the failed pipe line, even after the isolation of the steam generator, where the failure occurred.

As we have assumed an error-free operator and nominal (no failures) behaviour of hardware components and control instrumentation (Table 4.7), the first minutes of accidents are sufficient for the operator to develop the correct diagnosis of the accident and plan the initial isolation and regulation of mass flow according to the procedures.

The table of the salience and the graph of the diagnosticity are not reported as they show very clearly the convergence of the operator decision.

The study of the human-machine interaction can be developed from the analysis of the dynamic behaviour of the physical variables relative to SG_1 and SG_2 (Figures 4.49 and 4.50).

With reference to Figure 4.49 the following comments and observations can be made, concerning SG_1:

- As the amount of water that escapes from the steam generator is relatively small. The level (L_{SG1}) decreases very slowly, even after the isolation of the steam generator and the consequent cut off of the feed-flow.
- The feed-flow to SG1 (Q_{SG1}) increases rapidly during the first 5 minutes (at $t = 300\ s$, $Q_{SG1} \approx 40\ Kg/s$), as a consequence of:
  a) The fully open regulation valves on the feed-lines (by design of the initial automatic control); and
  b) The de-pressurisation of the steam generator (due to the failure of the steam-line).

  As soon as the diagnosis is performed and the isolation is completed, the feed-flow is completely cut off.
- The pressure in SG_1 is strictly related to the feed-flow, and decreases rapidly as a reaction to the pressure drop generated by the break on the steam-line and to the considerable amount of water fed into the steam generator by the AFWS pumping system. As soon as the feed-flow to SG_1 is cut off, the negative derivative of the pressure is reduced and, even if the pressure is not stabilised, the decrease rate becomes very small and much more controllable.

Figure 4.50 contains the results of the time dependent evolution of the physical variables of SG_2. The following comments and observations can be made:

- The behaviour of the level (L_{SG2}), mass flow (Q_{SG2}), and pressure (p_{SG2}) of SG_2 are similar to those of SG_1. However, considering that this steam generator does not present failures, these quantities change very smoothly after the initiating event and after the regulation of the inlet mass flow.

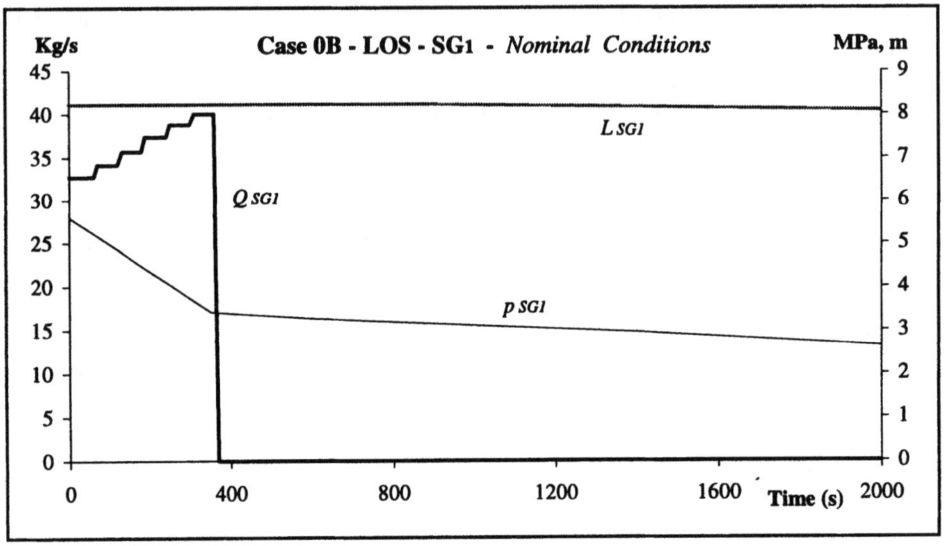

Figure 4.49. Case 0B - LOS -SG_1. Physical variables of SG_1. Mass Flow, Q_{SG1} (Kg/s); Pressure, p_{SG1} (MPa); Level, L_{SG1} (m).

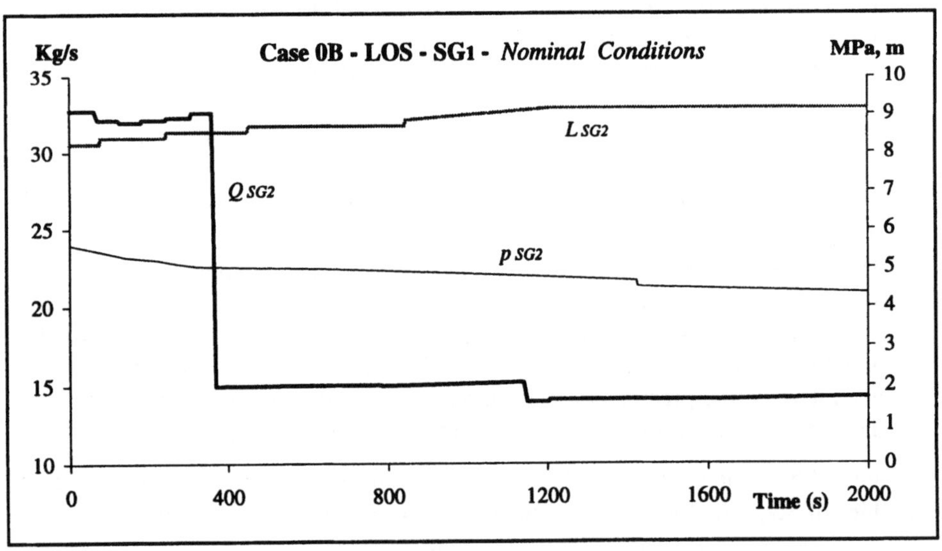

Figure 4.50. Case 0B - LOS - SG_1. Physical variables of SG_2. Mass Flow, Q_{SG2} (Kg/s); Pressure, p_{SG2} (MPa); Level, L_{SG2} (m).

- The considerable amount of water fed in SG_2 by the initial setting of the regulation valves (fully open by design): at $t = 0\ s$, $Q_{SG2} \approx 33\ Kg/s$ causes a decrease in the pressure and a corresponding increase of water level.
- However, as soon as the diagnosis and initial regulation are performed, and the feed-flow to the available steam generators is set to *13.9 Kg/s* (*50 t/h*), according to the procedures, the physical variables of steam generator 2 stabilise at the expected values.
- As the exchanged power decreases in time, some further reduction of feed-flow will be necessary (at time $t > 2000\ s$) as soon as these variables, specially the level, diverge from acceptable boundaries.

This case study has shown the robustness of the emergency procedures in the case of Loss Of Steam-flow accident. As no operator error nor component failure were foreseen, the accident has been immediately identified and controlled within the first few minutes.

Case Study Loss of Off-site Power - AFWS

The case study *Loss of Off-site Power* (*LOP*) (*Case 0C - LOP*) has been performed under the assumption that all hardware components and control instrumentation respond according to design specification ("nominal state") and the operator follows the procedures correctly ("error-free behaviour") (Table 4.7).

In this case, the reactor is shut-down as a consequence of a loss of external electrical power. However, the power to the reactor components, and, thus, all pumps and commands is maintained.

All four steam generators are available and, as there is perfect thermal and hydraulic equilibrium, the feed-water flow to all of them and their physical performances are very similar.

The operator behaviour follows the procedure, and no remarkable differences exist between expected and perceived values and trends of the physical quantities of the steam generators. No sudden alarm appears on the display panels. Consequently, *salience* and *diagnosticity* remain relatively low and no steam generator failure is, correctly, diagnosed.

The diagnosis of *Loss of Off-site Power* is not performed at 300 seconds, as the operator waits and continues to verify values and trends of physical quantities for another 25 minutes, before making the decision. This process ascertains that the accident is the *LOP* and that no other event or failure has affected any of the steam generators.

At this time the regulation takes place according to procedures and the mass flow of all four *SG*s are set to *30 t/h* (*8.33 Kg/s*). The accident is considered under control.

The dynamic behaviour of the physical variables of SG_1, taken as example for analysing the performance of the steam generators, are shown in Figure 4.51. In particular the following remarks can be made:

- The feed-flow to SG_1 starts ($t = 0$) at *33 Kg/s* with the regulation valves fully open. This flow of water (Q_{SG1}) is too high for the heat transferred from the Primary Circuit. Therefore, the temperature of the secondary side, and consequently the pressure (p_{SG1}) start to decay progressively with respect to their nominal values.
- Similarly, as more water than necessary is fed into the steam generator, the level (L_{SG1}) increases.
- However, no critical values of pressure, mass flow and level are reached. The operator's attention is not attracted to any particular steam generator. No threshold values of alarm are attained. The operator can continue the observation of values and trends for the first half hour of accident.
- When the diagnosis of the event is finally made (t = *1800 s*) the feed-water flows in all steam generators are set to *8.33 Kg/s*. The pressures and levels begin to respond by levelling their values in accordance to a reduced water flow, which is now regulated at an adequate level to cool the reactor from the "decay" heat still generated in the nuclear core.

The case study Loss of Off-site Power, in nominal conditions for operator and component behaviour, has shown that the reactor is controlled very efficiently. However, the diagnosis of the accident and the regulation of the steam generator could have been performed much earlier than simulated, as alarms and indicators allow the identification of the initiating event well within the first 5 minutes and the remaining time, up to 30 minutes, is used only for confirming the initial hypothesis.

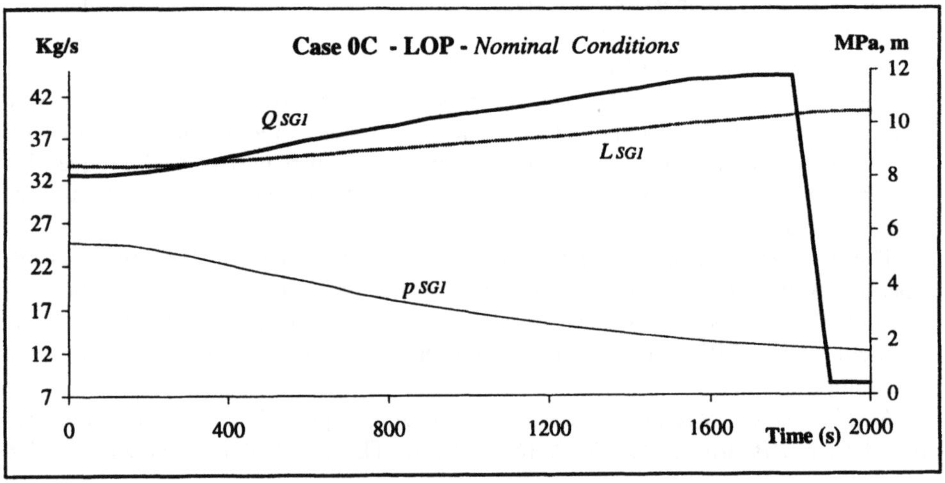

Figure 4.51. Case 0C - LOP. Physical variables of SG_1.
Mass Flow, Q_{SG1} (Kg/s); Pressure, p_{SG1} (MPa);
Level, L_{SG1} (m).

4.6 Critical Review of Case Studies CVCS and AFWS

The process of development of a human-machine interaction simulation for safety analysis has been discussed in two specific case studies: the Chemical and Volume Control System, and the Auxiliary Feed-Water System of a Nuclear Power Plant.

These two examples have been accurately developed, especially for the human simulation. The theories and algorithms discussed in the preceding Chapters have been put into practice.

These two case studies have been developed with two objectives in mind:

1. To show how to develop simulations of human-machine interaction systems in different cases, depending on the objectives of the analyst.
2. To demonstrate the scope and the type of results that can be obtained from the application of such methods.

These two case studies present a number of *similarities* and *differences*, precisely aimed at enhancing the procedures for developing simulations and the scope of safety analysis.

<u>Similarities between CVCS and AFWS Simulations</u>

The *similarities* between the CVCS and the AFWS case studies are associated with the reference points or milestones of the simulation architecture.

These similarities can be summarised in the following three main points:

1. *The DYLAM methodology for the Interaction Model.*
   The use of DYLAM for the Interaction Model ensures the necessary flexibility and conservation of the logic relationships between the different models, i.e., the human and machine models, and, within each model, between different components.
   Moreover, DYLAM contains a completely developed model for the evaluation of error/failure and recovery/repair probabilities, by which it is possible to perform a reliability analysis. Therefore, it is possible, with DYLAM, to carry out Probabilistic Safety Assessment studies, provided that the necessary data (failure rates, error probabilities, repair rate, etc.) are available.
2. *The methods for Machine Model.*
   The Machine Model for both case studies has been developed following the same conservation principles, same simplifying assumptions and solution techniques. The reason for this similarity is that the focal point of the Human-Machine simulation is associated with the human component. Thus, a simple machine simulation is usually considered acceptable, as long as it is able to represent at least the control instrumentation and the response of the machine to operator actions.
   In the CVCS and AFWS case studies, the hydraulic problems were solved with almost the same algorithms. However, the thermal and dynamic problems were rather different, and therefore specific solutions had to be identified for each case.

3. *The dynamics of human interactions.*
   The dynamic analysis of the Human-Machine Interactions has allowed the evaluation of the robustness of emergency procedures and protection systems. To do so, both case studies identify similar human behaviours induced by failures of the information display panels and possible random human errors.
   This common feature is, however, only formal, i.e., relative to the manifestation of behaviour and to the type of malfunctioning of components. In reality, as we will see in the next paragraph, the two cognitive models and simulations differ quite substantially, even though their outcomes are similar as far as the manifestations of behaviour are concerned.

Differences between CVCS and AFWS Simulations

The *differences* between the two case studies are very significant mainly in relation to way in which the operator's behaviour and the control procedures have been considered.

In particular, these differences focus on three important aspects: 1) *The cognitive model and the simulation*; 2) *The procedure and tasks*; and 3) *The errors and recoveries.*

The CVCS case study.

In the case study of the Chemical and Volume Control System these three main aspects have been tackled as follows.

1. *The cognitive model and the simulation* adopted to represent the operator are extremely simple: they are based on a representation of the emergency procedures, as they have been developed by the designer.
   The context and environment affect the operator's behaviour through very simplistic stress and recovery functions, which depend on some important physical and contextual quantities. These functions are responsible for identifying and introducing human errors, as well as recoveries from previous errors, during the dynamic interaction.
   All models and functions have been described by simple analytical expressions.
2. *The procedure and tasks* are performed in sequence during the operations of the CVCS. The control process lasts for several hours. The operator interacts with the CVCS during short periods, according to the procedure and in relation to the CVCS performances.
   More than 4 hours of interaction are simulated.
3. *Errors and recoveries*, as well as initial failures of components, are selected by the safety analyst and are then generated and managed in the simulation by DYLAM in relation to the stress and recovery function. If probabilities of errors and failures were assigned, the DYLAM could also provide the human probabilistic analysis, related to the CVCS management in the case of accidents followed by the reactor shut-down.

The cognitive simulation developed for the case study CVCS could be defined as an "engineering oriented model", because it is focused on the manifestations of behaviour and on the consequences of human errors on system management and emergency responses and vice versa. While little attention has been paid to the cognitive functions and processes that govern decision making.

In any case, the analysis of many accident scenarios has been carried out and certain critical issues have been highlighted, specially in relation to the design of some emergency procedures and to the management of the system in the case of multiple failures and human errors.

The AFWS case study.
The model selected for the case study of the Auxiliary Feed-Water System is based on quite different characteristics, specially with respect to the same three main aspects of the operator behaviour discussed above:

1. *The cognitive model and the simulation* of the operator behaviour for the AFWS is much more sophisticated than the one for the CVCS, especially for what concerns the cognitive functions and cognitive processes. A theoretical model of reference has been chosen. The four fundamental cognitive functions (*Perception, Interpretation, Planning,* and *Execution*) have been simulated, by logical expressions and fuzzy logic. This detailed modelling has been coupled to the equally accurate simulation of the AFWS control display system.
   In particular, the simulation of indicators and actuators has supported the implementation of the cognitive functions *perception* and *execution.*
2. *The procedure and tasks* for the AFWS have been very accurately analysed for what concerns the initial diagnosis, the isolation, and the regulation processes. These initial operations are performed between the 5th and 35th minute of the accident. The rest of the operating time of the AFWS, up to 6-7 hours, is less demanding in terms of cognitive activity. Consequently, the analysis has been focused on the first crucial half hour of accident.
3. *Errors and recoveries* have not been managed by DYLAM. This is because the attention was concentrated on the interactions between faulty information displays (false alarms) and "nominal" operator behaviour. In these cases the DYLAM method has only been utilised for combining logical events, rather than generating faults. However, the method has been very useful to ensure the logical and synchronised combination of human and machine simulations.

The case study AFWS revolves around a simulation of cognitive functions and processes, where the sensorial components of the operator simulation, i.e., perception and execution of control actions, are governed by mental functions and processes, i.e., interpretation and planning, and by the context.

The cognitive model and simulation for the AFWS case differs quite substantially from the previous one developed for the CVCS, even if both approaches generate the same type of results. In simple terms, it can be argued that the former approach is more suited for the validation of procedures and interfaces, while the latter is more appropriate and applicable to study operators' cognitive activity.

4.7 Summary

In this Chapter three major subjects have been presented and discussed in detail, namely:

1. The models and simulation methods for developing the Machine component of the Human-Machine Architecture (§ 4.2).
2. The models and methods for the Interaction Model, which sustains the communication and synchronisation of the human and machine simulations (§ 4.3).
3. The development of the human-machine interaction simulation for safety analyses in two specific case studies: the Chemical and Volume Control System (§ 4.4); and the Auxiliary Feed-Water System of a Nuclear Power Plant (§ 4.5).

<u>Models and Simulations of Machines</u>
The discussion on models and simulations of Machines has been very concise as it has been assumed that the theoretical and numerical approaches, normally applied for the simulation of hardware components, control instrumentation and physical variables are well known and developed.

<u>Interaction Model</u>
The methods and techniques proposed for the Interaction Model have instead demanded a more specific and detailed analysis and evaluation.

The main requirements of an Interaction Model have been identified in the ability to:

- Manage and control the time evolution and synchronisation of the machine and human simulations.
- Evaluate and maintain the dynamic logical interactions.
- Evaluate the reliability of machine components and human behaviour.

The application of "classical" dynamic reliability approaches for the implementation of the Interaction Model was considered. The complexity of the theoretical problem and the need to combine logic and probabilistic theories with data management in a dynamic simulation environment have also been considered.

This process led to the identification of the DYLAM methodology as a very appropriate approach. DYLAM was, then, described in some detail, particularly for what concerns the handling of probabilities and the mechanism for the time sequence management.

Even if other methods exist for managing the interaction between simulations of humans and machines, the DYLAM methodology is particularly suited for the Human-Machine Interaction architecture that has been proposed and has been applied to both sample case studies developed in the final part of this Chapter.

<u>Sample Case Studies</u>
These two sample case studies refer to quite complex real auxiliary and emergency systems of a nuclear power plant: the Chemical and Volume Control System, and the

Auxiliary Feed-Water System. All the steps of model and simulation development have been described in as much detail as possible, so as to show how the theories and algorithms developed in the previous Chapters may be put into practice.

In particular, in both cases the procedure for developing models and simulation of operators has been followed, from the definition of the problem boundaries to the task analysis, to the model development, numerical simulation and computerised implementation.

A similar procedure has been followed for the Machine model and for the application of DYLAM as an Interaction Model.

These two cases present two rather different operator models and simulations. This was done on purpose, so as to show:

1. The potential results and scope of application that can be obtained with human behaviour models of different nature.
2. The detailed procedure and the importance of including the simulation of human behaviour in any safety study of complex systems.

The variety of results obtained from the scenarios analysed and the consequences of the interactions, in case of human errors and hardware or control system failures, have clearly indicated that the complete overview of an accident can only be achieved when all the actors of the system are called into play.

These main actors include operators, machines, procedures, interfaces, emergency and protection systems, which need to be modelled and simulated in a combined architecture.

A reference Human-Machine Interaction Architecture, has been discussed and presented all throughout this book and all its major elements have been the subject of each individual Chapter.

ACRONYMS AND ABBREVIATIONS

AFWS	Auxiliary Feed-Water System
AI	Artificial Intelligence
AoR	Allocation of Resources
BB	Blackboard
BCCS	Boron Concentration Control System
BRS	Boron Recycle System
BTRS	Boron Thermal Regeneration System
CAMEO	Cognitive and Action Modelling of an Erring Operator
CBB	Control Blackboard
CCS	Component Cooling water System
CES	Cognitive Environmental Simulation
CIF	Currently Instantiated Frame
COCOM	Contextual Control Model
COSIMO	Cognitive Simulation Model
CREATE	Cognitive Reliability Assessment Technique
CTA	Cognitive Task Analysis
CVCS	Chemical and Volume Control System
DB	Design Basis
DBB	Domain Blackboard
DI	Direct Inference
DS	Direct Search
EAM	Entity Attribute Matrix
ET	Event Tree
FG	Frequency Gambling
FMEA	Failure Mode and Effect Analysis
FT	Fault Tree
FT	Flow Transducer
FWM	Focal Working Memory
GMTA	Goals-Means Task Analysis
HHI	Human-Human Interaction
HHIf	Human-Human Interface
HMI	Human-Machine Interaction

HMIA	Human-Machine Interaction Architecture
HMIf	Human-Machine Interface
HMS	Human-Machine System
HPS	Human Problem Solving
IE	Inference Engine
IPS	Information Processing System
KB	Knowledge Base
KBF	Knowledge Based Frames
LOCA	Loss Of Coolant Accident
LOF	Loss of Feed-water accident
LOP	Loss of Off-site Power accident
LOS	Loss Of Steam-flow accident
LT	Level Transducer
LTM	Long Time Memory
MAD	Méthode Analytique de Description
MIDAS	Man Machine Integration Design and Analysis System
MTM	Medium Time Memory
NPP	Nuclear Power Plant
OCDM	Observer/Controller/Decision Model
OCM	Optimal Control Model
OFM	Operator Function Model
PAF	Performance Adjustment Factor
PID	Proportional, Integral, and Derivative (regulation)
PIPE	Perception, Interpretation, Planning and Execution
PROCRU	Procedure Oriented Crew model
PRT	Pressure Relief Tank
PSA	Probabilistic Safety Assessment
PWM	Peripheral Working Memory
PWR	Pressurised Water Reactor
RBF	Rule Based Frames
RCTD	Reactor Coolant Drain Tank
RHRS	Residual Heat Removal System
RMC	Reference Model of Cognition
RMWS	Reactor Make-up Water System
ROC	Receiver Operating Characteristic graph
RWST	Refuelling Water Storage Tank
SBO	Station-Black-Out
SG	Steam Generator
SHEL	Software, Hardware, Environment, Liveware
SL	Step Ladder
SM	Similarity Matching
SRK	Skill, Rule, Knowledge
SSA	System Simulation Architecture
STM	Short Time Memory

SYBORG	Simulation System for Behaviour of an Operating Group
THERP	Technique for Human Error Rate Prediction
VCT	Volume Control Tank
VSMoC	Very Simple Model of Cognition
WGC	Waste Gas Compressor
WM	Working Memory

REFERENCES

Aldemir, T. (1987). Computer-assisted Markov failure modelling of process control system. *IEEE Transaction on Reliability,* **R-36**, 133-144.

Aldemir, T., N. Siu, A. Mosleh, P. C. Cacciabue, and B. G. Göktepe (Eds.) (1994). *Reliability and Safety Assessment of Dynamic Process Systems*, Springer-Verlag, Hidelberg, Germany.

Amalberti, R. (1996). *La coduite de systèmes à risque*. Presse Universitaire de France, Paris.

Amalberti, R., and F. Deblon (1992). Cognitive modelling of fighter aircraft process control: A step towards an intelligent onboard assistance system. *International Journal of Man-Machine Studies, IJMMS*, **36**, 639-671.

Amalberti, R., C. Valot, and J. P. Menu (1987). Etude préliminaire dans le cadre du développement d'aides intelligentes au pilotage. Report CERMA, 87-31, Brétigny sur Orge, F.

Amendola, A. (Ed.) (1986). System Reliability Benchmark Exercise, EUR 10696 EN. CEC-JRC, Ispra, Varese, Italy.

Amendola, A., and G. Reina (1981). Event sequences and consequence spectrum: a methodology for probabilistic transient analysis, *Nuclear Science and Engineering*, **77**, 297-315.

Alty, J. L., and G. Guida (1985). The rule-based system technology in design of human machine systems. Proceedings of 2nd IFAC/IFIP/IFORS/IEA Conference on Analysis, Design and Evaluation of Man-Machine Systems. Survey Lecture. Varese, Italy, 10-12 September. pp. 23-38.

Anderson, J. R. (1983). *The architecture of cognition.* Harvard University Press, Cambridge, MA.

Andrews, J. D., and T. R. Moss (1993). *Reliability and Risk Assessment.* Longman Scientific & Technical, Harlow, UK.

Ashby, W. R. (1956). *An introduction to cybernetics*. Chapman and Hall, London.

Bagnara, S., C. Di Martino, B. Lisanti, G. Mancini, and A. Rizzo (1989). A Human Error Taxonomy Based on Cognitive Engineering and on Social Occupational Psychology. EUR 12624 EN. CEC-JRC, Ispra, Varese, Italy.

Bainbridge, L. (1989). Development of skill, reduction of workload. In L. Bainbridge, and S. A. Ruiz Quintanilla (Eds.), *Developing skill with Information Technology,* Chichester, UK: J. Wiley & Sons. pp. 87-116.

Baron, S., G. Zacharias, R. Muralidharan, and R. Lancraft (1980). PROCRU: a model for analyzing flight crew procedures in approach to landing. NASA CR-152397.

Barr, A., P. R. Cohen, and E. A. Feigenbaum (1989). *The Handbook of Artificial Intelligence - Volume IV*. Addison-Wesley, Reading, MA.

Bersini, H., (1989). La modélisation cognitive dans l'étude des systèmes homme-machines. Dissertation de Doctorat en Sciences Appliquées. (in French). Université Libre de Bruxelles. Bruxelles, Belgium.

Bharucha-Reid, A. T., (1960). *Elements of the Theory of Markov Processes and Their Applications*. McGraw-Hill, New York.

Billings, C. E. (1997). *Aviation Automation: The Search for a Human-Centered Approach*. Lawrence Erlbaum Associates, Mahwah, New Jersey.

Bobrow, D. G. (Ed.) (1984). *Qualitative Reasoning about Physical Systems*. North-Holland, Amsterdam, The Netherlands. Reprint from *Artificial Intelligence*, **24**.

Bonar, J., J. Collins, K. Curran, R. Eastman, D. Gitomer, R. Glaser, L. Greenberg, S. Lajoie, A. Lesgold, D. Logan, M. Magone, V. Shalin, A. Weiner, R. Wolf, and L. Yengo (1986). Guide to cognitive task analysis. Learning Research and Development Center (LRDC), University of Pittsburgh, Pittsburgh, OH.

Broadbent, D. E., A. Baddeley, and J. Reason (Eds.) (1989). *Human Factors in Hazardous Situations*, Proc. of a Royal Society Discussion Meeting 28-29 June, Clarendon Press, Oxford.

Cacciabue, P. C., (1994). Affidabilità Dinamica e Fattori Umani in Sistemi Nucleari. European Commission, Joint Research Centre Report. EUR-15988 IT, EC Bruxelles (in Italian).

Cacciabue, P. C., (1997). A Methodology for Human Factors Analysis for System Engineering: Theory and Applications. *IEEE-System Man and Cybernetics, IEEE-SMC*, **27** (3), 325-339.

Cacciabue, P. C., A. Carpignano, and C. Vivalda (1992). Expanding the Scope of DYLAM Methodology to Study the Dynamic Reliability of Complex Systems: the Case of Chemical and Volume Control in Nuclear Power Plants. *Reliability Engineering and System Safety*, **36**, 127-136.

Cacciabue, P. C., A. Carpignano, and C. Vivalda (1993). A Dynamic Reliability Technique for Error Assessment in Man-Machine Systems. *International Journal of Man Machine Studies*, IJMMS, **38** (3), 403-428.

Cacciabue P. C., and G. Cojazzi (1994). A human factor methodology for safety assessment based on the DYLAM approach. *Reliability Engineering and System Safety*, **45**, 127-138.

Cacciabue P. C., G. Cojazzi, E. Hollnagel, and S. Mancini (1992). Analysis and Modelling of Pilot-Airplane Interaction by an Integrated Simulation Approach. Proceedings of 5th IFAC/IFIP/IFORS/IEA Symposium on Analysis, Design and Evaluation of Man-Machine Systems. The Hague, The Netherlands, 9-11 June. Pergamon Press, Oxford, UK. pp. 227-234

Cacciabue, P. C., F. Decortis, B. Drozdowicz, M. Masson, and J. P. Nordvik (1992a). COSIMO: A Cognitive Simulation Model of Human Decision Making and Behaviour in Accident Management of Complex Plants. *IEEE Transaction on Systems, Man and Cybernetics, IEEE-SMC*, **22** (5), 1058-1074.

Cacciabue, P. C., G. Guida, and A. Pace (1987). An expert system for modelling operators behaviour in control of a steam generator. In M.C. Majumdar, D. Majumdar and J.I. Sackett (Eds.), *AI and Other innovative Computer Applications in the Nuclear Industry*. Plenum Press, New York. pp. 313-320.

Cacciabue, P. C., and E. Hollnagel (1995). Simulation of Cognition: Applications. In J. M. Hoc, P. C. Cacciabue, and E. Hollnagel (Eds.), *Expertise and Technology: Cognition and Human-Computer Interaction.* Lawrence Erlbaum Associates, Hillsdale, New Jersey. pp. 55-73.

Cacciabue, P. C., G. Mancini, and H. Bersini (1990). A Model of Operator Behaviour for Man-Machine System Simulation. *Automatica*, **26**, (6), 1025-1034.

Capper, D. (1994). *C^{++} for Scientists, Engineers and Mathematicians.* Springer-Verlag, London, UK.

Carbonell, J. R. (1966). A Queuing model of many-instruments visual sampling. *IEEE Transactions on Human Factors in Electronics, IEEE- HFE*, **4**, 157-164.

Card, S. K., T. P. Moran, and A. Newell (1983). *The Psychology of Human Computer Interaction.* Lawrence Erlbaum Associates, Hillsdale, New Jersey.

Carpignano, A. (1989). Studio affidabilistico del sistema CVCS di un impianto PWR. Graduation Thesis in Nuclear Engineering (in Italian). Politecnico di Torino, Italy.

Chandrasekaran, B., and R. Milne (1985). Reasoning about structure, behavior and function. *ACM Sigart Newsletter*, **93**, 4-9.

Cohen, P. R., and E. A. Feigenbaum (1986). *The Handbook of Artificial Intelligence - Volume III.* Addison-Wesley, Reading, MA.

Cojazzi, G., and P. C. Cacciabue (1994). The DYLAM Approach for the Reliability Analysis of Dynamic Systems. In T. Aldemir, N. Siu, A. Mosleh, P. C. Cacciabue, and B. G. Göktepe (Eds.). *Reliability and Safety Assessment of Dynamic Process Systems*, Springer-Verlag, Berlin, Germany. pp. 8-23.

Cojazzi, G., P. C. Cacciabue, and P. Parisi (1993). DYLAM-3. A Dynamic Methodology for Reliability Analysis and Consequences Evaluation in Industrial Plants. EUR 15265 EN. EC Bruxelles.

Conant, R. C., and W. R. Ashby (1970). Every good regulator of a system must be a model of that system. *International Journal of Systems Science*, **1** *(2)*, 89-97.

Corker, K. M., and G. M. Pisanich (1995a). Analysis and modelling of flight crew performance in automated air traffic management systems. Proceedings 6th IFAC/IFIP/IFORS/IEA Symposium on Analysis, Design and Evaluation of Man-Machine Systems, Cambridge, MA, 27-29 June. pp. 629-634.

Corker, K. M., and G. M. Pisanich (1995b). A multiple agent model of human performance in automated air traffic control and flight management operations. Proceedings of 5th International conference on Human-Machine Interaction and Artificial Intelligence in Aerospace, Toulouse, France, 27-29 September.

Corker, K. M., and B. Smith (1993). An architecture and modeling for cognitive engineering simulation analysis: application to advanced aviation analysis. AAIA Conference on Computing in Aerospace, San Diego, CA.

Degani, A., C. M. Mitchell, and A. R. Chappell (1995). Task models to guide analysis: Use of the Operator Function Model to represent mode transitions. Proceedings of the 8th International Symposium on Aviation Psychology. Columbus, OH, 24-27 April. Volume 1 pp. 210-215.

Degani, A., and E. L. Wiener (1994a). On the design of flight-deck procedures. NASA Contractor Report 177642. NASA, Ames Research Center, Moffett Field, CA.

Degani, A., and E. L. Wiener, (1994b). Philosophy, policies, procedures and practice: The four "P"s of flight deck operations. In N. Johnston, N. McDonald, and R. Fuller (Eds.). *Aviation Psychology in Practice,* Avebury Technical, Aldershot, UK. pp. 68-87.

De Keyser, V., and D. D. Woods (1989). Fixation errors in dynamic and complex systems. In A.G. Colombo, and R. Misenta (Eds.), *Advanced Systems Reliability Modelling*, Kluwer Academic, Dordrechts, The Netherlands.

De Kleer, J., and J. S. Brown (1984). A qualitative physics based on confluences. *Artificial Intelligence,* **24**, 7-83.

DoD, Department of Defence (1984). System safety program requirements. MIL-STD-882B. DoD, Washington DC.

Douglas, J. F., J. M. Gasiorek, and J. A. Swaffield (1983). *Fluid mechanics*, Pitman International, New York.

Dubois, D., and H. Prade (1980). *Fuzzy sets and systems: Theory and applications.* Academic Press, New York.

Edwards, E. (1972). Man and machine: Systems for safety. In Proceedings of British Airline Pilots Association Technical Symposium. British Airline Pilots Association, London. pp. 21-36

Edwards, E. (1988). Introductory overview. In E. L. Wiener, and D. C. Nagel (Eds.), *Human Factors in Aviation*, Academic Press, San Diego, CA. pp. 3-25.

El Wakill, M. M. (1971). *Nuclear Heat Transport.* Int. Textbook Company, New York, USA.

Engelmore E., and T. Morgan (1988). *Blackboard Systems.* Addison-Wesley, Workingham, England, UK.

Ericsson, K. A., and H. A. Simon (1984). *Protocol Analysis, Verbal Reports as Data.* The MIT Press, Cambridge, MA.

Feigenbaum, E. A., and H. A. Simon (1984). EPAM-like models of recognition and learning. *Cognitive Science.* **8**, 305-336.

Feller, W. (1968) *An introduction to probability theory and its application.* Volume 1. John Wiley and Sons, Singapore.

Firley, M., and D. Hellens (1991). *Knowledge elicitation. - A practical handbook.* Prentice-Hall, London, UK.

Fitts, P. M. (1954). The information capacity of the human motor system in controlling the amplitude of movement.. *Journ. of Experimental Psychology,* **47**, 381-391.

Forbus, K. D. (1984) Qualitative process theory. *Artificial Intelligence,* **24**, 85-168.

Fujii, M., T. Ogino, and M. Tanaka (1988). Application of a cognitive model to Man-Machine Interfaces for nuclear power plant. International ENS/ANS Conference on Thermal Reactor Safety "NUCSAFE 88", Avignon, France, 2-7 October. pp 39/1-39/10.

Fujita, Y. (1989). Improved annunciator system for Japanese Pressurised Water Reactors. *Nuclear Safety*, **30** (2), 210-221.

Fujita, Y. (1992). Human Reliability analysis: a human point of view. *Reliability Engineering and System Safety, RE&SS,* **38** (1-2), 71-79.

Fujita, Y., M. Tohyama, I. Yanagisawa, T. Ida, and H. Arikawa (1991). Designing a knowledge-based operator support system for practical applications. *Nuclear Technology*, **95**, 116-128.

Fujita, Y., I. Yanagisawa, J. Itoh, N. Yamane, K. Nakata, R. Kubota, and M. Tani (1993). Modelling operator with task analysis in mind. ANS Top. Meeting on NPP Instrumentation, Control and Man Machine Interface Technologies, Illinois.

Goodstein, L. P., H. A. Andersen, and S. E. Olsen (Eds.) (1988). *Tasks, Errors and Mental Models.* Taylor & Francis, London, UK.

Grant, S., and T. Mayes (1991). Cognitive Task Analysis ? In G. R. S. Weir, and J. L. Alty (Eds.), *Human-Computer Interaction and Complex Systems.* Academic Press, London. pp. 147-167.

Green, D. M., and J. A. Swets (1966). *Signal Detection Theory and Psychophysics.* Peninsula Publishing, Los Altos, CA.

Guida, G., (1985). Reasoning about physical systems: Shallow versus deep models. In J. Efstathiou, and A. Mamdani (Eds.), *Expert Systems and Optimisation in Process Control,* Gower Technical Press, Aldershot, UK. pp. 135-159.

Guida, G., P. Baroni, G. Cojazzi, L. Pinola, and R. Sardella (1996). Preliminary requirements for expert judgement acquisition in safety assessesment. In P. C. Cacciabue and I Papazoglou (Eds.), Proceedings of International Conference on *Probabilistic Safety Assessment and Management.* Crete, Greece, 24-28 June. pp. 491-498.

Hansen, J. P., and K. Itoh (1995). Building a cognitive model of dynamic ship navigation on basis of verbal protocols and eye-movement data. In L. Norros (Ed.), Proceedings of 5th European Conference of Cognitive Science Approaches to Process Control, Espoo, Finland, August 30 - September 1. pp. 325-337.

Hassan, M., and T. Aldemir (1990). A Data Base Oriented Dynamic Methodology for the failure analysis of closed loop control systems in process plants, *Reliability Engineering and System Safety, RE&*SS, **27**, 275-322.

Hayes-Roth, B. (1985). A blackboard architecture for control. *Artificial Intelligence*, **26** (2), 251-321.

Hayes-Roth, F., D. A. Waterman, and D. B. Lenat (Eds.) (1983). *Building expert systems.* Addison-Wesley, Reading, MA.

Helander, M. (Ed.) (1988). *Handbook of Human-Computer Interaction.* Elsevier, Amsterdam, The Netherlands.

Henley, E. J., and H. Kumamoto (1981). *Reliability Engineering and Risk Assessment.* Prentice-Hall, Englewood Cliffs, N.J.

Hewitt, C. (1977). Viewing control structures as patterns of passing messages. *Artificial Intelligence.* **8** (3), 323-364.

Hewitt, C., P. Bishop, and R. Steiger (1973). A universal modular actor formalism for artificial intelligence. Proceedings of the Third International Joint Conference on Artificial Intelligence (IJCAI-73). pp.235-245.

Hick, W. E. (1952). On the rate of gain of information. *Quart. Journ. of Experimental Psychology.* **4**, 11-26.

Hofstede, G. (1991). *Cultures and Organisations: software of the mind.* McGraw-Hill, Maidenhead, UK.

Hollangel, E. (1991). The phenotype of erroneous actions: implications for HCI design. In G. R. S. Weir, and J. L. Alty (Eds.), *Human-Computer Interaction and Complex Systems.* Academic Press, London. pp. 73-121.

Hollnagel, E. (1993a). *Human Reliability Analysis: Context and Control.* Academic Press, London.

Hollnagel, E. (1993b). Requirements for dynamic modelling of man-machine interaction. *Nuclear Engineering and Design*, **144**, 375-384.

Hollnagel, E. (1997). Personal communications

Hollnagel, E. (1998). *Cognitive Reliability and Error Analysis Method.* Elsevier, London.

Hollnagel, E., and P. C. Cacciabue (1991). Cognitive Modelling in System Simulation. Proceedings of Third European Conference on Cognitive Science Approaches to Process Control, Cardiff, UK, September 2-6.

Hollangel, E., G. Mancini, and D. D. Woods (Eds.) (1986). *Intelligent Decision Support inProcess Environments.* NATO ASI Series, Springer-Verlag, Berlin, Germany.

Hollangel, E., and P. Marsden (1996). Further development of the phenotype-genotype classification scheme for the analysis of human erroneous actions. European Commission, Joint Research Centre Report. EUR-16463 EN. EC Bruxelles.

Høyland, A., and M. Rausand (1994). *System reliability theory: Models and statistical methods.* John Wiley and Sons, New York.

Hudson, P. T. W., J. Reason, W. Wagenaar, P. Bentley, M. Primrose, and J. Wisser (1994). Tripod Delta: Proactive approach to enhanced safety, *Journal of Petroleum Technology*, **46**, pp. 58-62.

Itoh, K., and T. Enkawa (1991). Analysis and evaluation of cognitive task analysis using an information transition model. In H. J. Bullinger (Ed.), *Human aspects in computing: Design and use of interactive systems and work with terminals.* Elsevier, Amsterdam, The Netherlands. pp. 646-650.

Itoh, K., J. P. Hansen, and F. R. Nielsen (1998). Cognitive modelling of ship navigation based on protocol and eye-movement analysis. To appear in *Le Travail Humain.*

Iwasaki, Y., and H. A. Simon (1986). Causality in device behavior. *Artificial Intelligence,* **29**, 3-32.

Jeong, K. S., S. H. Chang, and T. W. Kim (1987). Development of the dynamic fault tree using markovian process and supercomponents, *Reliability Engineering and System Safety, RE&SS*, **19** 137-160.

Johannsen, G. (1992) Towards a new quality of automation in complex Man-Machine Systems. *Automatica,* **28** (2), 355-373.

Johnston, N., N. McDonald, and R. Fuller (Eds.) (1994). *Aviation Psychology in Practice.* Avebury Technical, Aldershot, UK.

Just, M. A., and P. A. Carpenter (1980). A theory of reading: From eye fixation to comprehension. *Psychological Review,* **87**, 329-354.

Kalman, R. E. (1960). A new approach to linear filtering and prediction problems. *J. Basic Engineering, Trans. ASME,* **82D**, 33-45.

Kelley, C. R. (1968). *Manual and Automatic Control.* Wiley, New York.

Kleinman, D. L., S. Baron, and W. H. Levison (1970). An optimal control model of human response, Part I. *Automatica,* **6**, N.3, 636-640.

Kok, J. J., and H. G. Stassen (1980). Human operator control of slowly responding systems: Supervisory control. *Journal of Cybernetics and Information Science*, **3**, 123-174.

Kosko, B. (1991). *Neural networks and fuzzy systems.* Prentice-Hall, New Jersey.

Kuhlmann, A. (1986). *Introduction to Safety Science.* Springer-Verlag, New York.

Kuipers, B. (1984). Common sense reasoning about causality: deriving behavior from structure. *Artificial Intelligence,* **24**, 169-203.

Kuipers, B. (1986). Qualitative simulation. *Artificial Intelligence,* **29**, 289-338.

Marsden, P. (1987). The actual frequency of encounter of American Presidents. Department of Psychology, University of Manchester.

Masson, M. (1991). Understanding, reporting and preventing human fixation errors. In T. W. Van der Schaaft, D. A. Lucas and A. R. Hale (Eds.), *Near Miss reporting as a safety tool.* Butterworth-Heinemann, Oxford, UK. pp. 35-50.

Masson, M. (1994). Prevention Automatique des Erreurs de Routine. PhD Thesis (in French). University of Liege, Faculty of Psychology and Educational Sciences, Liege, Belgium.

Matsuoka, T., and M. Kobayashi (1988). GO-FLOW: a new reliability analysis methodology, *Nuclear Science and Engineering*, **98**, 64--78.

Maurino, D. E., J. Reason, N. Johnston, and R. B. Lee (1995). *Beyond Aviation Human Factors.* Avebury Aviation, Aldershot, UK.

McRuer, D. T, D. Graham, E. Kredel, and W. Reisener (1965). Human pilot dynamics in compensatory systems - theory, models and experiments with controlled elements and forcing function variations. Report AFFDL-TR65-15, Wright Patterson AFB, Ohio.

McRuer, D. T., and H. R. Jex (1967). A review of quasi-linear pilot models. *IEEE Transactions on Human Factors in Electronics, IEEE-HFE,* **4**, 231-249.

McRuer, D. T., and E. S. Krendel (1957). Dynamic response of human operators. Wright Air Development Center. Tech. Rep. 56-524.

Miller, C. O. (1965). Safety and semantics. Alumni Review. University of Southern California, Aerospace Division, Los Angeles.

Minsky, M. (1975). A framework for representing knowledge. In P. Winston (Ed.), *The psychology of Computer Vision.* McGraw-Hill, New York.

Mitchell, C. M. (1987). GT-MSOCC: A Domain for Research on Human-Computer Interaction and Decision Aiding in Supervisory Control Systems. *IEEE Transaction on Systems, Man and Cybernetics, IEEE-SMC,* **17** (4), 553-572.

Mo, J. (1997). Contribution de l'Informatique à la Sûreté de Functionnement d'un Système Socio-Technique. LAAS-CNRS, Doctorate Thesis (in French). Institut National Polytechnique de Toulouse, France.

Moray, N. (Ed.) (1979). *Mental Workload its Theory and Measurement,* NATO Conference Series, III : 8, Plenum Press, N.Y.

Moray, N. (1988). Mental workload since 1979, *International reviews of ergonomics*, **2**, 123-150.

Moray, N., W. R. Ferrell, and W. B. Rouse (Eds.) (1990). *Robotics, Control and Society*. Taylor and Francis, Bristol, PA, USA.

Nagel, D. C. (1988). Human Error in Aviation Operations. In E. L. Wiener and D. C. Nagel (Eds.), *Human Factors in Aviation*. Academic Press, San Diego, CA. pp. 263-303.

Neisser, U. (1967). *Cognitive Psychology*. Appleton-Century-Crofts, New York.

Neisser, U. (1976). *Cognition and Reality: Principles and Implications of Cognitive Psychology*. W. H. Freeman, San Francisco.

Newell, A., and H. A. Simon (1963). GPS, a program that simulates human thought. In E. A. Feigenbaum and J. Feldman (Eds.), *Computers and Thought*. McGraw-Hill, New York.

Newell, A., and H. A. Simon (1972). *Human Problem Solving*. Prentice-Hall, Englewood Cliffs, N.Y.

Nii, H. P. (1986) Blackboard systems (Part 1), *AI Magazine*, **7** (2), 38-53.

Ordys, A. W., A. W. Pike, M. A Johnson, R. M. Katebi, and M. J. Grimble (1994). *Modelling and Simulation of Power Generation Plants*, Springer-Verlag, London.

Pace, A. (1987). Uso di tecniche di modellizzazione qualitativa per la simulazione del ragionamento dell'operatore di un impianto industriale. Graduation Thesis (in Italian). Electronic Engineering. Politecnico di Milano, Italy.

Papenhuijzen, R. (1994). Towards a human operator model of the navigator. Doctoral Thesis, Technical University of Delft, TUD, Delft, The Netherlands.

Papenhuijzen, R and T. Dijkhuis (1995). Evaluation of two human operator models of the navigator's behaviour. Proceedings 6th IFAC/IFIP/IFORS/IEA Symposium on Analysis, Design and Evaluation of Man-Machine Systems, Cambridge, MA, 27-29 June. pp. 741-748.

Papenhuijzen, R., and H. G. Stassen (1987). On the modelling of the behaviour of a navigator. Proceedings of 8th Ship Control Systems Symposium, The Hague, The Netherlands. pp. 2.238-2.254.

Parry, G. W. (1994). Critique of Current Practice in the Treatment of Human Interactions in Probabilistic Safety Assessment. In T. Aldemir, N. Siu, A. Mosleh, P. C. Cacciabue, and B. G. Göktepe (Eds.), *Reliability and Safety Assessment of Dynamic Process Systems*, Springer-Verlag, Hidelberg, Germany. pp. 156-165.

Paulin, J. H., and K. Itoh (1995). Building a cognitive model of dynamic ship navigation on basis of verbal protocols and eye-movement data. Proceedings of 5th European Conference on Cognitive Science Approaches to Process Control. August 30 - September 1, Espoo, Finland. pp. 325-337.

Payne, S. J., and T. R. G. Green (1986). Task-action grammars: A model of the mental representation of task languages. *Human Computer Interaction*, **2**, 93-133.

Pedrali, M. (1993). Modelli e tassonomie di errori umani per l'analisi di sicurezza nell'aviazione civile. Il caso studio delll'incidente di Zurigo: ricerca delle cause e indagine prospettica. Graduation Thesis in Aeronautical Engineering (in Italian). Politecnico di Milano, Italy.

Pedrali, M. (1996). Vers un environnement multimédia pour l'analyse vidéo des causes d'erreurs humaines. Application dans les simulateurs d'avions. Thèse de Doctorat en Informatique (in French). Université Toulouse 1, Toulouse, France.

Pedrali, M., G. Cojazzi, and P. C. Cacciabue (1995). A methodology for retrospective analyses of accidents involving human factors. Proceedings of the 8th International Symposium on Aviation Psychology. Columbus, OH, 24-27 April. Volume 2 pp. 1278-1283.

Peterson, J. L. (1981). *Petri Net Theory and the Modelling of Systems.* Prentice-Hall, Englewood Cliffs, NJ

Peterson, W. W., T. G. Birdsall, and W. C. Fox (1954). The theory of signal detectability. *Transactions IRE Professional Group on Information Theory,* **PGIT-4**, 171-212.

Pew, R. W., and S. Baron (1983). Perspectives on Performance Modelling. *Automatica,* **19**, No. 6, 663-676.

Pew, R. W., S. Baron, C. E. Feehrer, and D. C. Miller (1977). Critical Review and Analysis of Performance Models Applicable to Man-Machine-Systems Evaluation, Report **BBN 3446**, Cambridge, MA.

Pisanich, G. M., and K. M. Corker (1997). A cognitive system model for human/automation dynamics in Airspace management. 13th Triennial Congress of the International Ergonomics Association, June 29 - July 4. Tampere, Finland.

Pople, H. E. Jr. (1977). The information of composite hypotheses in diagnostic problem solving: An exercise in synthetic reasoning. Proceedings of the Fifth International Joint Conference on Artificial Intelligence (IJCAI-77). pp.1030-1037.

Pople, H. E. Jr. (1985). Evolution of an Expert System: from Internist to Caduceus. In I. De Lotto, and .M. Stefanelli (Eds.), *Artificial Intelligence in Medicine*, Elsevier Science (North Holland), New York. pp. 179-208.

Ralli, M. (1993). *Fattore Umano* (in Italian). Libreria All'Orologio. Roma, Italy.

Rasmussen, J. (1974). The human data processor as a system component. Risø-M-1722, Risø National Laboratory, Roskilde, Denmark.

Rasmussen, J. (1976). Outline of a hybrid model of the process operator. In T. B. Sheridan and G. Johannsen (Eds.) *Monitoring Behavior and Supervisory Control.* Plenum Press, New York.

Rasmussen, J., (1983). Skills, Rules and Knowledge: signals, signs and symbols; and other distinctions in human performance model. *IEEE Transactions on Systems, Man, and Cybernetics, IEEE-SMC*, **13**, 3, 257-267.

Rasmussen, J. (1986). *Information processes and human-machine interaction. An approach to cognitive engineering*. Elsevier - North Holland, Amsterdam, The Netherlands.

Rasmussen, J., O. M. Pedersen, C. Carnino, M. Griffon, G. Mancini, and P. Cagnolet (1981). Classification System for Reporting Events Involving Human Malfunction. Risø-M-2240, EUR-7444 EN. Risø National Laboratory, Roskilde, Denmark.

Rasmussen, J., A. M. Pejtersen, and K. Schmidt (1990). Taxonomy for cognitive work analysis. In J. Rasmussen, B.Brehmer, M. de Montmollin, and J. Leplat (Eds.) Proceedings of the 1st MOHAWC Workshop, Liege, May 15-16. Vol. 1 ESPRIT Basic Research Project 3105, European Commission, Brussels, Belgium. pp. 1-153.

Rayner, K., and G. W. McConkie (1976). What guides a reader's eye movement? *Vision Research*, **16**, 829-837.

Reason, J. (1986). Recurrent errors in process environments: some implications for the design of Intelligent Decision Support Systems. In E. Hollnagel, G. Mancini and D. D. Woods (Eds.), *Intelligent Decision Support in Process Environment*. NATO ASI Series, Springer-Verlag, Berlin, FRG, 1986. pp. 255-270

Reason, J. (1987a). Generic Error Modelling System (GEMS): A cognitive framework for locating common human error forms. In J. Rasmussen, K. Duncan and J. Leplat (Eds.), *New technology and human error*. J. Wiley, London, UK. pp. 63-83.

Reason, J. (1987b). Cognitive aids in process environments: prostheses or tools? *International Journal of Man-Machine Studies*, IJMMS, **27**, 5&6, 463-471.

Reason, J. (1990). *Human error*. Cambridge University Press, Cambridge UK.

Reason, J. (1992). Personal Communications.

Redding, R. E., and T. L. Seamster (1994). Cognitive Task Analysis in air traffic controller and aviation crew training. In N. Johnston, N. McDonald, and R. Fuller (Eds.), *Aviation Psychology in Practice*, Avebury Technical, Aldershot, UK. pp. 190-222.

Reynard, W. D., C. E. Billings, E. Cheaney, and R. Hardy (1986). The development of the NASA Aviation Safety Reporting System (ASRS), NASA R.P. 1114.

Rouse, W. B. (1978a). A model of human decision making in a fault diagnosis task. *IEEE Transaction on System, Man, and Cybernetics*, *IEEE-SMC*, **8** (4), 258-271.

Rouse, W. B. (1978b). A model of human decision making in a fault diagnosis task. *IEEE Transaction on System, Man, and Cybernetics*, *IEEE-SMC*, **8** (5), 357-361.

Rouse, W. B. (1979). A model of human decision making in fault diagnosis tasks that include feedback and redundancy. *IEEE Transaction on System, Man, and Cybernetics*, *IEEE-SMC*, **9** (4), 237-241.

Rouse, W. B. (1980). *Systems Engineering Models of Human-Machine Interaction*. Elsevier North Holland, New York, N.Y.

Rouse, W. B. (1983). Models of human problem solving: Detection, diagnosis, and compensation for system failures, *Automatica*, ***19*** (6), 613-625.

Rouse, W. B. (1990). Human Resource Issues in System Design. In N. Moray, W. R. Ferrell & W. B. Rouse (Eds.), *Robotics, Control and Society*. Taylor and Francis, London, UK. pp. 177-186.

Rouse, W. B., and S. H. Rouse (1983). Analysis and classification of human error. *IEEE Transaction on System, Man, and Cybernetics*, *IEEE-SMC*, **13** (4), 539-549.

Rubin, K. S., P. M. Jones, and C. M. Mitchell (1988). OFMspert: Inference of Operator Intentions in Supervisory Control Using a Blackboard Architecture. *IEEE Transaction on Systems, Man and Cybernetics*, *IEEE-SMC*, **18** (4), 618-637.

Sasou, K., K. Takano, and S. Yoshimura (1993). Development of a team behaviour simulation model. Proceedings of 4th European Conference of Cognitive Science Approaches to Process Control, Grundtvigs Højskole, Frederiksborg, Denmark, August 25-27.

Scapin, D. L., and C. Pierret-Golbreich (1990). Toward a method for task description: MAD. In L. Berlinguet, and D. Berthelette (Eds.) *Work with display units 89*, Elsevier, Amsterdam, The Netherlands. pp. 371-380.

Schank, R. C., and R. P. Abelson (1977). *Scripts, Plans, Goals, and Understanding: An Inquiry into Human Knowledge Structures*. Erlbaum Associates, Hillsdale, N.J.

Searle, J. R. (1983). *Intentionality: An essay in the philosophy of mind.* Cambridge University Press, Cambridge, UK.

Sebillotte, S. (1995). Methodology guide to task analysis with the goal of extracting relevant characteristics for human-computer interfaces. *International Journal of Human-Computer Interaction, IJHCI,* **7** (4), 341-363.

Sebillotte, S., and D. L. Scapin (1994). From user's task knowledge to high-level interface specification. *International Journal of Human-Computer Interaction, IJHCI,* **6** (1), 1-15.

Senders, J. W. (1964). An investigation of the visual sampling behavior of human observers. NASA-CR-434.

Senders, J. W., and N. P. Moray (1991). *Human Error: Cause Prediction and Reduction.* Lawrence Erlbaum Associates, Hillsdale, N.J.

Shannon, C. E., and W. Weaver (1949). *The Mathematical Theory of Communications.* University of Illinois Press, Urbana, IL.

Sheridan, T. B. (1970). How often the supervisor should sample. *IEEE Transaction on Systems Science and Cybernetics, IEEE-SSC,* **6**, 140-145.

Sheridan, T. B. (1985). Forty-five years of Man-Machine systems: history and trends. Keynote Address. Proceedings of 2nd IFAC/IFIP/IFORS/IEA Symposium on Analysis, Design and Evaluation of Man-Machine, Varese, Italy, 10-12 September. pp. 5-13.

Sheridan, T. B. (1992). *Telerobotics, Automation and Human Supervisory Control.* The MIT Press, Cambridge, MA.

Sheridan, T. B., and W. R. Ferrell (1974). *Man-Machine Systems: Information, Control and Decision Models of Human Performance.* MIT Press, Cambridge, MA.

Siu, N. (1994). Risk assessment for dynamic systems: An overview. *Reliability Engineering and System Safety, RE&SS,* **43**, 1, 43-73.

Siu, N., and C. Acosta (1991). Dynamic event tree analysis - an application to SGTR. In G. Apostolakis (Ed.) *Proceedings of the International Conference Probabilistic Safety Assessment and Management (PSAM),* Beverly Hills, California, USA, February 4--7, Elsevier Science Publishers, London. pp. 539-541.

Skinner, B. F. (1957). *Verbal Behaviour.* Appleton-Century-Crofts. New York.

Stassen, H. G. (1988). Human supervisor modelling: Some new developments. In E. Hollnagel, G. Mancini, and D. D. Woods (Eds.), *Cognitive Engineering in Complex Dynamic Worlds.* Academic Press, London, UK. pp. 159-164.

Stassen, H. G. (1990). Supervisory control behavior modelling: The challenge and necessity. In N. Moray, W. R. Ferrell, and W. B. Rouse (Eds.), *Robotics, Control and Society.* Taylor and Francis, Bristol, PA, USA. pp. 105-117.

Stassen, H. G., G. Johannsen, and N. Moray (1990). Internal representation, internal model, human performance model and mental workload. *Automatica,* **26** (4), 811-820.

Stokes, A. F., and C. D. Wickens (1988). Aviation Displays. In E. L. Wiener, and D. C. Nagel (Eds.), *Human Factors in Aviation,* Academic Press, San Diego, CA. pp. 387-431.

Swain, A. D., and H. E. Guttmann (1983). Handbook on Human Reliability Analysis with Emphasis on Nuclear Power Plant Application. NUREG/CR-1278. SAND 80-0200 RX, AN. Final Report.

Takano, K., K. Sasou, and S. Yoshimura (1995). Simulation System for behaviour of an operating group (SYBORG). XIV European Annual Conference on Human Decision Making and Manual Control. Delft, The Netherlands, June 14-16.

Taylor, R. K., and D. A. Lucas (1991). Signal passed at danger: Near miss reporting from a railway perspective. In T. W. van der Schaaf, D. A. Lucas, and A. R. Hale (Eds.) (1991). *Near miss reporting as a safety tool.* Butterworth-Heinemann, Oxford, UK. pp. 79-92.

Valot, C., R. Amalberti, M. Bataille, F. Deblon and J. M. Paignay (1989). Metaknowledge for time and reliability: luxury or necessity. Proceedings of 2nd International Conference on Cognitive Science Approaches to Process Control. Siena, Italy, 24-27 October. pp 81-92.

Valot, C., R. Amalberti, and J. Y. Cibiel (1991). AIDE, A model of cognitive activity features of reasoning for time and resource management. In B.Brehmer, and J. Leplat (Eds.) Proceedings of the 4th MOHAWC Workshop, Bamber, Germany, October 22-23, 1991. Vol. 2 ESPRIT Basic Research Project 3105, European Commission, Brussels, Belgium. pp. 28-41.

Van Meter, D., and D. Middleton (1954). Modern statistical approaches to reception in communication theory. *Transactions IRE Professional Group on Information Theory,* **PGIT-4**, 119-141.

Van Orman Quine, W. (1959). *Methods of Logic.* Henry Holt and Co., New York.

Vesely, W. E., and F. F. Goldberg (1977). Time dependent unavailability analysis of nuclear safety systems, *IEEE Transaction on Reliability,* **R-26 4**, 257-260.

Vivalda, C. (1989). Modellizzazione e controllo del CVCS in situazioni incidentali. Graduation Thesis in Nuclear Engineering (in Italian). Politecnico di Torino, Italy.

Vosniadou, S., and A. Ortony (Eds.) (1989). *Similarity and Analogical Reasoning.* Cambridge University Press, Cambridge, UK.

Westrum, R. (1995). Organisational dynamics and safety. In N. McDonald, N. Johnston, and R. Fuller (Eds.), *Applications of Psychology to the Aviation System.* Proceedings of the 21st Conference of the European Association of Aviation Psychology, Avebury Aviation, Aldershot, UK. pp. 75-80.

Wickens, C. D. (1984). *Engineering psychology and human performance.* Charles Merrill, Columbus, OH.

Wickens, C. D., and J. M. Flach (1988). Information Processing. In E. L. Wiener and D. C. Nagel (Eds.), *Human Factors in Aviation.* Academic Press, San Diego, CA, pp. 111-155.

Wiener, E. L., B. Kanki, and R. L. Helmreich (1993). *Cockpit resource management.* Academic Press, Inc. San Diego, CA.

Wiener, N. (1948). *Cybernetics.* MIT Press, Cambridge, MA.

Woods, D. D., E. M. Roth, and H. E. Pople Jr. (1987). Cognitive Environment Simulation: An Artificial Intelligence System for Human Performance Assessment. US-Nuclear Regulatory Commission, Technical Report NUREG-CR-4862. US-NRC, Washington DC.

Woods, D. D., H. E. Pople Jr., and E. M. Roth (1990). The Cognitive Environment Simulation: as a tool for modelling for Human Performance and Reliability. US-Nuclear Regulatory Commission, Technical Report NUREG-CR-5213. US-NRC, Washington DC.

Woods, D. D., E. M. Roth (1995). Symbolic AI computer simulations as tools for investigating the dynamics of Joint Cognitive Systems. In J. M. Hoc, P. C. Cacciabue, and E. Hollnagel (Eds.). *Expertise and Technology: Cognition & Human-Computer Cooperation.* Lawrence Erlbaum Associates, Hillsdale, N. J., USA. pp. 75-90.

Yoshida, H., M. Takeda, Y. Hayashi, and E. Hollnagel (1997). Cognitive control behaviour of the operator in the emergency. Proceedings of the 13th Triennial Congress of the International Ergonomics Association, June 29 - July 4. Tampere, Finland. Vol. 3 pp. 82-84.

Yoshikawa, H. (1989). Human Factors study on Man-Machine Interface form cognitive psychological aspect. Second European Conference on Cognitive Science Approaches to Process Control, Siena, Italy, 24-27 October. pp.283-292

Zadeh, L. (1965). Fuzzy sets. *Information and Control,* **8**, 338-353.

Zadeh, L. (1973). Outline of a new approach to the analysis of complex systems and decision processes. *IEEE Transaction on Systems Man and Cybernetics, IEEE-SMC,* **3**, 28-44.

SUBJECT INDEX

T

V

W

AUTHOR INDEX

MIX
Papier aus verantwortungsvollen Quellen
Paper from responsible sources
FSC® C105338

If you have any concerns about our products,
you can contact us on
ProductSafety@springernature.com

In case Publisher is established outside the EU,
the EU authorized representative is:
Springer Nature Customer Service Center GmbH
Europaplatz 3, 69115 Heidelberg, Germany

Printed by Libri Plureos GmbH
in Hamburg, Germany